Endorsements for *Maintenance Benchmarking and Best Practices*

Top leaders of manufacturing must understand the true value of maintenance. They must have an equivalent Chief Maintenance Officer (CMO) at or near the top. This book can help top leaders with profit optimization or better customer service or both. My experience as the Worldwide VP for PC Manufacturing at IBM proved the key points found in this book. The author's worldwide experience is shared very well in this book for you to apply in your own manufacturing operation. A must read for all in maintenance and top leaders as well.

RICHARD L. "DICK" DAUGHERTY
Former Vice President of IBM Worldwide PC Manufacturing
Research Triangle Park, NC

Maintenance excellence is mission-essential for our military forces and for public school systems at all levels across the USA. This excellent book will be your guide as a leader. It is based on maintenance excellence experiences of the author during his 26 years of service in the North Carolina Army National Guard in infantry and engineer units at all levels plus his concurrent professional career. The high cost of gambling with poor maintenance can be deadly to your organization. Maintenance excellence should be a priority goal for facilities and fleets. Maintenance excellence is an ongoing critical need in any organization. This book will help you as a leader at any level to take charge of your maintenance program in order to promote maintenance excellence.

BG KENNETH R. NEWBOLD
Retired from North Carolina Army National Guard
(Commanded 30 Separate Mechanized Infantry Brigade)
Retired Superintendent of Greensboro Public Schools

Other McGraw-Hill Maintenance Books

PALMER • *Maintenance Planning and Scheduling Handbook*

BLOOM • *Reliability Centered Maintenance: Implementation Made Simple*

BORRIS • *Total Productive Maintenance*

BAGADIA • *Computerized Maintenance Management Systems Made Easy*

Maintenance Benchmarking and Best Practices

A Profit- and Customer-Centered Approach

Ralph W. "Pete" Peters

McGraw-Hill

New York Chicago San Francisco Lisbon London Madrid
Mexico City Milan New Delhi San Juan Seoul
Singapore Sydney Toronto

The McGraw·Hill Companies

CIP Data is on file with the Library of Congress.

1 2 3 4 5 6 7 8 9 0 DOC/DOC 0 1 3 2 1 0 9 8 7 6

ISBN 0-07-146339-9

The sponsoring editor for this book was Kenneth P. McCombs, the editing supervisor was David E. Fogarty, and the production supervisor was Pamela A. Pelton. It was set in Century Schoolbook by International Typesetting and Composition. The art director for the cover was Anthony Landi.

Printed and bound by RR Donnelley.

This book was printed on acid-free paper.

McGraw-Hill books are available at special quantity discounts to use as premiums and sales promotions, or for use in corporate training programs. For more information, please write to the Director of Special Sales, McGraw-Hill Professional, Two Penn Plaza, New York, NY 10121-2298. Or contact your local bookstore.

Contents

Part 2 Determining Where You Are as a Profit- and Customer-Centered Maintenance Operation: The Scoreboard for Maintenance Excellence

Preface

Maintenance is forever and the prime objective for this book is to help you maximize the value of your current maintenance operation. It is to provide very practical and proven benchmarking and best practice tools for supporting either profit or customer service in your organization or both. It is about benchmarking maintenance from four key levels that can lead you toward improvement for "your business of maintenance." The tools in this book help you to start right now and "Just do it" as Nike says in commercials.

We will talk repetitively about the positive attitude, the foundation philosophies, and the proactive practices you must have for both *profit-centered maintenance and customer-centered maintenance* (PCCM). And you will understand the important synergistic multiplying effect both can have. Your impact to the bottom line as a leader, manager, or craftsperson is not mission-critical. Your work is mission-essential and a core requirement for total operations success. This need remains with us forever, just as the need for maintenance is forever. As an incentive for you to read this book, here is the deal. Whoever is first to provide the correct number of times the word *profit* is used, wins a free admission to our four-day workshop on this book's focused topics for achieving measurable results.

This book is for top leaders, maintenance leaders, and crafts leaders (as I will make frequent reference to generically) at all levels from both large and small operations. It is to give you proven tools for benchmarking that you can use right now! It is not about planning for the industrial tourism type of benchmarking. This book is definitely not another "leaner" as I call it with more current buzzwords from Harvard Business Review. There will be only just a few new ones that I have had to use to name some new things (as you just saw from PCCM earlier) and some others like CRI, OCE, ACE, and PRIDE in Maintenance.

My desire is for this material to really help the small maintenance operations of less than 5 or 10 crafts in, say, a small manufacturing plant or a small facilities complex. Small shops face the same technical and maintenance management challenges as faced in the 100 to 200 or more total craft employee operations. Many times the small shop feels they can not afford the outside support from a "consultant." The science and basic technologies of physical asset maintenance are common all across the world. Only the scope and location provide the unique differences for the profession and practice of maintenance.

This book has really universal application to all size operations. It can apply to almost all types of maintenance operations. We will not go into the deep end of the maintenance science/technology pool. But rather we will go to the real deep end of the other pool for leadership of profit- and customer-centered maintenance operations. There are countless books from McGraw-Hill and others, and articles that focus on the art and science for improving many specific maintenance functional areas. I encourage you to buy and read every book like this that you can afford from the deep end of the maintenance science and technology pool.

This book provides extensive case studies accumulated over the past 36 years. Many are personal lessons learned about the true value of maintenance. They come from having direct P/L responsibilities for two manufacturing operations (Crescent-Xcelite and Channel Master), fleet maintenance operations (U.S. Army ranging from detachment, company, separate company up to battalion levels over 26 years of service), and from facilities management as facilities director for all state buildings in my home town and our three state aquariums. They come from the plant maintenance side as a plant manager and as a corporate manager of industrial engineering as in-house consultant/industrial engineer to include maintenance. And last but not least, they come from what some may term as an "outhouse consultant for maintenance excellence" such as George Smith and Jack Burgess. There will be some positive topics in this book, so all will not be on the dark side for sure.

From my experiences in the maintenance trenches (that also includes highway road maintenance and golf course maintenance), what I outline in this book is what I have tried to do when I figuratively "owned the company." You will find here the complete step-by-many-steps strategy that we have used at The Maintenance Excellence Institute and with my previous employer; Tompkins Associates, Cooper Tools Group and the NC Department of Transportation. Of all the many steps *toward* maintenance excellence you must take the most important step we all can take, that is, the first step of defining that line in the sand as to "where we are right now."

And in summary, this book is also about most of all PRIDE in maintenance (second Buzz) which is *people really interested in developing excellence* in maintenance. If you are not a top leader reading this, my goal is for you to become one. If you are a maintenance leader or craft leader, I want this book to help you today on that journey toward, and to your definition of, maintenance excellence. We have helped over 5000 organizations in well over 70 countries around the world with free information at www.PRIDE-in-Maintenance.com. So as you chose then so let it be that you read it all, use what you agree with, gain the commitment and courage (you got to be brave to be in maintenance) and then go ahead and "Just do it!"

RALPH W. "PETE" PETERS
Raleigh, NC; Oak Island, NC

Acknowledgments

Personal thanks to Joyce Peters, my best friend and wife of over forty years. She has helped me focus on completing this book with her encouragement and with her stern insistence that I get back to work on it as I drifted off course on many other things. Thanks to Ralph (Sr.) and Wilhelmina Peters, my parents who helped me understand and focus on the really important things in this life and beyond. And thanks to our sons Jay and Brian who have created the joy of grandchildren and in-laws with their two families.

Thanks to the friends and associates who provided encouragement as well as material for this book and who will be noted later with their respective contributions. Thanks to Bob Gaskins, Ben Gibbs, Larry Peters, Jane Davison, Remy Kolb, Jim Ross PE, Lee Peters PE, Carla Reed, Roland Newhouse, Jack Burgess, Sergio Rossi, Tom O'Kelley PE, Ricky Smith, Sandy Dunn, Mick Windsor, Philip Slater, Lindsay Homenuik, Doug Smith, Tom Rhyne, Jim Tompkins, Myra Swartz, and last but not least, Lewis Burns.

This book is dedicated to maintenance professionals: crafts employees, storeroom and MRO procurement staff, and Maintenance Leaders at all levels who perform, manage, and lead the important profession of maintenance. May this group continue to get the proper recognition and appreciation for the valuable work that they do. This book is also dedicated to five Chief Maintenance Officers and one special Top Leader that I have known from the US Army; Mr. Berry (1970), Mr. Donaldson (1971), Mr. Deardorf (1971), Mr. Cheek (1975), Mr. McGee (1980), Mr. Metcalf (1989), and BG Kenneth Newbold (1980). These five Chief Warrant Officers are true maintenance professionals with BG Newbold being a true Top Leader and a role model for maintenance excellence. All were important contributors to my philosophies about mission-essential maintenance. A very special acknowledgement is made to Mr. George Smith, a World War II veteran, Navy pilot, and a maintenance consultant who has contributed inspiration and knowledge to my own professional journey.

Final thanks goes to the McGraw-Hill staff, to Kenneth McCombs, Senior Editor, who got it all started and to Gita Raman and other ITC staff in Noida, India who got it all finalized. We look forward to more projects together to support the important area of maintenance.

Introduction

A Plant Maintenance Focus but Facilities Maintenance Application

We will focus on plant maintenance but all the material is also for those involved with pure facilities maintenance work as well. Applying the tools presented will help you, the maintenance leader, from both the public and private sectors to benchmark your current operation against today's best practices We will approach it from a profit-centered and customer-centered strategy for improvement. There are numerous definitions of benchmarking but in its simplest it is "making improvement by learning from others." Benchmarking also is the process of identifying, understanding, and adapting today's outstanding maintenance practices to help your organization improve its performance.

It Is Not about Lean Maintenance or Pokey Yokey

One thing should be made very clear. This book is not another "leaner" a collection on "lean maintenance," pokey yokey, jishu kanri, kaizen, or other new/emerging buzz from the island nation east of China. I personally cringe when I hear the term "lean maintenance" after being around well over 1000 manufacturing plants, hospitals, university campuses, facility complexes, and fleet maintenance operations. Most people do not really understand the real meaning of productivity (which we will review in terms of craft productivity in Chap. 13) much less lean maintenance. And then add on the negative connotation that just the term *lean maintenance* brings to the mind of a top leader who has no basic knowledge of the science and technology of maintenance. It makes it very hard to listen to the maintenance messenger when they report "we can not meet our core requirements for maintenance." However, we will define briefly the basics of *total preventive maintenance* (TPM) and key elements that originated within the United States.

This book is absolutely not about the current buzzword style of benchmarking where one does research on companies with superior performance, designs previsit questionnaires, and makes numerous plant visits to analyze/analyze in order to find where your own gaps might be in terms of your current performance. For maintenance operations, there already exists a wealth of knowledge as to currently accepted maintenance best practices. There are numerous books and countless seminars on the major best practice areas such as planning and scheduling, preventive and predictive maintenance, maintenance engineering, reliability centered maintenance, maintenance storeroom and MRO materials management, and CMMS applications. For the most part I really feel that the true maintenance leader knows what they are. There will be a few new terms/buzzwords that I have coined for some new methodologies as I mentioned in the Preface.

But even these few buzzwords are the accumulation of 36 years including direct responsibilities for manufacturing operations management (Crescent-Xcelite and Channel Master), fleet maintenance management (U.S. Army ranging from detachment, company, separate company up to battalion levels), facilities management director for a major North Carolina state agency, plant maintenance as a plant manager, and last but not least as in-house consultant/industrial engineer providing about every type of maintenance including highway road maintenance and golf course maintenance. If more buzzwords are needed we have a special pocket size buzzword generator for you to also use for 1331 more for special occasions shown in Fig. 1.

Maintenance benchmarking does not take weeks and weeks of analysis paralysis with a travel agent for industrial tourism for visits and more data from other organizations or "benchmark partners." They may

The Maintenance Excellence Institute—Buzzword Generator 4.0: Pocket Card					
0	Profit	0	Centered	0	Maintenance
1	Customer	1	Productive	1	Leadership
2	Continuous	2	Reliability	2	Improvement
3	Team-based	3	Accounting	3	Management
4	Total	4	Financial	4	Success
5	Virtual	5	Throughput	5	Technology
6	Physical	6	Asset	6	Logistics
7	Reliable	7	Continuous	7	Planning
8	Supply	8	Operations	8	Quality
9	Demand	9	Capacity	9	Service
10	Service	10	Based	10	Productivity
Pick 3 numbers each from 0 to 10 for 1,331 Buzzwords					

Figure 1 Buzzword generator 4.0 pocket card.

not be even close to yours when comparing "benchmark statistics" to begin with. Put the analysis time where it will do the most good: into doing, trying new things, and implementation. Maintenance benchmarking recognizes that no organization whether public or private is exceptional at every best practice that we all read about. Benchmarking must be an ongoing process and is not a one-shot event. Implementing maintenance best practices is not a one-time "kaizen event," it is a profit- and customer-centered attitude and philosophy to ensure you have the basics firmly in place and continuously more up the scale with more advance technologies that apply.

Benchmarking requires that you constantly search for better solutions never becoming satisfied with the status quo. If you continuously search for and implement maintenance best practices from the best firms around the world, then you will become the leader of an exceptional maintenance operation.

Every process and task of your total organization can be benchmarked, from production, to marketing, to purchasing, to information technology management, to customer service, and certainly maintenance. We will cover the type of benchmarking termed as "best practices benchmarking" or "process benchmarking" to help you improve the total maintenance process.

Personal Lessons Learned and Case Studies

Throughout this book, I will share a number of personal lessons learned as well as many case study examples, some as chapters and some to illustrate key point. We can definitely "make improvement by learning from others" by sharing of ideas through professional contacts, current literature, seminars, and trade shows. As I have seen in my travels around the world we can find good ideas and learn from our counterparts about the business of maintenance. What is does take is making sure you have the basic foundation right and then having the courage to continuously improve, that is, kaizen and industrial engineering. Maintenance benchmarking at the strategic or global level is about "determining where you are now" and "where you want to go" to improve your maintenance business processes. It is also internal benchmarks that get down to "your shop level" for measurable results of implementation.

You Must Think Global but Start Local

A benchmark is a point of reference for a measurement. This book will allow you to begin right now to define your own benchmark statistics and current baselines with these four very important benchmark measurement tools:

1. *The Scoreboard for Maintenance Excellence* for maximizing overall best practices
2. *The CMMS benchmarking system* for optimizing your information technology investment
3. *The Maintenance Excellence Index* for validating bottom line results
4. *ACE team benchmarking process* for developing reliable planning times

Each of these benchmark tools when used can become your point of reference for measurement of mission-essential maintenance from four distinct points of reference. In surveying, benchmarks are landmarks or objects of a reliable and a precisely-known location that are unlikely to change (survey monuments) and set permanently into the earth. This book is a hands-on guide for improving your maintenance business processes. It must start with a personal philosophy of continuous improvement. Benchmarking must also start with a firm commitment to implement needed best practices after you have "determined where you are."

The four key benchmarking tools presented in this book can help you start right now and enable you to turn improvement opportunities into visible profit- and customer-centered results. If you are the maintenance leader of a private or public sector facilities maintenance operations it will be service-centered results. We will provide the details for developing and applying these four proven benchmarking tools that have been part of a proven approach from *The Maintenance Excellence Institute* (TMEI) for many years. Profit optimization can be enhanced and well underway with these essential benchmarking tools in place to benchmark "where you are" with today's best practices. And these benchmarking tools will allow you to measure the results of "where you want to go" with your plan of action for improvement.

Often our toughest competitor is our own organization that does not understand the true value of maintenance, the high cost of deferred maintenance, or the dangers of gambling with a run-to-failure strategy of continuous reactive maintenance. It is important to realize that there are two basic forms of benchmarking, internal and external, and to understand why both are essential during the journey to maintenance excellence.

Global Maintenance Best Practices Serve as External Benchmarks

There are many maintenance best practices that can serve as the global external benchmarks. External benchmarking is about gaining the knowledge and understanding of best practices and then applying them within the maintenance operation to help pursue and gain a manufacturing or service edge. Today's best maintenance practices are in areas such as:

- Preventive/predictive maintenance
- Continuous reliability improvement
- Reliability centered maintenance
- Maintenance parts/materials control
- Maintenance storeroom operations
- Work order and work control
- Maintenance planning/scheduling
- Maintenance budget and cost control
- Operator-based maintenance
- Team-based continuous improvement
- Improving and measuring equipment effectiveness and reliability
- Craft skills development
- Maintenance performance measurement
- Computerized maintenance management systems
- Continuous maintenance improvement

Effective benchmarking should start locally with a total evaluation of current maintenance practices and procedures and then lead to the development of plan of actions that lead to successful implementation. Plan of actions may be at various levels: strategic, tactical, or operational levels and very, very, very short term; "Do it now!" ones. All provide the road map of varying map scale for applying maintenance best practices through a long-term process of continuous maintenance improvement. Benchmarking at its best is when maintenance effectively measures its level of services and practices, and develops its own unique benchmarking criteria with high standards for maintenance excellence.

External benchmarking within maintenance allows for taking the global view of identifying best practices and determining how they can be transferred and applied successfully within your own unique maintenance operation. External benchmarking provides for developing broad-based comparisons with other maintenance operations in terms of best practices, standard operating procedures, and industry-wide practice. As we will see later in Chap. 4, "The Scoreboard for Maintenance Excellence" is today's most comprehensive guide for external benchmarking. It covers 27 best practice categories and 300 evaluation criteria.

We must "think global but start local" with maintenance benchmarking: Internal benchmarking starts locally within the maintenance operation at the shop floor. It focuses on measuring the successful execution of best practices such as CMMS, preventive and predictive maintenance, maintenance planning and scheduling, and effective maintenance storeroom operations.

Internal benchmarking is about developing specific internal metrics or performance indicators. It is about determining progress from an internal baseline or starting point and measuring the progress toward a performance goal specific to your type of maintenance operation. For example, an internal benchmark could be the current level of maintenance related downtime hours for a critical asset or the maintenance cost per unit of output such as cost per ton, cost per carton, or cost per equivalent standard hour if a standard cost system is in place. As we will see in Chap. 27, The Maintenance Excellence Index will provide an important means to measure the results of our maintenance improvement process.

Personally, I feel it is much more important to begin measuring where you are now with key performance indicators to define your current baseline, than to worry about the published performance statistics of other maintenance organizations. Industry standards can be useful, but your progress from your current baseline toward preestablished performance goals is the most important issue.

Internal benchmarking is also about how well we are using our existing CMMS to enhance maintenance best practices. The CMMS benchmarking system in Chap. 11 is an important internal benchmarking tool to help gain maximum value from an existing CMMS or from the implementation of a new system for computerized maintenance management. The following provides a brief look at each of the six key parts of this book and the underlying objectives for each one as a cornerstone for maintenance excellence and profit optimization.

Part 1: Maximizing Maintenance Best Practices with a Profit- and Customer-Centered Strategy

Here is your guide for a profit- and customer-centered strategy and the *key requirements for profit- and customer-centered maintenance* as the foundation for your profit- and customer-centered maintenance strategy. Part 1 brings all to the realization that maintenance within manufacturing operations has a major impact on profit, throughput, and quality in many ways. That impact can easily be negative if maintenance is cut indiscriminately and true maintenance requirement are not being met. But with focused investments and continuous reliability improvement, the impact of maintenance can have a very positive impact on the bottom line and profit optimization. It also strives to help all leaders understand the importance of managing their maintenance and physical asset management operations as a profit-center and that maintenance truly is forever!

Part 2: Determining Where You Are? As a Profit-Centered Maintenance Operation: The Scoreboard for Maintenance Excellence

Here we will get down to the detailed level of "determining where you are?" with actually applying today's best practices for maintenance. It introduces "The Scoreboard for Maintenance Excellence" as today's most complete benchmarking tool to assess your current operation. It addresses 27 major evaluation categories with 300 very specific evaluation criteria. "The Scoreboard for Maintenance Excellence" provides the first of four benchmarking tools introduced in this book and is the major one that benchmarks where you are with applying external best practices that other successful maintenance operations recognize and use. This part defines how you can also develop your own unique "Scoreboard for Maintenance Excellence," outlines steps for conducting a self-assessment and how to use it to continuously evaluate progress on your journey toward maintenance excellence. This part also defines how your unique "Scoreboard for Maintenance Excellence" can provide a strategy for a multiple site operation. Developed originally as "The Scoreboard for Excellence" in 1981, this external benchmarking process has evolved from over 25 years of successful application to many different types of public and private organizations

Part 3: Developing Your CMMS/EAM as a True Maintenance Business Management System

In 1999, I edited and was primary author, for working/living book entitled, *Guide to Computerized Maintenance Management Systems*, published and sent to subscribers by Alexander Communications Group as a chapter of the month on CMMSCMM. This was later acquired by *Scientific American* and later died a painless death. Part 3 brings to you very important strategies for evaluating, selecting, implementing, using, and improving return on this important information technology investment for maintenance. Fully integrated information technology systems to manage the business of maintenance are now essential business management tools almost like the Crescent wrench. Part 3 provides five extensive chapters to support making the right choices on CMMS and how to improve its value to your operation.

Part 3 introduces the second benchmarking tool and the improvement process for getting maximum use from current information technology. *The CMMS benchmarking system* is introduced as a means to evaluate the effective use of your current CMMS, to define functional gaps, and to define how to enhance current use. Results from using *the CMMS benchmarking system* will also help to develop and justify a replacement strategy if that is needed. *The CMMS benchmarking system* is easily adaptable and should be specifically tailored to your existing CMMS and to its intended application. This tool is an internal benchmarking guide that is a model process for benchmarking effective utilization of CMMS for maximum value.

Part 4: The Profit- and Customer-Centered Maintenance Operation

Part 4 with 11 chapters provides a recommended strategy and profit- and customer-centered philosophy toward maintenance operations that has been used successfully by TMEI. It applies to both public and private sector maintenance operations and to major contract maintenance providers where in-house maintenance has already been privatized or contracted out.

Part 4 starts with how to define, measure and improve craft labor productivity in Chap. 13 to support productivity and profit of the total operation. In Part 4 we will review direct and indirect savings from increased craft productivity as gained value. The *overall craft effectiveness* (OCE) factor for craft productivity is introduced in Chap. 13 as an important companion metric to OEE (asset productivity) and production operator productivity as three of the key elements contributing to total operations productivity. Part 4 also details the step-by-step method for using the ACE team benchmarking process for developing reliable planning times for your planning and scheduling function, the third benchmarking tool. A complete application guide is provided for using this new method for maintenance work measurement in Chap. 15. The remainder of Part 4 comprises chapters summarizing well-recognized functional best practices, with some of MEI proven strategies and many helpful case studies/lessons learned from the real world of maintenance.

Part 5: Validating Results with Your Maintenance Excellence Index

The goal is profit optimization and exceeding expectations for customer service. It not a report with benchmark evaluation details from "The Scoreboard for Maintenance Excellence" neatly bound in a report. The goal is successful implementation of prioritized improvement opportunities from the benchmarking self-assessment or assessment recommendations from a third party service such as TMEI. It is to help improve all internal maintenance resources do a better job for production operations or the tenant/customer within a pure facilities maintenance operation, the patients, and staff in a healthcare facility or visitors to Disney World or McDonalds. Part 5 links closely with Part 4 to introduce the fourth benchmarking tool, the *Maintenance Excellence Index* (MEI). This part covers the process of defining and gaining consensus on very specific key performance indicators related to the total manufacturing and maintenance operation. This has application to absolutely all maintenance operations regardless of the size or scope. This section helps validate maintenance contribution to profit and the return on maintenance investments. It covers a listing of possible metrics, a recommended set of internal benchmarks or metrics for today's leader with the purpose for each, where they traditionally can be found in the CMMS (or financial system), how to calculate each one, and how to determine your current baseline. Each element for developing and calculating your own MEI is covered. Most importantly, this section recommends an attainable performance goal and how your own uniquely developed Maintenance Excellence Index will validate results and *return on investment* (ROI) for maintenance operations.

Part 6: The Journey Toward Maintenance Excellence

Our summary chapters are to support growth of commitment, courage and cooperation needed within every maintenance operation to continue or start a journey toward your definition of maintenance excellence. It is also about collaboration needed from others: top leaders, maintenance leaders, crafts leaders, and operations leaders. Part 6 covers four key topics:

1. Nontraditional ROI for improving maintenance ROI
2. Developing and Implementing a profit-centered action plan
3. Maintenance excellence begins with *pride* in maintenance
4. The journey toward *maintenance excellence*? Where do you go now?

The new millennium view toward maintenance and physical asset management must see maintenance helping to maximize profits and customer service. The strategy defined in this book has been proven for application to a multitude of different types of maintenance and physical asset management operations within both the public and private sectors. The approach is simple but powerful in terms of achieving results and validating return on investment. Organizations that clearly understand that "maintenance is forever" will find the keys to balancing all maintenance resources toward optimum total operations success. These are the organizations that will succeed in the remainder of twenty-first century.

Part

1

Maximizing Maintenance Best Practices with a Profit-Centered Approach

Part

1

Maximizing Maintenance Shop Practices with a Profit-Centered Approach

Chapter 1

A Profit- and Customer-Centered Maintenance Strategy

Maintenance

Does maintenance and physical asset management operate as a profitable in-house business within your organization? This might seem like a strange question. But, if your current maintenance practices, leadership philosophies, and organizational culture do not allow you to manage maintenance like a profitable internal business, you could be in trouble or heading toward serious trouble. An organization that still views maintenance as a "cost center" and continuously "squeezes blood from the maintenance turnip" is on the road to major problems with physical asset management. This attitude has resulted in catastrophic failures in airlines, refineries, ships at sea, and many other operations. Maintenance requirements exist everywhere and the need for effective maintenance is continuous because maintenance is forever!

The high cost of gambling with maintenance

There can be a very high cost of gambling with maintenance and most operations lose when they gamble with their maintenance chips. There is an extremely high cost to bad maintenance on the shop floor, in combat, in the skies, and everywhere the maintenance process fails in the proper care of physical assets. If you have not invested wisely in maintenance to the point that you can manage it as a profitable internal business, then you also could be a potential takeover target for contract maintenance. Many operations have lost heavily due to indiscriminate cuts to a core requirement—the resources necessary for effective physical asset management and maintenance. Quantum leaps backwards will occur for the organizations that fail to view maintenance as a core business requirement. Indiscriminate downsizing and "dumbsizing" of maintenance is finally being recognized as a failed business practice. Where are your maintenance chips being stacked? Do not view maintenance as a cost center. View it with a profit- and customer-centered mentality and with an attitude that promotes initiative, customer service, profit optimization, and ownership.

Profit- and customer-centered contract maintenance

You might say that profit- and customer-centered maintenance is not possible for an in-house maintenance operation. But a profit-centered strategy does exist in the thousands of successful businesses that provide contract maintenance services everywhere we look. Maybe we invest heavily in profit-centered contract maintenance providers who are truly in the maintenance business for profit and for serve their customers. They will multiply even further if organizations continue to give up on in-house maintenance operations. Third-party maintenance will continue to be a common practice in organizations that have continually gambled with maintenance costs and have lost. For some of the maintenance operations that I have seen, as a result of hundreds of maintenance benchmarking assessments, the best answer for survival is a partnership with a contract-maintenance provider. It is often a hard choice, especially when it is tempered with all the relentless pressure from unions. But, for some operations, quality service from qualified maintenance service providers is unfortunately the best choice available. But we should not give up on in-house maintenance when contract maintenance could be just as bad, operating within our current organizational culture.

This is not a scare tactic, which advocates third-party maintenance in total for an organization. It positively and unequivocally does not support the dumbsizing of in-house maintenance to provide lean maintenance, which fails to meet the actual maintenance requirements. Dumbsizing of maintenance and reengineering without true engineering has failed. Third-party maintenance in specialty areas, or areas where current maintenance skills or competencies are lacking will be needed and be a growing practice. It provides real profit to the maintenance provider and savings to the customer.

More third-party maintenance of all types is occurring and will continue to occur in operations where maintenance is not treated as an internal business opportunity. It will obviously occur where the maintenance operations have deteriorated to the point that a third-party service is more effective and less costly than in-house maintenance staff.

Where Is Your Chief Maintenance Officer?

We now have more "C-positions" than we know the terms for: CEO, COO, CFO, CIO, CPO, and CRO. But, there is an important one that is missing, that is, the *chief maintenance officer* (CMO). The evolution of the CMO position must occur to provide leadership to physical asset management within large manufacturing operations. I sincerely believe that the real maintenance leaders will begin to emerge as CMOs in the business world. This new staff addition of a CMO is desperately needed and the smart organizations will have someone near the top who is officially designated to ensure that physical assets are properly cared for. I believe that the CMO will join the ranks of the CEO, COO, CFO, and CIO in large multisite manufacturing operations to manage physical assets. The real CMOs will manage, and most importantly lead maintenance forward as a "profit center." A good CMO with "profitability" will be in place to lead maintenance forward to profitability. A good CMO will help the CFO take the "right" fork in the road as it relates to physical asset management and profitability. Regardless of the size of the operation, every manufacturing operation needs a CMO. For smaller operations, it might be a CMO equivalent, a maintenance manager, or a maintenance supervisor who can really manage maintenance as a business and as an internal profit center.

Profit ability and leadership

Leadership ability is an important personal attribute, but being in the "maintenance for profit business" also requires an important new type of ability—"profitability." To lead maintenance forward we must learn from the leaders within the third-party maintenance business. There are many good ones out there, but one that I personally know about is Viox Services Inc.—a Cincinnati-based organization. Viox Services is a facility maintenance and management company that customizes and bundles services depending on the client's needs. Viox, a third-generation family business, focuses on increasing the client's operational efficiencies, reducing the client's costs, and achieving continuous improvement. They have the ability to focus on customer service, to develop skilled craft people to do the work, and to produce measurable profits to both Viox and their customers. This is an example of true profitability. Having a good CMO adds the missing link to achieving total operations success. Maintenance has rapidly evolved into an internal business opportunity and can now be stated as a true profit center. The change from a "run to failure" strategy into a proactive, planned process for asset management requires a CMO with demonstrated technical and personal leadership. Plan on becoming the CMO within your operation regardless of your organization's size and your current level in your organization.

A Fundamental Lesson Learned

I learned one very important lesson about maintenance and CMOs during my career spanning the past 35 years. CMOs do exist for sure in the U.S. Army Corps of Engineers and are quite effective. In this book I will tell you about five of the finest CMOs you could find and one extraordinary *chief executive officer* (CEO). I found out very quickly about the importance of a CMO during my tour of duty in Vietnam in 1970–71 with direct responsibility for 72 Ford and General Motors Corporation (GMC)tandem, 20-ton dump trucks of 1969 and 1970 vintage. Twenty-four were driven by Vietnamese drivers and forty-eight by our men. Combat engineering and construction assets require very extensive and effective maintenance. These "Yellowbirds," as they were called, were big old yellow commercial trucks in a sea of army green road construction equipment. So as a 20-something First Lieutenant, I talked to and listened very intently to our CMO at breakfast, noon, and night. My company commander talked to our CMO and me at breakfast, noon, and night about maintenance and our support to the road-building mission of the 20th Engineer Brigade—our highest headquarters. My commander and I listened, took notes, and then took action on our CMO's technical advice for maintenance of our engineer equipment—the "Yellowbirds." We did what was required for maintenance excellence on engineer construction equipment and the mission of building South Vietnam's primary road system working six and a half days per week. That half day was Sunday morning before noon church services, and our CMO was there looking at each and every truck just as he did on the other six days, to ensure maintenance excellence of our fleet.

The CMO in our case was our chief warrant officer, CW3 Berry, a career maintenance professional and soldier in the U.S. Army, Corps of Engineers. We called him by either of his two first names: either *mister* or *chief*. When we called for him we did not have to shout because he was always close making sure mission-essential maintenance stayed as our top priority for survival and for building roads. Mister Berry could unleash a reign of terror in a lieutenant or platoon sergeant's mind for bad maintenance. He was a true CMO and highly respected by all in the unit. He was one of those top people really interested in developing excellence in maintenance, and he did spread the attitude of *pride* NO PRIDE in maintenance and *pride* in ownership to all officers, *noncommissioned officers* (NCOs), and most importantly to each Yellowbird driver. We completed an important engineering mission and survived building roads in Vietnam due to having one of the U.S. Army's best maintenance professionals as our CMO. Many other American soldiers in many other wars have lived through combat due to good maintenance, while some have died due to bad maintenance and not by the hand of the enemy.

Later on, during that same tour I had the job as Company Commander of a direct support maintenance company in our battalion that provided maintenance to all other equipment. We maintained dozers, scrapers, graders, 5-ton military dump trucks, rock crushing plants, asphalt plants, and the paving equipment in our engineer battalion—everything except the Yellowbirds that Chief Berry and crew maintained. Here again I was blessed with not one but two CMOs, Chief Donaldson and Chief Deardorff, with the same passion for maintenance excellence as CMO Berry. Maintenance responsibilities came very early for me, but those three CMOs really made it happen along with a 6 ft 8 in. First Sergeant of about 50 years old. Some sight—a 6 ft 4 in. First Lieutenant saluting a 6 ft 8 in. giant of a man.

No matter how bad something is, it can always be used as an illustrative bad example

We all learn lessons either the hard way or the easy way. Therefore, bad examples are not wasted. I think we can learn important lessons about maintenance the easy way by having an effective CMO. I think that a new breed of corporate officers will evolve. An effective CMO will be a firm requirement for organizational success. CMOs will take their place near the top with the CEOs, COOs, CFOs, CIOs, and the corporate quality gurus. I think we will start to listen closely to the maintenance messenger—our CMOs. We will not and we should not shoot the maintenance messenger. Many have been seriously wounded when they have tried to state the "true state of maintenance" within an organization. Manufacturing plant managers, CFOs, COOs, and VPs of manufacturing operations—who do not understand the true value of maintenance—will continue to be the bad examples. The CMOs of successful organizations will have an important and unprecedented role in the success of their total operation.

The successful manufacturing company will have true maintenance leaders, not managers of the status quo. The true maintenance leaders and CMOs of these successful companies will know the contribution to profit that their maintenance operations provide. They will view maintenance improvements as value-adding investments that provide a measurable return on investment. They will measure the results of the maintenance process whether it is internal or outsourced maintenance. They will validate the investments they have made just as they try to validate other return on investments. The CMO will be the maintenance messenger!

The true CMO will also be the maintenance leader who understands how to operate the total maintenance process as an internal business within a business. They will be able to turn in-house maintenance into a profit center comparable to contract service providers. All true corporate leaders must strive to understand current and future trends, take action, and proactively plan for the future of maintenance within their total operation.

There must be a maintenance champion. The real maintenance leader readily accepts the role as champion for maintenance excellence. Likewise, integrity of purpose and the integrity of the maintenance champions must set an example for others in the organization to follow. Ralph Waldo Emerson said it very well when he remarked,"What you are thunders so loudly, I cannot hear a word you say to the contrary." Leadership by example and "walking your talk" is essential for the maintenance champion and all company leaders.

The maintenance champion as the CMO understands and can communicate the true cost of deferred maintenance as well as the cost of inadequate preventive/predictive maintenance. The CMO is prepared to provide proactive leadership and support to the company's compliance to regulatory issues. The real CMO must be prepared to take bad news about the true state of maintenance to the company leaders with courage, confidence, and most importantly with credibility.

The effective CMO utilizes a true teaming process to bring maintenance, operations, and operators together to detect, solve, and prevent maintenance problems. The effective CMO will take the lead for implementing best practices such as an effective *computerized maintenance management system* (CMMS). They will work closely with information services staff and CMMS vendors over the long term to make it work to enhance the business of maintenance.

The CMO and PRIDE in Ownership

The true CMO also encourages pride in ownership with equipment operators and maintenance staff as they do their part to fix and prevent maintenance problems through a cooperative team effort of operation-based maintenance. The CMO has the integrity and inspires individual integrity to the point that all employees do their jobs as if they too owned the company. Individual integrity includes pride in one's work no matter what the task. An effective CMO can help your operation go beyond the bottom line to ensure long-term total operations success of the company and the maintenance process. What if the following headline and news clip appeared in a future *Wall Street Journal* or *Harvard Business Review* (HBR)?

"The Evolution of the Corporate Chief Maintenance Officer"

"The time has come for CMOs to be a force in today's global economy and sit with the C-position team at the top. CEOs, COOs, CFOs and CIOs come and go with the musical chair game played by the board of directors. Chief Maintenance Officers are sitting at the right (or to the left) of the chief at annual stockholders' meetings. By the Year 2030, it is projected that CMOs will be leading 18 percent of the manufacturing companies within the Fortune 500. Small operations even have a CMO equivalent. What has sustained this evolution of the CMO to the center of attention? Is it *asset management* or *physical asset management*? Or is it the fact that *Maintenance is Forever*? The latest Fox News polls provide the answers to these questions which are..."

Have You Ever Had Future Shock?

HBR would be shocked if they had not gotten the inside scoop on this Orson Welles type headline. Is this headline really a hoax, an addendum to Alvin Tofler's 1970 "Future Shock" or another NO ZAP the book zap by Zig Siglar ? This hypothetical news clip is simply a description of what should be present in all major/minor corporations. But what do we see when we do word searches for the HBR archive of articles, not free ones, but all those for sale? There is nothing on real maintenance. You will find much on asset management of dollars however. So, it is very interesting to note that maintenance is still available as a hot new business topic for the HBR. Would like HBR reference to stay!

There was not even one single reference to the word *maintenance* in article titles from July 1995 to October 1999 when I first searched their (HBR) archives. I recently checked in August and again no luck. Back in 1991 there was (HBR) an article entitled "Northwest Airlines Financing Maintenance Facility in Minnesota." But in March to April 1968 there was an article by John J. Wilkinson on "How to Management Maintenance." I still have the reprint version I bought and keep it in my stack of 1960s maintenance references. HBR does have a good article (for sale) on "How High Is Your Return on Management" ... but nothing on your return on maintenance investment—the ROMI of all the physical assets that produce all goods and services in the world for asset management of profit and dollars. That is strange to me, and of course there are no references at all (now) to a CMO.

Take action on this question

As a maintenance leader, you must take action on this key question: "If I owned my maintenance operation what would I do different to make a profit? Another question could be "How high is your return on maintenance management." If you begin to think like the CMO, you will get others to think this way too. You will get more people thinking "profit- and customer-centered." As each crafts person feels they own part of the business, you will experience a groundswell of profitability. One key part of this answer will be to get maximum value received from your information technology tools, your CMMS, as we will see in Part III. For every benchmark assessment I have done over many years CMMS is not being fully used in 95.6 percent of my personal observations.

Effective In-House Maintenance Plus High-Quality Maintenance Contractors

Profit- and customer-centered in-house maintenance in combination with the wise use of high-quality contract maintenance services will be the key to the final evolution that occurs. There will be revolution within organizations that do not fully recognize maintenance as a core business requirement and establish core competencies for it. The bill will come due for those operations that have subscribed to the "pay me later syndrome" for deferred maintenance. It will be revolution within those operations that have gambled with maintenance and have lost with no time left before profit- and customer-centered contract maintenance provides the best financial option for a real solution. Back in September 1999 I was asked to submit responses to eight really good questions from the editor of *IMPO Magazine* related to *maintenance at the millennium.* Take a look at my responses in App. K because I think that traditional thinking about maintenance has changed and will change dramatically in the future. Whereas maintenance was once considered to be a necessary evil, it is now being viewed as a key contributor to profit in a manufacturing- or service-providing operation. My goal for this book is to build your case clearly within your organization.

Where is the profit in maintenance really?

You might ask yourself after reading this far: where is the profit in maintenance *really* for an in-house operation trying to keep its head above water? We will cover in detail many areas later where there are profit opportunities available—in Chaps. 12, 14, and 24 very specifically with case studies. But for one example: what if the net profit ratio of an operation is 4 percent? What does a 4 percent net profit ratio mean in terms of the amount of equivalent sales needed to generate profits? A net profit ratio of 4 percent requires $25 of equivalent sales for each $1 of net profit generated.

Therefore, when we view maintenance in these terms we can readily see that a small savings in maintenance can mean a great deal to the bottom line, *equivalent to pure sales revenue.* Maintenance as a profit center is illustrated in Table 1.1 showing that only a $40,000

TABLE 1.1 Maintenance as a Profit Center

Maintenance savings that impact net profit ($)	Equivalent sales required for generating net profit ($)
1	25
1000	25,000
10,000	250,000
20,000	500,000
30,000	750,000
40,000	1,000,000
80,000	2,000,000
120,000	3,000,000
200,000	5,000,000

savings is required to translate into the equivalent of $1,000,000 in sales revenue. As we will discuss in later chapters, there are many more areas such as the value of increased asset uptime, increased net capacity and just-in-time throughput, increased product quality, and increased customer service that contribute to the bottom line and subsequently to profit.

Investments in maintenance that successfully implement the practices summarized in this book (and detailed in many books focused on functional maintenance processes) can achieve results that are comparable to the following:

- 15 to 25 percent increase in critical capacity constraining equipment uptime
- 20 to 30 percent increase in maintenance productivity of the craft workforce
- 25 to 30 percent increase in planned maintenance work
- 10 to 25 percent reduction in emergency repairs
- 20 to 30 percent reduction in excess and obsolete inventory
- 10 to 20 percent reduction in maintenance repair costs

Other improvements can include:

- Improved product quality
- Improved utilization of equipment operators
- Improved equipment productivity (OEE) and production throughput capacity
- Improved equipment life lower *life cycle cost* (LCC)
- Improved productivity of the total operation and pure profit

The foregoing results can be achieved by maintenance organizations that have committed to continuous maintenance and reliability improvement (or to what terms you use such as maintenance benchmarking and best practices implementation). Your organization must realize that there are no easy answers and no "quick fixes." Organizations that have invested in maintenance over the long term, have realized a tangible return on that investment. Consider what would happen if your numbers were used in the following very basic examples:

- *Maintenance craft productivity increase of 20 percent.* Net improvement in craft productivity of 20 percent (craft utilization, craft performance, and craft service quality) would be 20 × 40 Craftsmen × $35,000/year = $280,000/year.
- *Increased equipment uptime of 25 percent.* Downtime reduced 25 percent from 8 to 6 percent means that the value of increased uptime would be 0.25 × baseline $800,000 value of downtime = $200,000/year.
- *Inventory reduction in maintenance storeroom of 25 percent.* Reduction of 25 percent from $1,000,000 to $800,000 would be $200,000 × 0.30 inventory carrying costs = $60,000/year.
- *Improved pricing from suppliers of 1 percent.* Direct price saving of 1 percent (not high cost of low-bid buying) would be 0.01 × $1,000,000 purchase volume/year = $10,000/year.
- *Reduction in net maintenance repair costs of 10 percent.* Reduction of 10 percent would be 0.10 × $750,000 annual repair cost = $75,000/year, if all required maintenance requirement were being met.
- *Improved product quality of 1 percent.* Reduction of 1 percent in equipment-related scrap, rework returns, waste, and better yields would be 0.01 × $2,000, 000 value of production at standard cost = $20,000/year.
- *Improved equipment life of one-half year.* One-half year longer productive asset life would be 0.5 years × $10-million capital investment × 0.10 expected capital ROI = $500,000 minus additional $200,000 additional maintenance cost = $300,000/year.

These examples all contribute to the bottom line either directly or indirectly. They illustrate briefly that tangible return on investment (ROI) can be significant, depending on the size of the maintenance operation and the type of organization being supported. Maintenance leaders must be able to gain support for continuous maintenance improvement by developing valid economic justifications. Take the time to evaluate the potential savings and benefits that are possible within your own organization. Gain valuable support and develop a partnership for profit with operations. Include all other key departments who will receive benefits from improved maintenance. The application of today's best maintenance practices will provide the opportunity for maintenance to contribute directly to the bottom line. However, the pursuit of maintenance excellence requires leadership.

Core Requirement versus Core Competencies for Maintenance

The *core requirement* for good maintenance never ever goes away because "Maintenance Is Forever!" There will always be a need to maintain. Maintenance of our physical bodies, minds, souls, cars, computers, and all physical assets providing products or services will always be required. Maintenance, gravity, extinction, and change are truly forever. Yet some organizations today have neglected maintaining their core competencies in maintenance to the point that they have lost complete control. The core requirement for good maintenance will always remain (forever), but the *core competency* to do good maintenance can be missing. In some cases we know that the best and often the only solution is value-added outsourcing. Maintenance is a core requirement for profitable survival and total operations success. If the internal core competency for maintenance is not present it must be regained with internal leadership of the CMO. Neglect of the past can either be overcome internally or externally. The core requirement for maintenance can be reduced, but it can never go away.

Neglect of Past Can Be Overcome, Start Your New Millennium Now

The neglect of the past will be overcome externally with a growing number of profit- and customer-centered maintenance providers that clearly understand providing value-added maintenance service at a profit. It will also be overcome internally by a growing number of internal CMOs who can lead maintenance forward to profitability as if they owned the internal maintenance business. The new millennium strategy for maintenance excellence will be an effective CMO leading the physical asset management process with attitudes and actions toward profit- and customer-centered maintenance.

Chapter

2

Key Requirements for Profit- and Customer-Centered Maintenance

As a maintenance leader, how would you answer the key question, if I owned my maintenance operation as a business, what would I do differently? This question is one that we will help you answer throughout this book. It is important that today's maintenance leader operate maintenance with a strategy and an attitude that maintenance is indeed an internal business. To do this there are a number of fundamental principles and practices that provide the foundation upon which to develop improvements. It is important to understand the key requirements for profit- and customer-centered maintenance because they in turn provide the measurable benefits for maintenance and the total operation.

Fundamental principles and best practices become the cornerstone for achieving, maintaining, and continually improving the maintenance process. Organizations that are establishing best practices for a profit- and customer-centered maintenance strategy will be actively pursuing the following key requirements for profit- and customer-centered maintenance. Parts of the following appeared as part of a Tompkins Press book: *The Future Capable Company*, by Jim Tompkins.* This publication at www.tompkinsinc.com has important points that every maintenance leader should read to understand the customer (the operations) side of the total operation better. Top leaders should read it too for better understanding of total operations and how the maintenance piece is a mission-essential element of total operations success, profit, customer service, and much more.

---

*Paraphrased with permission from Tompkins Press.

1. *View maintenance as a priority and as an internal business opportunity.* The process of performing maintenance and managing physical assets must be recognized as a top priority. It must also be viewed as an internal business and not as a necessary evil. It will be viewed as an area that contributes directly to the bottom line when a profit- and customer-centered strategy and continuous maintenance improvement are adopted. The successful maintenance leader will understand today's best practices and will have identified top priority areas for improvement based upon on a total benchmark evaluation of their maintenance operation. Investments will be made to implement best practices that provide a measurable return on investment.
2. *Develop leadership and technical understanding.* Maintenance leaders must understand the challenges of maintenance and provide effective maintenance leadership to operate maintenance as an internal business. Maintenance leadership must continually develop the skills, abilities, and attitudes to lead maintenance into the future. Maintenance leaders must completely understand the key requirements for profit- and customer-centered maintenance. Maintenance leaders must create a better understanding within the organization about the value of maintenance and develop a vision of continuous maintenance and reliability improvement shared throughout the organization.
3. *Develop pride in maintenance.* Maintenance operations will experience fundamental improvements in work ethic, attitude, values, job performance, and customer service to achieve real pride in maintenance excellence. Tangible savings and improvements will occur as a result of continuous maintenance improvement. The successful maintenance operations will experience other fundamental improvements that develop *pride* in maintenance, where *pride* is "*People Really Interested in Developing Excellence in Maintenance.*" Successful maintenance operations will have leadership that instills *pride* in maintenance starting with the craftswork. Maintenance leaders with a clear vision of maintenance excellence will create inspiration, cooperation, and commitment throughout the organization. Figure 2.1 illustrates an important point you will see throughout—a sincere personal knowledge that the foundation of maintenance excellence begins with *pride* in maintenance at all levels in the organizations: top leader, maintenance leaders, and your *most valuable people* (MVP)—craft leaders doing the work.
4. *Recognize importance of the maintenance profession.* The profession of maintenance will gain greater importance as a key profession for success within all types of organizations as the role of the *chief maintenance officer* (CMO) becomes well established. Maintenance

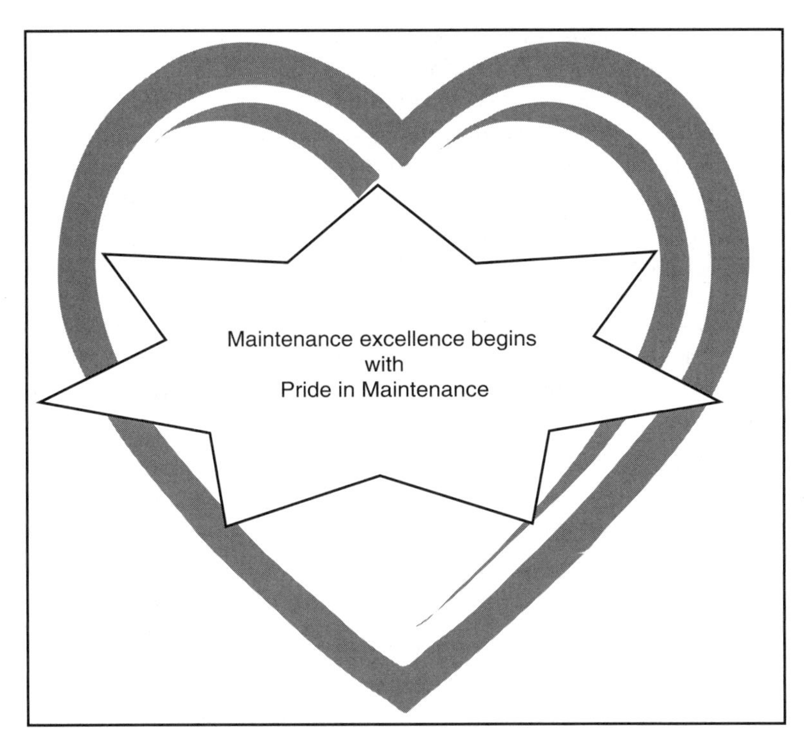

Figure 2.1 Most of all your MVPs must have *pride* in maintenance.

leaders will be recognized as critical resources that are absolutely necessary for the success of the total operation. The CMO within large multisite operations will create and promote standard best practices. The complexity and importance of the maintenance and physical asset management will continue to grow. New technologies and added responsibilities will require a higher level of technical knowledge and skills.

5. *Increase core competencies of your maintenance personnel.* A significant upgrade in the level of personnel involved with maintenance will take place to keep pace with new technologies and responsibilities. Maintenance operations will achieve a significant upgrade in the skill level of maintenance craftspeople in order to keep pace with new technology and responsibilities. Successful maintenance operations will continually upgrade the skill level of craftspeople through more effective recruiting with higher standards and more effective craft training programs. Pay increases will be more directly linked to performance and demonstrated competency levels in required craft skills.

6. *Initiate craft skills development to enhance people resources.* Successful maintenance operations will continually assess craft training needs and provide effective skills development through modern technical learning systems and competency-based development of required skills. A complete assessment of craft training needs will be accomplished to identify priority areas for skill development. Skill development will be competency based to provide demonstrated technical capabilities for each craft skill. The successful maintenance operations will develop an ongoing program for craft skill development. Continuous maintenance education based on modern technical learning systems will be viewed as a sound investment and an important part of continuous maintenance improvement as shown in Fig. 2.2.

7. *Develop adaptability and versatility.* The maintenance crafts workforce will become more versatile and adaptable by gaining value with new technical capabilities and multicraft skills. The development of more craftspeople with multiskills will occur to provide greater versatility, adaptability, and capability from the existing workforce. Multiskilled personnel will have added value and will be compensated according to well-defined policies. Craftspeople will become

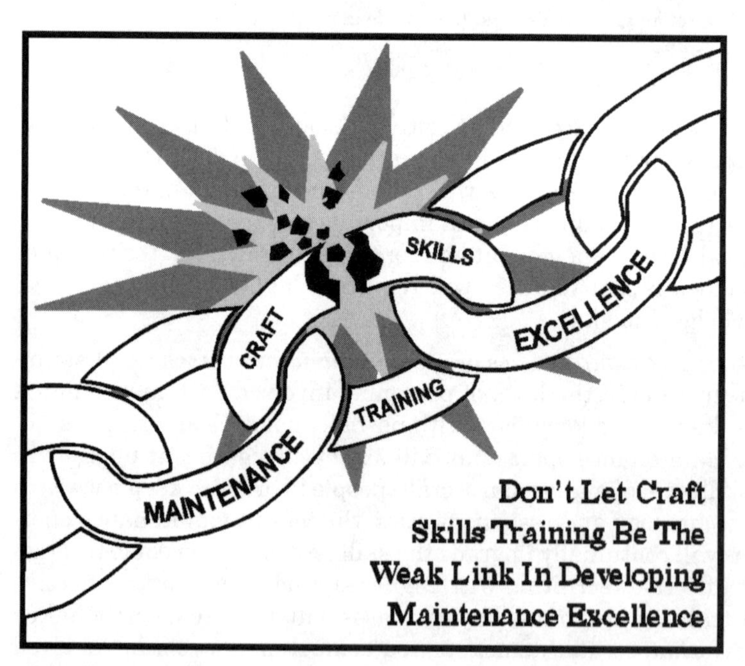

Figure 2.2 Craft training can be your weak link.

more adaptable, versatile, and valuable as a result of ongoing programs for craft skill development.

8. *Promote teamwork as a profit- and customer-centered strategy.* Maintenance will need team players who maintain a leadership-driven, team-based approach to continuous maintenance improvement. It is not about a self-managed team without leadership and focus. True team-based processes can create a revolution as we will see in some of our case studies. Maintenance leadership will accept its role as a top priority operation and will set the example as team players within the organization. The strategy for continuous maintenance improvement will be an approach that captures the knowledge, skills, and ideas of the entire maintenance workforce. Cross-functional teams with representatives from maintenance, operations, engineering, and the like, will be formally chartered to address improvements to equipment effectiveness, reliability, and maintainability.

9. *Establish effective maintenance planning, estimating and scheduling.* Maintaining customer satisfaction and the utilization of available craft time will improve through more effective planning and scheduling systems. The development of more effective planning and scheduling systems will be a top priority for a profit- and customer-centered strategy. As reductions in breakdown repairs occur through more effective preventive/predictive maintenance, the opportunity to increase planned maintenance work will result. Maintenance and operations work closely to schedule repairs at the most convenient time. Maintenance will become more customer oriented and focus on achieving greater customer satisfaction by completing scheduled repairs on time. The utilization of craft time will increase as levels of planned work increases and as the uncertainties and inefficiencies associated with breakdown repairs are reduced. Figure 2.3 illustrates key elements of an effective planning, estimating, and scheduling workflow.

10. *Maintenance and manufacturing operations: a partnership for profits.* Maintenance and manufacturing operations will become integrated and function as a supportive team through improved planning, scheduling, and cooperative team-based improvement efforts. Operations will be viewed as an important internal customer. Improved planning and scheduling of maintenance work will provide greater coordination, support, and service to manufacturing-type operations. Maintenance and manufacturing operations of all types will recognize the benefits of working together as a supportive team to reduce unplanned breakdowns, to increase equipment effectiveness and to reduce overall maintenance costs. Manufacturing will be viewed as an important internal customer. Manufacturing will gain

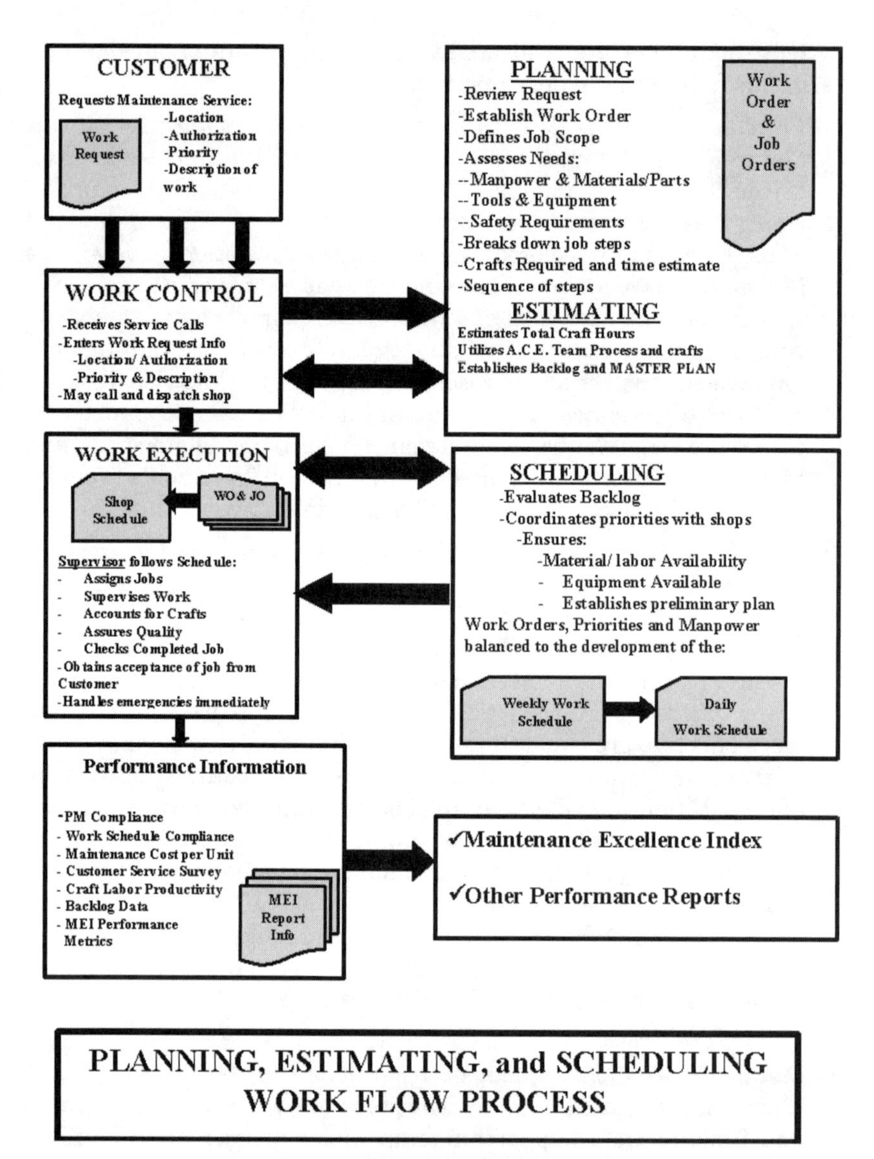

Figure 2.3 Key elements of an effective planning, estimating, and scheduling process.

greater understanding of the key requirements for profit- and customer-centered maintenance and accept its important partnership role in supporting maintenance excellence.

11. *Develop pride in ownership.* Equipment operators and maintenance will develop a partnership for maintenance service and prevention and take greater pride in ownership through operator-based maintenance.

Equipment operators will assume greater responsibility for cleaning, lubricating, inspecting, monitoring, and making minor repairs to equipment. Maintenance will provide training support to operators to achieve this transfer of responsibility and to help operators with early detection and prevention of maintenance problems. Operators will develop greater pride in ownership of their equipment with their expanded responsibilities.

12. *Improve equipment effectiveness.* A leadership-driven, team-based approach will be used by maintenance and manufacturing operations to totally evaluate and subsequently improve all factors related to equipment effectiveness. The goal is maximum availability of the asset for performing its primary manufacturing function. Continuous improvement of equipment effectiveness will address major losses due to equipment breakdowns, setup/adjustments, idling/minor stoppages, reduced speeds, process defects, and reduced yields. Maintenance operations within successful companies will use a total team effort—by operators, engineering, operations staff, and maintenance—to identify and resolve root causes of equipment problems.

13. *Maintenance and engineering: a partnership for profitable technology application.* Maintenance and engineering will work closely during systems specification, installation, start-up, and operation to provide maintenance with the technical depth required for maintaining all assets and systems. Engineering will provide technical resources and support to ensure that maintenance has the total technical capability to maintain all equipment and systems. Engineering will play a key support role with maintenance in improving the effectiveness of existing equipment. Maintenance and engineering will work closely in developing specifications for new equipment. During installation and start-up, maintenance and engineering will also work closely to ensure that operating specifications are achieved.

14. *Continuously improve reliability and maintainability.* Machines and systems will be specified, designed, retrofitted, and installed with greater reliability and ease of maintainability. Equipment design will focus on maintainability and reliability and not primarily on performance. Design for maintainability will become an accepted philosophy that fully recognizes the high cost of maintenance in the life-cycle cost of equipment. The causes for high life-cycle costs will be reduced through the application of good maintainability and reliability principles during design. Design will be focused on life-cycle reliability by identifying potential problems before they are designed into the equipment, or facility equipment and facility design will include a higher level of internal diagnostic capabilities and provide

for greater use of expert systems for troubleshooting. Maintenance will work closely with equipment designers to share information about problems with existing equipment. They will provide possible maintenance-prevention solutions during the design and/or specification process for new equipment.

15. *Manage life-cycle cost and obsolescence.* The life-cycle costs of physical assets and systems will be closely monitored, evaluated, and managed to reduce total costs. A profit- and customer-centered strategy will achieve significant reductions in total life-cycle costs through an effective design process prior to purchase and installation. During the equipment's operating life, systems will be developed to continually monitor equipment costs. Information to identify trends will be available to highlight equipment with high maintenance costs. Action can be taken to address critical high-cost areas in order to reduce future costs. A complete equipment history of repair costs will assist maintenance in making decisions on equipment replacement, equipment overhaul/retrofit, and overall equipment condition.
16. *Minimize uncertainty and eliminate root causes.* Uncertainty will be minimized through effective preventive/predictive maintenance programs and through continuous application of reliability-centered maintenance techniques and continuous monitoring systems. Effective preventive/predictive maintenance programs will be used to anticipate and predict maintenance problems in order to eliminate the uncertainty of expected breakdowns and high repair costs. Preventive/predictive maintenance tasks will be adequately planned based upon criticality of failure and will cover all major assets within the operation. Maintenance will maintain current technical knowledge and experience for applying a combination of preventive, predictive, and reliability-based technologies that are best suited for the specific asset. They will use techniques such as *continuous reliability improvement* (CRI) and *reliability-centered maintenance* (RCM). Chronic problems will be analyzed using tools such as statistical process control, graphs, process charts, and cause- and effect-analysis.
17. *Maximize use of computerized maintenance management and enterprise asset management.* Systems that support the total maintenance operation will improve the quality of maintenance and physical asset management and be integrated with the overall business system of the organization. *Computerized maintenance management systems* (CMMS) will provide greater levels of manageability to maintenance operations. CMMS will cover the total scope of the maintenance operation providing the means to improve the overall quality of maintenance management. *Enterprise asset management* (EAM) will provide a broader scope of integrated software to manage

physical assets, human resources, and parts inventory in an integrated system for maintenance management, maintenance, procurement, inventory management, human resources, work management, asset performance, and process monitoring. Vast amounts of data associated with maintenance tasks will come under computer control and be available as key information for planning, scheduling, backlog control, equipment history, parts availability, inventory control, performance measurement, downtime analysis, and so forth.

18. *Use maintenance information to manage the business of maintenance.* The maintenance information system and database will encompass the total maintenance function and provide real-time information to improve maintenance management. The implementation of CMMS and EAM provides the opportunity for improved maintenance information systems. With CMMS and EAM, the maintenance information system can be developed and tailored to support maintenance as a true "business operation." Information to support planning, scheduling, equipment history, preventive/predictive maintenance, storeroom management, and the like, can be established to improve decision making and overall maintenance management. Improved maintenance information will allow for an open information flow to exist between maintenance, operations, and all departments within the organization. Maintenance will become an important part of the overall information flow, and maintenance will be kept well informed about current operational plans and future plans. True information to manage the business of maintenance is needed, not a sea of more data as illustrated in Fig. 2.4

19. *Ensure an effective maintenance storeroom operation.* The maintenance storeroom will be orderly, space efficient, labor efficient, responsive, and provide the effective cornerstone for maintenance excellence. The maintenance storeroom for *maintenance repair operations* (MRO) items will be recognized as an integral part of a successful maintenance operation. Initial storeroom design or modernization will include effective planning for space, equipment, and personnel needs while providing a layout that ensures efficient inventory control and includes maximum loss control measures. It will be professionally managed and maintained in a clean, orderly, and efficient manner. The trend will be toward larger centralized storerooms with responsive delivery systems to eliminate craftspeople waiting or traveling to get parts. An effective maintenance storeroom catalog will be maintained to provide a permanent cross-reference of all storeroom items and to serve as a tool for identifying and locating items.

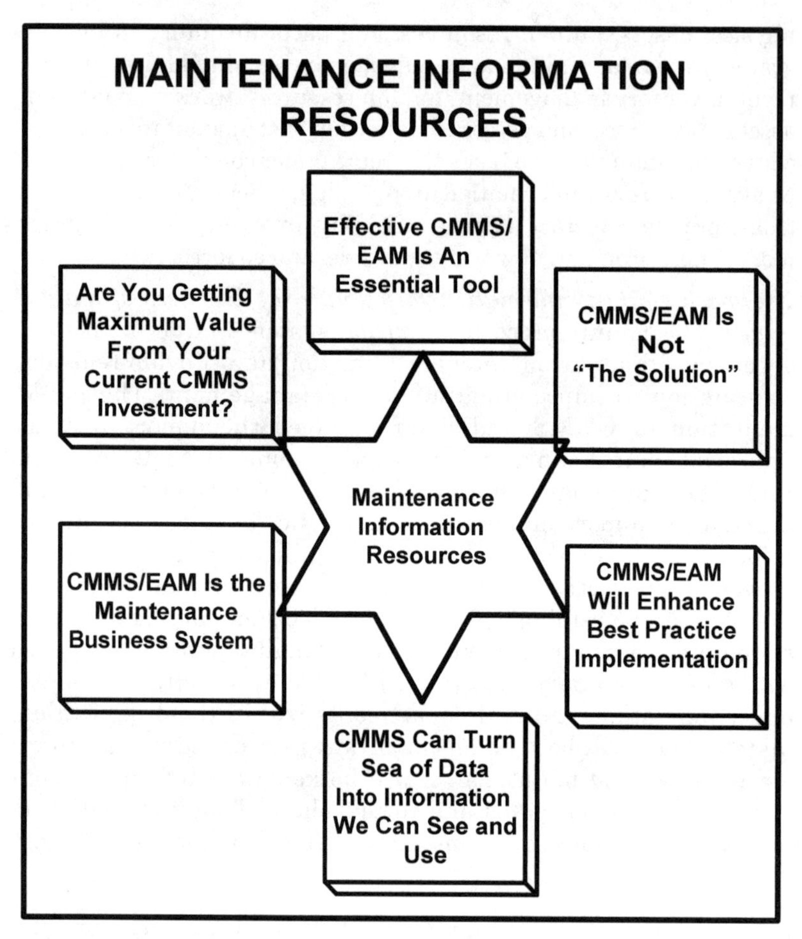

Figure 2.4 True information is needed, not a sea of more data.

20. *Establish the spare parts inventory as the cornerstone for effective maintenance.* The proper quantity of the spare parts will be on hand due to progressive MRO procurement, and internal storeroom controls, all to support maintenance excellence. The implementation of CMMS and EAM will include an inventory system that totally supports the requirements of maintenance and the storeroom. The maintenance inventory will be managed to ensure that the right part is available at the right time without excessive inventory levels. Information from all available sources will be used to determine optimum stock levels. A continuous review of stock levels will be made to eliminate excess inventory and obsolete parts. Inventory reductions will be achieved through more partnerships with suppliers and

vendors that establish joint commitments to purchase based on responsive service and fast delivery. Positions within MRO material management and procurement will increase in their importance and level of technical knowledge to perform effectively.

21. *Establish a safe and productive working environment.* Successful maintenance operations will be safe, clean, and orderly because good housekeeping is an indicator of maintenance excellence. Maintenance leaders will provide a working environment where safety is a top priority, which in turn allows maintenance to set the example throughout the organization. Good housekeeping practices in maintenance will provide the basic foundation for safety awareness. Maintenance will provide support throughout the organization to ensure that all work areas are safe, clean, and orderly.

22. *Aggressive support compliance to environmental, health, and safety requirements.* Maintenance must provide proactive leadership and support to regulatory compliance actions. Maintenance leaders must maintain the technical knowledge and experience to support compliance with all state and federal regulations. The issue of indoor air quality must receive constant attention to eliminate potential problems. Maintenance must work closely with other staff groups in the organization such as quality and safety to provide a totally integrated and mutually supportive approach to regulatory compliance.

23. *Continuously evaluate, measure, and improve maintenance performance and service.* Broad-based measures of maintenance performance and customer service will provide a continuous evaluation of the value of maintenance. CMMS will allow for a broad range of measurement for maintenance performance and service. Investment in best maintenance practices will require valid return on investment. Projected savings will be established and results will be validated. Measures will be developed in areas such as labor performance/utilization, compliance to planned repair and preventive/predictive maintenance schedules, current backlog levels, emergency repair hours, storeroom performance, asset uptime, and availability. Leaders of successful maintenance operations will continuously evaluate performance and service in order to manage maintenance as a business. They will adopt the philosophy of continuous maintenance improvement and have a method to measure progress.

24. *Benchmark where you are and improve maintenance in tota—not piecemeal.* Figure 2.5 illustrates a total approach to maintenance excellence and the 1993 version of the *Scoreboard for Maintenance Excellence* from Tompkins Associates. This was the second revision that evolved from the original *The Scoreboard for Excellence in Maintenance and Tooling Services* back in 1981 at The Cooper Tool Group.

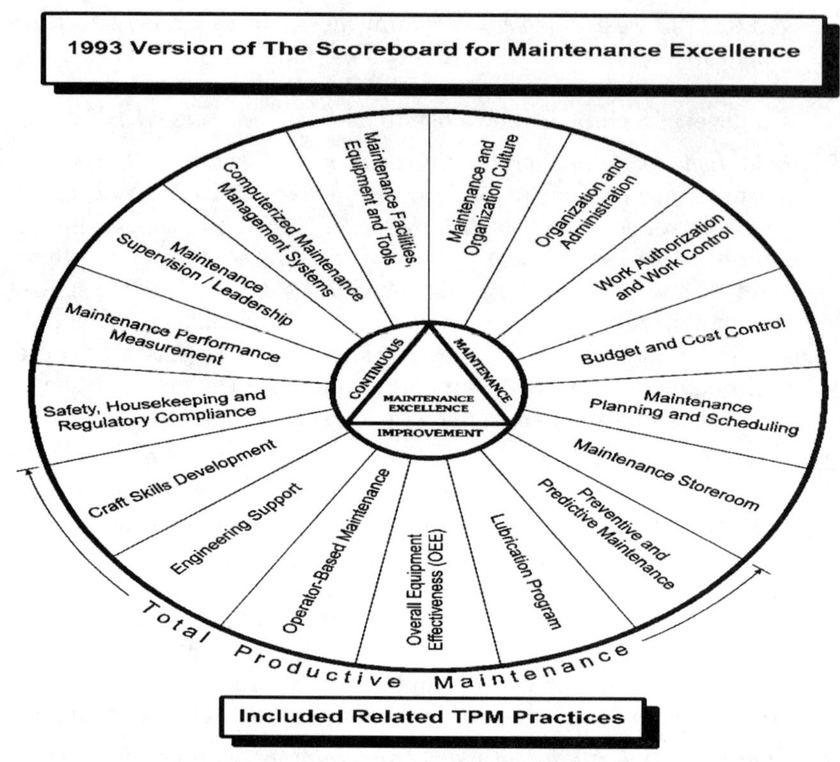

Figure 2.5 1993 version of The Scoreboard for Maintenance Excellence. (*Tompkins Associates 1993 to 2000*)

The key requirements for profit- and customer-centered maintenance provide the foundation for developing a profit- and customer-centered maintenance strategy. An effective maintenance process is essential to profitability. It all starts with the maintenance leader having a total commitment to improvement and with maintenance being managed as a business. It is the realization that maintenance is a key contributor to profit. It is the realization that maintenance best practices, plus people assets, plus MRO assets, and information technology asset all combine for the success and improvement of the total maintenance operation. The four benchmarking tools (illustrated in Fig. 2.6) are tools you can apply now, and they are discussed in the later chapters. These four well-proven and highly practical benchmarking tools can support the journey toward maintenance excellence in almost all types of maintenance operations if you do as Nike says: Just Do It!

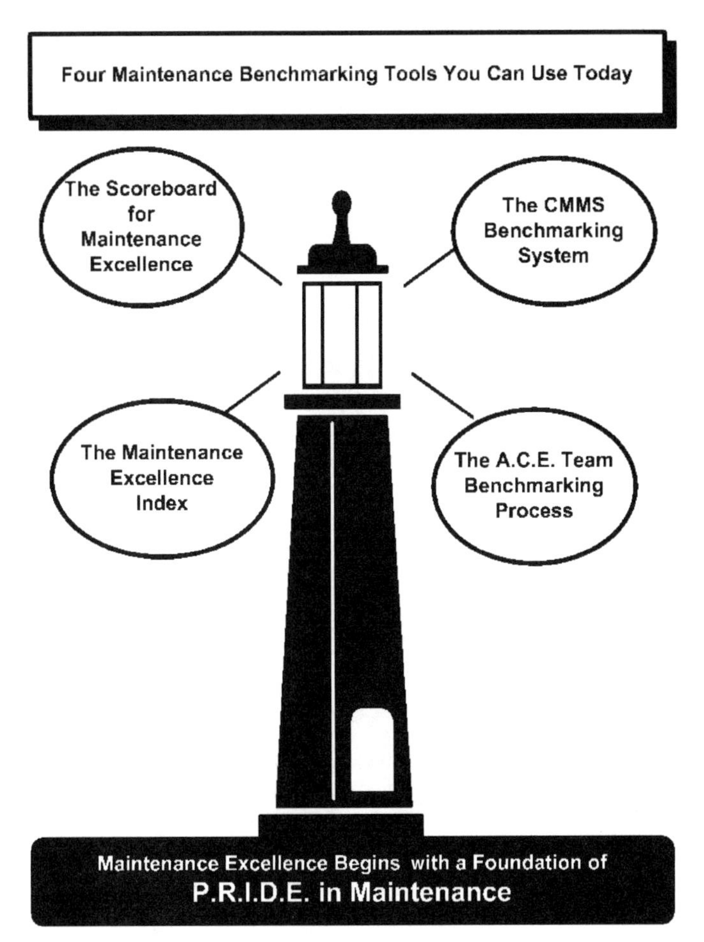

Figure 2.6 Just do it now with these four tools.

Chapter

3

Four Real Maintenance Challenges We All Face

We must remember that "Maintenance is forever!" No matter what we work on, we consider that maintenance plays an essential part. Our bodies, minds, souls, cars, and homes—and the entire infrastructure that makes up our daily environment—all require maintenance. We can outsource many things to other countries, but it is really hard to outsource maintenance of our physical assets overseas. (See Fig. 3.1.)

Four Real Maintenance Challenges

Regardless of the current size and operation scope, both plant maintenance and facilities management leaders are faced with four unique, but very interrelated challenges. These four challenges illustrated in Fig. 3.2 are:

- *Challenge one.* Maintaining existing production assets, equipment, and facilities in safe and sound condition.
- *Challenge two.* Improving, enhancing, and then maintaining existing assets and facilities to achieve environmental/regulatory standards, greater production capacity at better quality, and while using the best energy practices.
- *Challenge three.* Enhancing, renovating, and modifying/overhauling existing assets/facilities using capital funds or funds from tenant/customer, and then maintaining the new additions or enhancements.
- *Challenge four.* Commissioning new production assets or facilities. Assume increased scope of work to maintain the new assets. Be prepared to assume more work from challenges one, two, and three mentioned earlier, as production assets, equipment, and facilities get older and older.

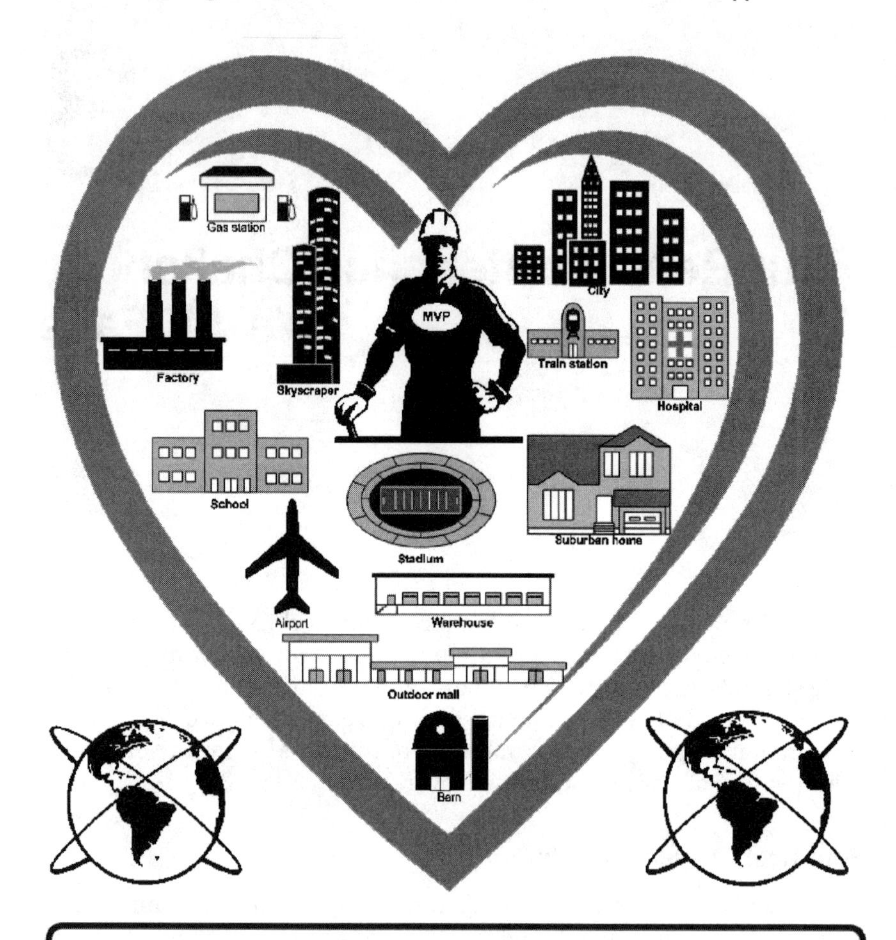

Figure 3.1 Maintenance is forever around the world.

Most maintenance operations must face these four challenges without additional craft resources. They are challenged by continuously trying to find ways to convince top leaders that additional resources to meet growing maintenance requirements are needed. And they are challenged by having to continuously improve their maintenance business process to keep up with growing maintenance requirements. Zero craft labor growth plus more work for challenges two, three, and four; (without craft productivity gains) when taken together results in more requirements. This simple fact can exponentially increase the volume of deferred maintenance and long-term costs. It is now "pay me

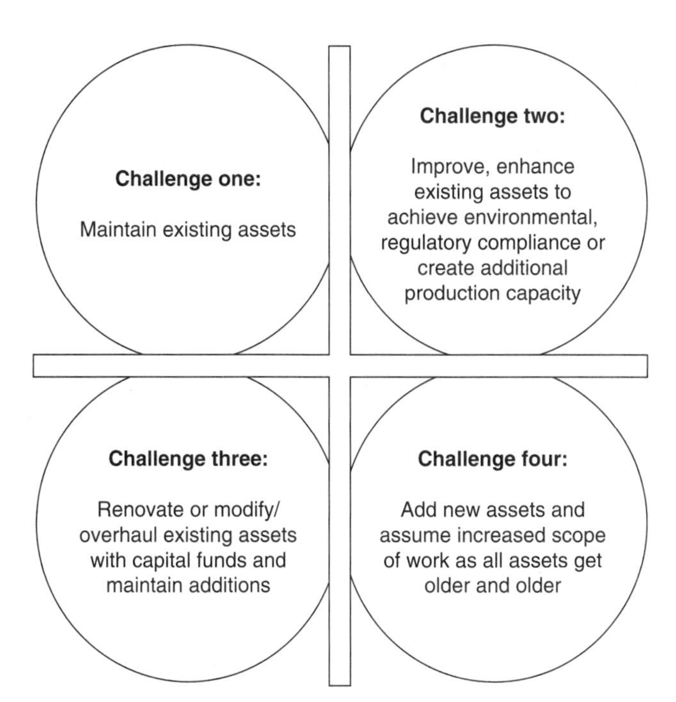

Figure 3.2 Maintenance leaders faced with four unique, interrelated challenges.

now or pay me more later" as we all have heard. The lesson to be learned is to emphasize the need to know as "precisely accurate" as possible your total maintenance requirements, so that if you are the maintenance messenger you will at least be precisely inaccurate when time comes to justify additional maintenance resources. Almost every benchmarking tool we discuss and each benchmark category can be a part of your justification process to add needed resources and best practices.

Lesson learned

Figure 3.2 and this chapter title evolved as my classical example of these four maintenance challenges within the facilities maintenance operation of large state university systems across the United States. The challenge of multiple campuses, major facilities, and infrastructure of all types exists on each campus. The U.S. Air Force's base maintenance challenge is another one as well, which we will discuss in Chap. 7. The scope of my example includes the facilities and physical plant operations within my home state of North Carolina. Here we have 16 major campus complexes and 59 community colleges across the state. This includes physical assets with over 110 million gross square feet valued at over $13.2 billion.

Billions added to challenge four—commissioning of new maintenance requirements

A statewide education bond issue was passed, providing $3.1 billion for new construction and renovation adding to challenge four. Each university campus will receive a designated amount for new construction, addition, and some replacements of major assets. Each complex also receives and uses state-appropriated funds supposedly at 3 percent of replacement value for maintenance and repair. These are dollars, hopefully sufficient for facility and infrastructure maintenance to maintain current facilities just for challenge one mentioned earlier. And hopefully enough to achieve each facility's primary function for the respective customer's/tenant's teaching or research mission.

Our universities also receive other pots of money in the form of trust funds from alumni gifts and funds that come from receipt-supported facilities such as dormitories, research facilities, basketball arenas, and football stadiums. The bottom line of our example here is not good. In the State of North Carolina, there is now a deferred facilities maintenance bill outstanding of nearly $1 billion and it is still growing rapidly! Our primary and secondary road maintenance needs in North Carolina represent another book entitled for example *Maximizing the Value of Your Gas Tax Funds for Road Maintenance Excellence.*

Challenge One

Often the appropriated (or budgeted) funds fail to meet existing facility needs for basic facility maintenance or for plant maintenance requirements in a manufacturing operation. This is compounded even further when overall deficits in state budgets occur (as in North Carolina in 2001 to 2004). So what is cut first (or never adequately included to begin with) is maintenance and repair. The vicious cycle continues year after year until the taxpayers have to pay up for government's addiction to the "high cost of gambling with maintenance costs." Then as in North Carolina we all pitch in and help out on challenge one with our tax dollars.

Determining existing maintenance needs

There are two foundational needs for an effective facility management or plant maintenance operation that need to be highlighted:

1. *Maintenance business process improvement.* Business process improvement is what this book strives to help and promote with a profit- and customer-centered strategy and related attitudes. If this is truly present, then the plant maintenance leader or facilities management

leader in governmental operations at least, has a chance to survive. However, regardless of the type of maintenance operation, they must be able to show top leaders that they really are maximizing all available maintenance resources and there is a true need for resources to address the next item—the basic maintenance requirements.

2. *Maintenance requirements for the physical asset.* Achieving the total maintenance requirement is the primary mission. It is executing the required maintenance while providing, maintaining, and improving the asset or facilities and related services for production operation of the tenants/customers. One word of caution is that we must always be aware of what the asset manufacturer wants for preventive maintenance and spares because it can be overkilled under the current operating condition of the asset. Maintenance leaders must achieve the basic maintenance requirements in addition to the many other activities that compete for engineering, craft, and administrative resources. Defining true maintenance requirements to top leaders is extremely important when all resources are maxed out and basic preventive maintenance is being neglected.

Challenge Two

Periodic determination of basic maintenance requirements and regulatory compliance issues can be determined relatively easily. But often there is so much to do that we cannot stop to do this important task. Basic maintenance requirements can be validated and reinforced for budgeting by periodic facility and asset condition evaluations by qualified professional engineering staff. Facility condition evaluation of existing facility components/systems are important to benchmark actual condition of primary and secondary electrical systems, major facility HVAC systems, elevators, roofing system, fire protection systems, other life safety conditions, *American's with Disabilities Act* (ADA) compliance, along with energy management opportunities. Condition assessment of production assets to perform primary function with quality output likewise is a valuable exercise.

In our example, facilities management leaders then figuratively have to "beg, borrow, and steal," with maximum creativity to get scarce appropriated state funds to accomplish their challenge one as well as their challenge two. To improve and enhance existing facilities to meet the environmental/regulatory standards and achieve energy management, the best practices are often a never-ending fiscal battle—to state it simply. Often when we save "here" we cannot invest it "there" for even higher return on investment (ROI) on initial savings. For governmental facilities leaders, they watch it revert back to the "big, big pot" and continue the vicious cycle for the next fiscal year. Most of the time this basic scenario applies to the plant maintenance leader as well: "If you save it, it will fly away" and not come back next year as available revenue/profit to be used for additional improvement. If you owned your maintenance operation as your business this may or may not occur.

Challenge Three Is Good News/Bad News

For challenge three: the receiving of capital funds, funds from the tenant/customer to enhance/renovate or add new production processes is both good news and bad news for the maintenance leader. The good news is that significant other special nonappropriated funds can and do evolve. Needed additions, renovations, and often major/minor tenant-/customer-funded construction and facilities enhancements can be achieved. Nonappropriated funds in North Carolina may even fund new craft positions, or if there is a large tenant population, then college student and football fan revenue can fund crafts positions. Others having deep pockets are those with continuous research grants, with trust fund gifts by alumni, and so forth. New production processes provide new products, greater throughput, improved quality for more profit, and hopefully better profit optimization choices. Profit optimization is good for all in a production environment.

But there is some real bad news

The real bad news side for challenge three is that the production asset additions, renovations, and facilities enhancements along with their major building system requirements all must be maintained . . . forever. More bad news is the fact that the budgeted craft positions used to construct or support these minor and major renovations at a university often may not be fully used to maintain work in the existing facilities they were originally "justified for." Pure maintenance and repair must come from "appropriated funds" paying for existing craft positions. The bad news and total maintenance costs exponentially compounds itself over time as "appropriated funds" then actually decrease for the enhanced existing facilities requirements. We very seldom see a bronze plaque in a building lobby with donor's names for someone providing the maintenance funds. Nor will you very often see a building or wing in a building named after someone who is paying for the maintenance. We the taxpayers do most of that for governmental and educational system facilities.

Challenge Four

The last but certainly not the least of the challenges for all types of maintenance leaders is major additions of production assets and new construction. Here the commissioning and the ongoing maintenance and operations of new assets and facilities systems without adequate technical craft additions is a common practice. I have personally witnessed this all across the United States, Canada, and all over the world. Often plant maintenance is faced with start-up of new production assets without adequate short- and long-term engineering support. To continuously assume increased scope of work to maintain new assets or new construction, be prepared to assume more work from challenges one, two, and three requiring every bit of maintenance business process improvement that can be mustered. Top leaders must understand and see that all available resources are being maximized (if they truly are) and that the new maintenance requirements are valid. These added requirements can come from any of the mentioned four challenges. It is very important here for the plant maintenance and facilities manager to have in place a way to document true requirements and to have an effective performance measurement process in place that we will discuss later in Part V.

Validate results of improvement. Performance measurement should cover multiple resource areas and be broad based to validate asset reliability and uptime, actual work accomplished, the backlog of work including deferred maintenance, the PM requirements and compliance, customer service metrics, time charged to work orders, and actual charge-backs to customer. These metrics along with maintenance repair and operations (MRO) materials management metrics and other metrics to validate craft utilization (wrench time) and even craft performance can be extremely helpful in documenting true resource needs. A contract maintenance provider sums the net positive impact of all these into two key areas: *profit and improved customer satisfaction.* And as we will see later, the *Maintenance Excellence Index* can provide the same basic process to manage the in-house maintenance operation as an internal business. In-house operations must validate results to top leaders and also provide for communications of positive results to all levels in their organization, especially down to our most important maintenance resource which is the craft workforce.

Ensure top leaders understand your total maintenance requirements. Maintenance leaders must ensure their top level leaders fully understand "the high cost of gambling with deferred maintenance costs" whether in a research facility or in a precision machining center making aircraft components. The growing labor resource needs for increasing maintenance requirements: challenges one, two, three, and four must come from somewhere. Labor resource needs can be offset by either new craft resources and/or greater productivity of existing craft resources.

The growing maintenance needs of an organization must continuously be highlighted to top level leaders. Just as important, the maintenance operation must continuously improve its operation and provide investments for implementing operational improvements that support a profit and customer-centered approach to maintenance.

A profit- and customer-centered approach is needed. Large maintenance operations can and must operate as true profit- and customer-centered maintenance organizations. This must encompass all aspects of their extensive business enterprise: administration, financial, design, maintenance, construction, planning/scheduling, procurement, and overall MRO materials management. Current constraints may continue from the public sector organizations, and attitudes that in-house maintenance is a "necessary evil" will die hard. But to survive in the twenty-first century, both private and public sectors must put in place effective performance-measurement processes that truly validate profit- and customer-centered results. Both must be able to show ROI when investment for improvements are received.

Do not kill the goose. Budget cuts often fall in the one place they can hurt the worst, and that is cutting of craft people, the technicians within all of the necessary trade's areas who are out there doing the real work, the PMs, the emergency responses, and the weekend service calls. The indiscriminate cutting of these scarce craft resources is a failed business practice of the twentieth century. Indiscriminate cutting is killing the goose that lays the golden egg. If an organization is not: (a) doing continuous reliability improvement; and (b) defining true maintenance requirements and achieving them, then cutting craft positions to meet budget is exactly like using blood letting as a new cure for a heart attack. It just will not work.

Another True Lesson Learned

An assessment was in progress at a plant for a major brand of tomato catsup owned by a famous person's wife when an interview with the manager of engineering and maintenance indicated the following.

A previous consultant had recommended that 11 craft positions could be eliminated if

a. Planning and scheduling is implemented with two planners

b. Current storeroom is modernized and centralized from satellite stores all across the plant

c. A new *computerized maintenance management systems* (CMMS) is installed

What did this plant do with that consultant's recommendation? They immediately eliminated the 11 craft positions and never completed any of the preceding three items. The two planners were on second shift as maintenance coordinators and parts chasers; parts were still scattered all across the plant and the CMMS was literally still in the box. This was a prime example of the dumbsizing of maintenance. So for all further write-ups of assessment results I would always put in something in the executive summary such as the following when projecting craft productivity gains.

> **Indiscriminate cuts in craft and technician labor must not occur as a result of these recommendations:** Significant opportunities to reduce the values of inventory via better materials management are available in the range of over $X. Better procurement practices can bring another $Y. Likewise with better shop level information, increased material availability, and effective planning, scheduling, and supervision, significant craft productivity gains are possible. These gains will be measured and actual craft productivity will be validated. The gained value of craft/technician time as result of this project is in the range of over $Z.

The core competency for doing maintenance may not be present and contracted skills from a service provider may truly be needed. But regardless of the operation's size or scope, *the core requirement for maintenance remains forever.* Dumbsizing of maintenance to match lean manufacturing trends can be fatal. It really will not work if one has truly maximized use of existing resources and valid maintenance requirements are not being accomplished. So do not kill the goose even for a chief finance officer (CFO) who wants a short term "golden egg."

Do not gamble, especially with maintenance costs

Do you know where you stand with applying today's best practices for maintenance and physical asset management? Do you have a baseline as to what is considered today's best practice and whether you have applied them effectively? Are there best practices that you have heard about that we now need to really consider? If not, you may very well be gambling with the long-term success of your total production operation. Effective maintenance and physical asset management add value. They support profit optimization whether at one site or multiple sites. There are some very important steps that you should take to gain maximum value from your maintenance operation. Parts II, III, IV, V, and VI of this book can now help you on your journey toward maintenance excellence.

When we view maintenance operations as a profit- and customer-centered process, we must view the cost of making improvements as an investment in maintenance and not as a traditional cost. Maintenance leaders must understand that investments require a ROI and they must be prepared to define potential benefits and put in place methods to validate the results of improvements. Maintenance and physical asset management operations within your organization can be true contributors to profit generation or increased service levels. The cost of external resources and internal support services can be a very good investment. The opportunities for measurable results in almost all organizations are significant as can be evidenced by the growing number of contract maintenance providers.

Successful contract maintenance providers clearly understand maintenance as a profitable business opportunity. Maintenance is forever, and as a result there will forever be new business opportunities for the true value-added contract maintenance provider. Contract maintenance providers clearly understand profit and the importance of overall craft effectiveness and quality service. Very simply their goal is to perform services equal to or greater than in-house maintenance at a profit and potential savings to you. The future will see third-party maintenance continue to replace in-house maintenance operations who have priced themselves out of the marketplace due to low craft labor productivity, poor service, and lack of internal leadership.

Who will step forward with the courage and the commitment necessary to bring the art and science of maintenance into being profit- and customer-centered within your operation? Will it be you or the vice president of ACME Technical Services Inc., who convinces your leadership that:

- We have the technical know-how and the leadership capabilities to operate your maintenance operation and take it to another level—a more reliable and profitable level.
- We will divest it completely from you and slowly sell its services back to you at a profit to us and a net savings to you now! We will validate immediate short-term benefits and continuously measure improvements and benefits to you over the long term.
- We will provide both the technical and personal leadership to avert certain failure of your business if you continue along your current path of no investments in maintenance best practices and craft skills development for your people assets.
- We will provide a measurable value-added maintenance service in the core business requirement that you have obviously given up on.
- When can we sign the contract and begin to create improved cash flow and more profits for your company?

Do not be a takeover target?

The in-house maintenance operation that continues with rising craft labor costs and no productivity gains and marginal at best customer service will be an easy takeover target. The in-house maintenance operations can be competitive with the proper leadership, equipment, tools, and application of today's maintenance best practices. It requires taking action to these questions: If I owned my maintenance operation what would I do different to make a profit? What will it take to regain my competitive edge and successfully bid against contracted services to keep work in-house or avoid a complete takeover? I have personally talked to many union shop stewards and we have discussed these two questions during assessments. The really smart ones can see the handwriting in the papers as more plants go south, east, and west. You can take action on these two basic questions right now without even going any further in this book. It is about "Just doing it" in most cases!

Part

2

Determining Where You Are as a Profit- and Customer-Centered Maintenance Operation: The Scoreboard for Maintenance Excellence

Chapter

4

The Scoreboard for Maintenance Excellence

Introduction

We will now get down to the detailed level of "determining where you are" with actually applying today's best practices for maintenance. This chapter introduces "The Scoreboard for Maintenance Excellence" focused on plant maintenance as your *global benchmark* for maintenance best practices. It has evolved since 1981 (with 10 benchmark categories and 100 benchmark criteria) into today's most complete benchmarking tool to define where you are on your journey toward maintenance excellence. Combined with our other three benchmarking tools, we feel sure that this book provides today's best process starting from global, external benchmarking and going down to the craft level with reliable benchmarks for craft repair times with the ACE Team Benchmarking Process included later in Chap. 15. The four levels of maintenance benchmarking are illustrated in Fig. 4.1.

Understanding the Types of Benchmarking

A *benchmark* is a point of reference for a measurement. The term presumably originates from the practice of making dimensional height measurements of an object on a workbench using a graduated scale or similar tool, and using the surface of the workbench as the origin for the measurements. In surveying, benchmarks are landmarks of reliable, precisely known altitude, and are often man-made objects such as features of permanent structures that are unlikely to change, or special-purpose "monuments," which are typically small concrete obelisks, approximately 3 ft tall and 1 ft at the base, set permanently into the

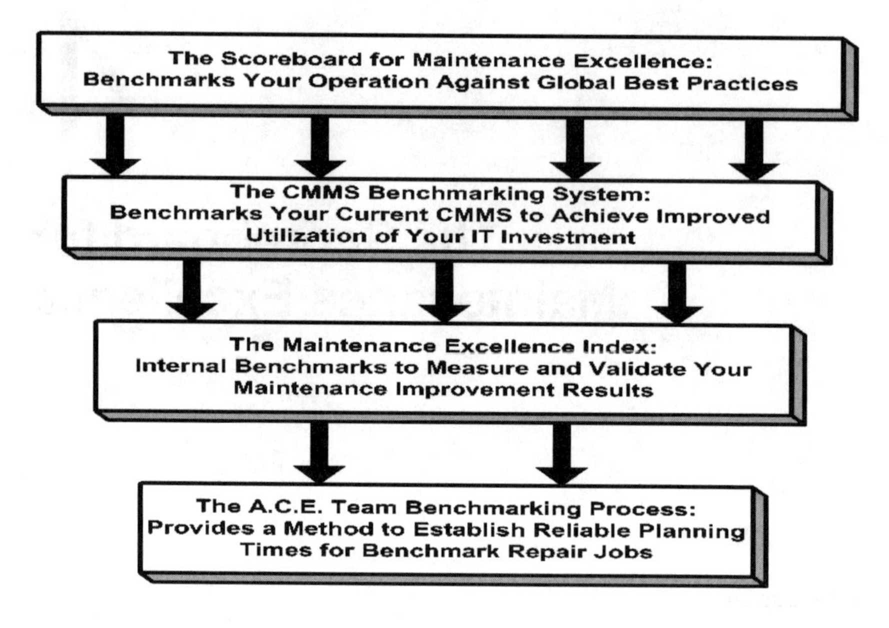

Figure 4.1 Maintenance benchmarking from four important levels.

earth. Do you have a current benchmark as to where you are with your maintenance process? If you desire to be committed to continuous reliability improvement and to be progressing toward a profit- and customer-centered operation, here is where we begin to help you get started.

The 2005 version of "The Scoreboard for Maintenance Excellence" within App. A of this book addresses 27 major benchmark categories with 300 benchmarking criteria. The benchmark categories are recognized maintenance best practices. Some are easier than others and are basically do-it-now items. Many are strategic opportunities requiring internal and external resources to implement. "The Scoreboard for Maintenance Excellence" provides the first of four benchmarking tools introduced in this book, a global one that benchmarks where you are with applying external best practices that other successful operations recognize and use.

Benchmarking is a very versatile tool that can be applied in a variety of ways to meet a range of requirements for improvement. The following is a summary of different terms used to distinguish the various ways of applying benchmarking and how benchmarking tools from The Maintenance Excellence Institute fit these summary definitions.

1. *Strategic benchmarking.* This improves an organization's overall performance by examining the long-term strategies and general

approaches that have enabled high-performers to succeed. It involves high-level aspects such as core competencies, developing new products and services, changing the balance of activities, and improving capabilities for dealing with changes in the culture of the organizational environment. This type of benchmarking may be difficult to implement and the benefits are likely to take a long time to materialize.

"The Scoreboard for Maintenance Excellence" is both *strategic benchmarking* and what we term *global benchmarking* as it applies to the total maintenance process and overall best practices for the "business of maintenance." Results from using "The Scoreboard for Maintenance Excellence" will include strategic, tactical, and operational level of difficulty improvement opportunities. There will almost always be *Do It Now* opportunities that can be implemented immediately.

2. *Performance benchmarking or competitive benchmarking.* A total organization considers its positions in relation to performance characteristics of key products and services compared to benchmarking partners (or competitors such as contract maintenance providers) from the same sector. In our scope this is physical asset management. In the commercial world, companies tend to undertake this type of benchmarking through trade associations or third parties to protect confidentiality. Very seldom will you see an in-house maintenance operation openly benchmark with a contract maintenance provider. Conversely, contract maintenance providers base their entire business case upon comparing/benchmarking *what they can do for your organization* compared *to what you are now doing or not doing.* As stated in several places in this book: you are almost always a takeover target for contract maintenance providers in one form or another.

3. *Process benchmarking.* This focuses on improving specific critical processes and operations. Benchmarking partners are sought from best practice organizations that perform similar work or deliver similar services. Process benchmarking invariably involves producing *process maps* to facilitate comparison and analysis. This type of benchmarking can result in both short- and long-term benefits.

   We feel very, very strongly that process benchmarking must go beyond analysis paralysis and lead directly to successful implementation and measured results. This gets down to the focused implementation of tactical and operational plans of actions based upon priorities defined from "The Scoreboard for Maintenance Excellence" benchmarking or assessment results. It may involve vendors of predictive technology equipment, integrated suppliers, or external consultant resources to help recruit, train, and install a planning function, to help install a *computerized maintenance management systems* (CMMS) or to help modernize a *maintenance repair operations* (MRO) storeroom. Process benchmarking is basically the complete business case analysis. Results of process benchmarking plus implementation must be nailed down with a valid measurement process that your financial folks understand and agree upon. Here is where "The Maintenance Excellence Index" comes into play as a key benchmarking tool at the shop level. Part V, "Validating Best Practice Results with the Maintenance Excellence Index," includes four chapters on this important topic for profit- and customer-centered maintenance.

4. *Functional benchmarking or generic benchmarking.* This is benchmarking with partners drawn from different business sectors, public/private operations, military organizations, and different maintenance environments to find ways of improving similar functions or work processes. This sort of benchmarking can lead to innovation and dramatic improvements.

   The various scoreboard versions include The Scoreboard for Maintenance Excellence (manufacturing plants), The Scoreboard for Facilities Management Excellence (pure facilities maintenance), The Healthcare Scoreboard for Maintenance Excellence (healthcare facilities), The Scoreboard for Fleet Management Excellence (equipment fleet operations), and The Golf Course Scoreboard for Maintenance Excellence (for the green industry and golf courses). These five represent unique maintenance processes all requiring the art and science of maintenance, but application within different environments. Often the sharing of practices and innovations across these maintenance practice areas can result in dramatic improvements or enlighten a practitioner as exposure to other types of maintenance environment is made via benchmarking.

5. *Internal benchmarking.* It involves seeking partners from within the same organization, for example, from business units located in different areas. The main advantages of internal benchmarking are (*a*) access to sensitive data and information and standardized data are often readily available; and (*b*) usually less time and resources are needed. There may be fewer barriers to implementation as practices may be relatively easy to transfer across the same organization. However, real innovation may be lacking and best-in-class performance may be found through external benchmarking outside the organization. Later we will see that the Air Combat Command desired to conduct external benchmarking against the private sector's facilities maintenance operation. They also had in place an existing internal benchmarking process for their civil engineers operations that was responsible for maintenance at their current air bases.

   "The Scoreboard for Maintenance Excellence" is ideal for this definition of internal benchmarking via comparison between maintenance operations within larger organization. Chapter 6, "A Strategy for Developing a Corporate-Wide Scoreboard," defines how multiple site operations can provide internal benchmarking with tools in this book. Case studies from Boeing Commercial Airplane Group, the U.S. Air Force's Air Combat Command, SIDERA (largest steelmaker in Argentina), BigLots, National Gypsum, and others have used "The Scoreboard for Maintenance Excellence" for internal benchmarking. (The Scoreboard is included in App. A.) All desired to define "where they were" with overall maintenance best practices at numerous sites and then develop plans of actions to improve high priority opportunities.

We also consider "The Maintenance Excellence Index" as an important internal benchmarking process, but this tool is used at the grass-roots level, the shop floor. It should also focus not just on maintenance, but rather on success of the total operations, whether a plant, a pure facilities complex, a fleet operations, or healthcare facility and the equipment maintenance process.

6. *External benchmarking.* Outside organizations that are known to be best in class are sought out to provide opportunities of learning from those who are at the leading edge. External benchmarking keeps in mind that not every best practice solution can be transferred to others. This type of benchmarking may take up more time and resources to ensure the comparability of data and information, the credibility of the findings, and the development of sound recommendations. External learning is also often slower because of the "not invented here" syndrome.

   "The Scoreboard for Maintenance Excellence" is also an external benchmarking tool defining "where you are" with overall maintenance best practices at a single site and also with plans of actions to improve high priority opportunities.

7. *International benchmarking.* This is used where partners are sought from other countries because best practitioners are located elsewhere in the world and/or there are too few benchmarking partners within the same country to produce valid results. Globalization and advances in information technology are increasing opportunities for international projects. However, these can take more time and resources to set up and implement, and the results may need careful analysis due to national differences.

   However, our work with many international maintenance operations has shown that we can learn or have our so called best practices reconfirmed when we see maintenance in other countries. For example, my experience with Wyeth Medica Ireland, SIDERA Argentina, Ford Argentina, Coke Argentina, Avon Brazil, and AO Smith Mexico reconfirms that a very simple best practice—*cleanliness in shops and plant areas*—is an important contributor to maintenance excellence and *pride* in maintenance.

An International Benchmarking Methodology Framework and Aide document can be downloaded from http://www.benchmarking.gov.uk/content/documents/intrntnl.doc.

Our Publications Are Free Except?

Our various scoreboards are the only proprietary items on our Web site at PRIDE-in-Maintenance.com. Our extensive collection of white papers, articles, and PowerPoint copies of speech offering material are all free to download. And we do not sell the scoreboard except as part of this book and our consulting support that we term as *Maintenance Excellence Services*. Our scoreboards all began originally as "The Scoreboard for Excellence in Maintenance and Tooling Services" in 1981 while I was the Industrial Engineering Manager for the Cooper Tool Group, the hand tool division of Cooper Industries. One of my many roles on a very small corporate staff was to improve maintenance processes at our seven plants: Cresent-Xcelite, Weller, Plumb, Nicholson File, Nicholson Saw, Wiss, and Lufkin. Ironically, after learning the manufacturing operations in each of these high-quality hand tool plants I got the chance to be plant manager of the one making the "knuckle buster"—*The Crescent Wrench*. I was hired at the corporate level, not only for my industrial engineering background but for my previous maintenance experience of helping Ben Gibbs (a contributor to this book) install a very extensive fleet maintenance management across 100 county shops and 15 division levels from 1973 to 1975 for the Division of Highway's Equipment Unit in the North Carolina Department of Transportation.

How Do You Get "There" if You Do Not Know Where "There" Is When You Start?

That was my first concern. A map is completely useless if you do not know your current location—"where you are right now on the ground." For the seven Cooper Group plants, my first task was to determine "where we are" with regards to current maintenance practices. So research began to put together an assessment tool to benchmark where we were with best practices. Tooling services were included in this first scoreboard because all plants had extensive tool rooms. Nicholson File in Cullman, Alabama, for example, made 80 percent of their repair parts for their custom-built file-making machines. It was a parts manufacturing business alongside the business of maintenance, both supporting the file making business. We began with a self-assessment which helped to get each plant to understand the basic best practices that were to come as we later got down to

TABLE 4.1 Evolution of "The Scoreboard for Maintenance Excellence"

Date	Scoreboard version	Benchmark categories	Benchmark criteria	Focus
1981	Scoreboard for excellence	10	100	Plant maintenance and tooling services
1993	Scoreboard for maintenance excellence	18	200	Plant maintenance
2003	Scoreboard for maintenance excellence	27	300	Plant maintenance
2003	Scoreboard for facilities management excellence	27	300	Facilities maintenance
2003	Scoreboard for healthcare maintenance excellence	27	300	Healthcare facilities maintenance
2004	Scoreboard for fleet maintenance	27	300	Fleet maintenance
2006	Scoreboard for golf course maintenance excellence	30	300	Golf course maintenance and other green industry operations

supporting each plant. The evolution of "The Scoreboard for Maintenance Excellence" is summarized in Table 4.1.

Then, while at Tompkins Associates Inc., the 1993 version of "The Scoreboard for Maintenance Excellence" shown in Fig. 2.5 evolved and was used until 2003 when the next enhancement was made.

Over 24 years of application and evolution

You can see how this external, global benchmarking process has evolved from over 24 years of successful application to many different types of public and private organizations. Its counterpart, "The Scoreboard for Facilities Management Excellence," was also developed for pure facilities maintenance operations. Appendix L provides a comparison of these two very similar benchmarking tools. The only basic difference is that "The Scoreboard for Facilities Management Excellence" replaces plant-specific categories with the following categories that are related to facilities:

a. Facilities condition evaluation program

b. Building automation and control systems technology

c. Utilities system management

d. Security systems and access control

e. Grounds and landscape operations

f. Housekeeping service operations

Later in Chap. 7, "Case Study: The Scoreboard Self-Assessment: Just Do It!," we will define how you can develop your own unique "Scoreboard for Maintenance Excellence" for continuously evaluating progress on your journey toward maintenance excellence.

Begin with a Very, Very Important First Step

Establish a maintenance excellence strategy team

The very first step of course is for the maintenance leader to have a firm commitment to improve the total maintenance operation. It is very important to avoid taking a piecemeal approach. Others within the organization must also share that commitment. One key element of success is having a commitment from top-level leaders across the organization and craft leaders. Establishing a maintenance excellence strategy team is highly recommended. This high-level leadership driven cross-functional team is made up from maintenance leaders, key operations leaders, shop-level maintenance staff, IT, engineering, procurement, operations/customer, financial, and human resources staff. A craft employee is always a recommended member of the *maintenance excellence strategy team*.

The mission of this team is to lead and facilitate the overall maintenance improvement process and to ensure measurement of results: top leader support and required resources are provided. This team can also sponsor other teams within the organization to support implementation of the recommended path forward. One of the very first things that this team should do is to sponsor a comprehensive evaluation of the total physical asset management and maintenance operation. Again this is the first step to help determine "where you are." A sample charter for a maintenance excellence strategy team is included in App. E. This example was used for a multisite operation that was also undergoing the installation of new CMMS, storeroom modernizations and maintenance planner selection and training.

How to determine "where you are"

The real goal is improvement of your total maintenance operation to support profit and customer service of the total operation better. Maintenance leaders must clearly understand—"your maintenance operation is not the tail wagging the dog." Within plant maintenance operations are included maintenance and repair of all production and facility assets, supporting infrastructure; overhaul and renovation activities; engineering support processes as well as all material management; and procurement of typical repair parts, supplies plus contracted services. You should benchmark your current operation against today's best practices for preventive maintenance, planning and scheduling, effective spare parts control, work orders, work management, the effective use of computerized systems for maintenance business management, and all other categories from "The Scoreboard for Maintenance Excellence" that we will review next. This step is important because it gives you a baseline as your starting point for making improvements and for validating results. It will help to ensure that you are taking the right steps for taking care of your mission-essential physical assets.

An independent evaluation, in most cases, helps to reinforce the local maintenance manager's desire to take positive action in the first place: to do something to improve the overall maintenance process. For multiple site operations, this provides a great opportunity for developing standard best practices that can be used across the corporation and for new sites.

The Scoreboard for Maintenance Excellence

Today's most comprehensive benchmarking guide, "The Scoreboard for Maintenance Excellence," will define "where you are" in terms of applying today's best practices for plant maintenance You can also develop your own "Scoreboard for Maintenance Excellence" and begin with a self-assessment as will be discussed in a later chapter. But we normally recommend getting help from a consulting resource with at least a pilot plant site. "The Scoreboard for Maintenance Excellence" provides a means to evaluate how we are managing our six key maintenance resources: people, technical skills, physical assets, information, parts and materials and our hidden resources, and the synergy of team efforts. Figure 4.2 illustrates how "The Scoreboard for Maintenance Excellence" includes these six key maintenance resources and how hidden resources can evolve from the synergy of team efforts.

Table 4.2 provides a summary of the 2005 "Scoreboard for Maintenance Excellence" for manufacturing plant maintenance operations.

The 2003 version of "The Scoreboard for Maintenance Excellence" was updated to its current 2005 version by adding nine important best practice categories:

1. Shutdown and major planning/scheduling and project management
2. Manufacturing facilities planning and property management
3. Production asset and facilities condition evaluation program
4. Process control and instrumentation systems technology
5. Energy management and control
6. Maintenance and quality control

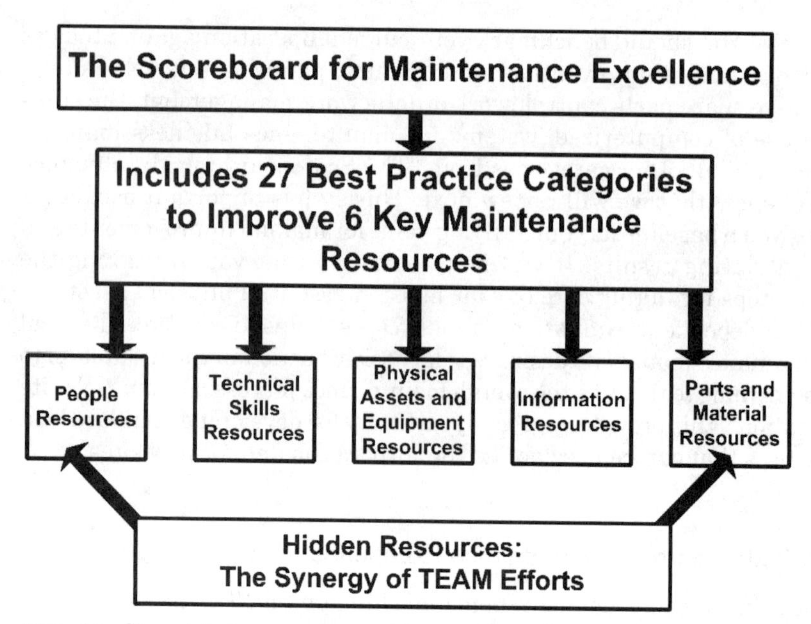

Figure 4.2 Six maintenance resources to improve maintenance in total, not piecemeal.

7. Critical asset facilitation and *overall equipment effectiveness* (OEE)
8. *Overall craft effectiveness* (OCE)
9. MRO materials management and procurement

The Scoreboard for Maintenance Excellence provides a means to evaluate how we are managing our six key maintenance resources: people, technical skills, physical assets, information, parts and materials and our hidden resources, and the synergy of team efforts. The following shows how the 27 evaluation categories are broken down across the six key maintenance resource areas:

People resources

- The maintenance organizational culture and *pride* in maintenance
- Maintenance organization, administration, and human resources
- Operator-based maintenance and *pride* in ownership
- Safety and regulatory compliance
- Shop facilities, equipment, and tools

Technical skill resources

- Craft skills development and *pride* in maintenance

TABLE 4.2 provides a summary of the 2005 "Scoreboard for Maintenance Excellence" for manufacturing plant maintenance operations.

Category	The scoreboard for maintenance excellence benchmark category descriptions	Benchmark items	Points
A.	The organizational culture and PRIDE in maintenance	6	60
B.	Maintenance organization, administration, and human resources	12	120
C.	Craft skills development and PRIDE in maintenance	12	120
D.	Operator based maintenance and PRIDE in ownership	6	60
E.	Maintenance supervision/leadership	9	90
F.	Maintenance business operations, budget, and cost control	12	120
G.	Work management & control: Maintenance and repair	12	120
H.	Work management & control: Shutdowns and major overhauls	6	60
I.	Shop level planning and scheduling	18	180
J.	Shutdown and major planning/scheduling and project management	9	90
K.	Manufacturing facilities planning and property management	9	90
L.	Production asset and facilities condition evaluation program	6	60
M.	Storeroom operations and internal MRO customer service	12	120
N.	MRO Materials management and procurement	12	120
O.	Preventive maintenance and lubrication	18	180
P.	Predictive maintenance and condition monitoring technology applications	15	150
Q.	Process control, building automation, and instrumentation systems technology	9	90
R.	Energy management and control	12	120
S.	Maintenance engineering support	9	90
T.	Safety and regulatory compliance	12	120
U.	Maintenance and quality control	9	90
V.	Maintenance performance measurement	12	120
W.	Computerized maintenance management system (CMMS) and business system	18	180
X.	Shop facilities, equipment, and tools	9	90
Y.	Continuous reliability improvement	15	150
Z.	Critical asset facilitation and overall equipment effectiveness (OEE)	15	150
ZZ.	Overall craft effectiveness (OCE)	6	60
	Total benchmark evaluation items and points	300	3000

- Maintenance engineering support
- Overall Craft Effectiveness (OCE)

Physical assets and equipment resources

- Production asset and facility condition evaluation program
- Preventive maintenance and lubrication
- Predictive maintenance and condition monitoring technology applications
- Process control and instrumentation system technology

- Energy management and control
- Maintenance and quality control
- Critical asset facilitation and OEE

Information resources

- Maintenance business operations, budget, and cost control
- Work management and control: *maintenance and repair* (M/R)
- Work management and control: major shutdowns and overhauls
- Shop-level maintenance planning and scheduling
- Shutdown and major maintenance planning/scheduling and project management
- Manufacturing facilities planning and property management
- Maintenance performance measurement
- CMMS and business system

Parts and material resources

- Storeroom operations and internal MRO customer service
- MRO materials management and procurement

Hidden resource: the synergy of *team* efforts

- Maintenance supervision/leadership
- Continuous reliability improvement

Your Global Benchmark

An assessment using "The Scoreboard for Maintenance Excellence" provides a "global, external" benchmark against today's best maintenance practices: it is very important to note that the overall total score for the assessment is not an absolute value. The assessment results and the overall total score, however do represent an important benchmark and a baseline value. Results will identify *relative strengths* within an operation and *opportunities for improvement* from among the 27 evaluation categories of best practices. The assignment of baseline values to each evaluation item *is not an exact science.* It should be based upon an objective assessment of the best practice item when observed at a point in time.

Benchmarking results

Overall assessment results fall into five possible overall rating levels as shown in Table 4.3: excellent, very good, good, average, and below average. Each of the five overall ratings levels represents at least a 9 percent point spread.

TABLE 4.3 Benchmark Rating Summary by Total Point Range

Total point range	Scoreboard for maintenance excellence: benchmark rating summary
90–100% of total points	Excellent: 2700–3000 points
80–89% of total points	Very good: 2400–2699 points
70–79% of total points	Good: 2100–2399 points
60–69% of total points	Average: 1800–2099 points
Less than 59% of total points	Below average: less than 1799 points

A summary of baseline value assignment for each rating level is as follows:

Baseline value 10: excellent. The practices and principles for this evaluation item are clearly in place, and this rating provides an example of world-class application of the practice. There is evidence that this practice has received high-priority support in the past to achieve its current level.

For an overall baseline rating of excellent, the range of baseline points is 2700 to 3000 points or 90 to 100 percent of total baseline points of 3000.

Baseline value 9: very good. This rating denotes that current practices are approaching world-class level as a result of high-priority focus on continuous improvement for this evaluation item. This practice continues to be a high priority for improvement within the organization.

For an overall very good baseline, the range of baseline points is 2400 to 2699 points or 80 to 89 percent of total baseline points of 3000.

Baseline value 8: good. This rating denotes that the practice is clearly above average and above what is typically seen in similar maintenance operations. There is a need for additional emphasis, to reassess priorities and to reconfirm commitments to improvement for this evaluation item.

For an overall good baseline, the range of baseline points is 2100 to 2399 points or 70 to 79 percent of total baseline points of 3000.

Baseline value 7: average. This rating represents the typical application of the practice as seen within a facilities maintenance operation that has had very little emphasis on improvement. It reflects an operation that typically is just maintaining the status quo.The organization should conduct a complete assessment of maintenance operations and review this practice in detail for improvement or for implementation if it is not currently in place.

For an overall average baseline, the range of baseline points is 1800 to 2099 points or 60 to 69 percent of total baseline points of 3600 .

Baseline value 6 or less: below average. This rating (a score of 6 or below) denotes that the application of this practice is below what is typically seen in similar maintenance operations. This practice may not be currently in place, and should be considered as part of the organization's continuous improvement process. Immediate attention may be needed in some areas to correct conditions having an adverse effect on customer service, safety, regulatory compliance, or facilities maintenance costs.

For an overall baseline of below average, the total baseline points are less than 1799 baseline points or less than 59 percent of total baseline points of 3000.

Table 4.4 provides comments for each of the Scoreboard for Maintenance Excellence benchmark rating levels.

Appendix A includes a complete "Scoreboard for Maintenance Excellence" formatted as shown in the following example in Fig. 4.3

- *Benchmark criteria 1.* This example also illustrates the very first benchmark evaluation criteria on the scoreboard which is related *to whether or not maintenance is a considered a priority in your operation.* Does your organization include maintenance within its plant goals as a key item?
- *Lesson learned.* As a plant manager of a high-quality hand tool manufacturing plant in the mid 1980s, I was fortunate to be a part of an operation that did have maintenance as a key part of our plant goals as shown in Fig. 4.4. In our case, effective *preventive maintenance* (PM) was one of our key operating goals for the plant and was one of the factors contributing to our quality and service level during a period of intense competition from foreign hand tool manufacturers. Our PM included not only heavy machining equipment, heat treat, plating, molding, and automated forging equipment but also all of the tooling fixtures, cutters, and dies used in a high-quality machining operation.
- *Never, never, never give up!* If you do not have a top leader that appreciates and values maintenance as a key contributor to profit and customer service, do not give up. As Jim Valvano and Winston Churchill both said: *Never, Never, Never Give Up!* Continue the good fight, get the facts for ROI, and continue to educate top leaders that *pride* in maintenance is a core requirement for success of total operations.

We will now take a look at the best practice evaluation criteria and the scoreboard benchmarking categories under each of the maintenance resource areas. The remainder of this chapter will discuss each criterion in this order, which is not the category order as shown previously in Table 4.2

TABLE 4.4 Rating Summary Comments for Results from a Scoreboard for Maintenance Excellence Assessment

Total point range	Scoreboard for maintenance excellence: rating summary comments
90–100% of total points	2700–3000 points—excellent: Practices and principles in place, for achieving maintenance excellence and world-class performance based on actual results. Reconfirm overall maintenance performance measures. Maintain strategy of continuous improvement. Set higher standards for maintenance excellence and measure results.
80–89% of total points	2400–2699 points—very good: Fine-tune existing operation and current practices. Reassess progress on planned or ongoing improvement activities. Redefine priorities and renew commitment to continuous improvement. Ensure top leaders see results and reinforce performance measurement process.
70–79% of total points	2100–2399 points—good: Reassess priorities and reconfirm commitments at all levels of improvement. Evaluate maintenance practices, develop, and implement plans for priority improvements. Ensure that measures to evaluate performance and results are in place. Initiate strategy of continuous reliability improvement.
60–69% of total points	1800–2099—average: Conduct a complete assessment of maintenance operations and current practices. Determine total costs/benefits of potential improvements. Develop and initiate strategy of continuous reliability improvement. Define clearly to top leaders where deferred maintenance is increasing current costs and asset life cycle costs. Gain commitment from top leaders to go beyond maintenance of the status quo.
Less than 59% of total points	Less than 1799 points—below average: Same as for average, plus, depending on the level of the rating and major area that is below average, immediate attention may be needed to correct conditions having an adverse effect on life, health, safety, and regulatory compliance. Priority to key issues, major building systems and equipment, where increasing costs and deferred maintenance are having a direct impact on the immediate survival of the business or the major physical asset. The capabilities for critical assets to perform intended function are being severely limited by current "state of maintenance." Consider immediate contract services as required for business survival and for achieving the core requirements for maintenance services if necessary investments for internal maintenance improvements is not going to be made.

A.	The Organizational Culture and PRIDE in Maintenance	
Item #	Benchmark Item Rating: excellent – 10, very good – 9, good – 8, average – 7, below average – 6, poor – 5 or less.	Rating
1.	The organization's vision, mission, and requirements for success include physical asset management and maintenance as a top priority.	

Figure 4.3 Example of format of benchmark evaluation criteria.

People resources—five benchmark categories

See Table 4.2.

1. The maintenance organizational culture and pride in maintenance
2. Maintenance organization, administration, and human resources
3. Operator-based maintenance and pride in ownership
4. Safety and regulatory compliance
5. Shop facilities, equipment, and tools

Plant Operating Goals: Crescent-Xcelite Plant
(Division of Cooper Industries)

1. Aggressive investigation and follow-through of potential new products and markets. Maintain a close working relationship with sales and marketing functions to enable Crescent-Xcelite to react to customer and market place changes.
2. New manufacturing technology will be integrated into our manufacturing methods and process. This will allow Crescent-Xcelite to be competitive in the basic areas of quality, price, and service.
3. Maintain technical competence or workforce (internal and external training).
4. Clean, orderly, and safe workplace.
5. Create atmosphere of pride in workmanship (quality awareness and quality improvement programs).
6. Balanced line product flow through manufacturing phases using hard automation or robotics when economically feasible.
7. Control all inventory levels using automated reporting and MRP system and capacity planning.
8. Preventive maintenance to monitor and adjust equipment to eliminate production disruptions.
9. Simplified quick change tooling to accommodate short runs, reduce EOQ.
10. Equipment and tooling monitored to achieve 100% first fun quality capability.
11. Create a no-crisis atmosphere for all employees.

Figure 4.4 Plant operations goals with maintenance clearly defined as top priority.

The maintenance organizational culture and *pride* in maintenance. This benchmark category is very important, but much more subjective than all others. Our on-site assessment includes a rating of this category after our time on-site. Ratings are based upon results of interviews, observations, and available information from throughout the total operation, not just within our host maintenance operation. For example, I always look at the plant bulletin boards. When we read a big sign in a plant with operations goals and reference to the words *maintenance or reliability* are missing, that is our first clue on rating the very first benchmark criteria on "The Scoreboard for Maintenance Excellence." That first impression or clue may or may not change the rating value at the end of the benchmark assessment. Benchmark items in this category include:

1. The organization's vision, mission, and requirements for success include physical asset management and maintenance as a top priority.
2. Senior management is visible and actively involved in continuous maintenance improvement and is obviously committed to achieving maintenance excellence.
3. The organization's strategy and plan for total operations' success are known to all in maintenance and include a strategy for maintenance improvement.
4. Maintenance is kept well informed of changing business conditions, strategies, and long-range plans.
5. The organization's culture and the maintenance environment result in innovation, *pride* in maintenance, trust, and an obvious spirit of continuous improvement.
6. Open communication exists within maintenance and the overall organization to ensure interdepartmental cooperation, idea sharing, and basic teamwork.

Maintenance organization, administration, and human resources. Elements related to this category, as you will see later, are intertwined within many other categories as well. Some of the items go below the evaluation, such as item number 1 below. We always look very carefully at how maintenance craft resources are deployed and try to answer these questions as well. Are maintenance crafts physically and organizationally deployed for maximum craft productivity? Does technical leadership of all maintenance processes exist from one maintenance leader or does assignment under others such as operations hinder total plant productivity and specifically craft productivity. Our position on maintenance leadership is very emphatic that there should be one person for the overall technical leadership of all maintenance processes in a plant. This is the *chief maintenance officer* (CMO, if you will) at plant level. Elements of maintenance resources can be divided and be effectively assigned *to focus on* specific production (or operations) departments, specific processes, zones, central shop or area shop, or satellite shops. But we feel very strongly that organizationally there is one overall technical leader of the *total maintenance business* supported by the appropriate number of supervisors and lead people at specific tasking levels at the site.

Benchmark items in this category include:

1. The maintenance organization chart is current and complete with fully defined areas of responsibility.
2. Clear-cut craft job descriptions have been developed that completely define job responsibilities and skill levels required for each craft.
3. Craft personnel are provided copies of their job descriptions and counseled periodically on job performance, job responsibilities, and craft skills development needs.
4. One single head of maintenance operations is supported by adequate clerical and technical staff of planners, first-line supervisors, stores personnel, maintenance engineering, and training support.
5. The maintenance department head has high visibility within the organization and reports to a level such as the plant manager.
6. The first-line supervisors are responsible for the performance of 12 to 15 crafts people.
7. A timekeeping system is in place to charge craft time to each job.
8. Monthly or weekly reports are available to show distribution of maintenance labor in critical categories: breakdown repairs, corrective work, PM work, and the like.
9. Monthly or weekly reports are available to monitor backlog status and priority of planned or project work, and so forth.
10. Backlog trend data are available to highlight the need for craft increases, scheduled overtime, or subcontracting.
11. Guidelines on the level of accepted backlog are established to determine the need for overtime or subcontracting as well as to identify potential problem areas.
12. Sufficient employee-hour data are available that allow valid decisions as to which jobs must be delayed if new jobs or projects are added to the schedule.

Operator-based maintenance and *pride* in ownership. Operator-based maintenance has been a best practice in America since our Revolutionary War and Civil War, where top leaders like George Washington and

Robert E. Lee were insistent upon operator-based maintenance of personal items (and the person), equipment, and horses/ mules used for personal and supply transportation. So I do not use any of the *total productive maintenance* (TPM) terms for this best practice that the U.S. Military has used for several centuries now. Chapter 19: "Pride in Ownership with Operator-Based Maintenance: Lessons Learned from the U.S. Army," can help you begin this proven best practice.

Operator-based maintenance is a best practice that every operation must consider. Maintenance leaders must understand both the risk and the benefits of this best practice. Operators can achieve *pride* in ownership and subsequently develop pride in maintenance. This can happen when operators of all types of equipment are properly trained, motivated, and paid according to the personal job responsibility description and wage payment plan for their respective organization; and if they can answer this question with a positive answer: how would I operate this piece of equipment if I owned it and was in business to sell back my output for a profit or to provide increased customer service?

Benchmark items in this category include:

1. Operators are responsible for cleaning their equipment and trained to perform selected levels of operator-based maintenance.
2. An initial cleaning to bring equipment to an optimal or "as new" status has been planned or has been completed for critical equipment.
3. Operators have been trained to perform periodic inspections on their equipment and report problems.
4. Operators have been trained and have proper tools and equipment to do selected lubrication, tighten bolts and fasteners, and to detect symptoms of deterioration.
5. The process of transferring maintenance tasks and skills to operators has been well coordinated between maintenance, operations, engineering, and human resources staff.
6. Operators have developed greater "pride in ownership" and understand their expanded role in detecting and preventing maintenance problems.

Safety and regulatory compliance. The 12 evaluation items for this critical category only scratch the surface of safety and regulatory compliance requirements. Most large organizations will have support to maintenance with internal resources devoted to safety. But the scope of maintenance work requires that the maintenance leaders and craft leaders and the craft workforce be extremely knowledgeable of all items in this category.

Benchmark items in this category include:

1. Maintenance leaders have created a broad-based awareness and appreciation for achieving a safe maintenance operation.
2. Maintenance employees attend at least one safety meeting per month.
3. Maintenance has shown a continual improvement in its safety record over the past five years.
4. All permits and safety equipment are available and prescribed for each job where they are required.
5. All cranes, hoists, lift trucks, and lifting equipment are inspected as part of the preventive maintenance program.
6. Good housekeeping within maintenance shops and storerooms is a top priority.
7. Maintenance tools, equipment, and leftover materials are always removed from the job site after work completion.
8. Maintenance continually evaluates areas throughout the operation where safety conditions can be improved.
9. The total scope of regulatory compliance issues within the organization has been defined, and a prioritized plan of action established.
10. Maintenance responsibilities related to regulatory compliance have been well defined.
11. Maintenance has the technical knowledge and experience to support the organization's regulatory compliance action.
12. Maintenance works closely with other staff groups in the organization for a totally integrated approach to regulatory compliance.

Shop facilities, equipment, and tools. This category is personally one of my first concerns. During my years as a Director of Facilities Management for the State of North Carolina, I saw a real need on my first day/week on the job. There were poor shop facilities in bad locations, a lack of special equipment, and some mobile shop vans and landscape trucks that looked like Fred Sanford trucks for hauling junk. Literally, our crews driving around our capital city of Raleigh looked very bad, which contributed to a less than average public image of a very dedicated craft workforce. Quickly. . . . the bottom line to this one is: *We must give our valuable craft resources the best possible shop facilities, equipment and tools that we can afford*! We can be penny wise and pound foolish as Ben Franklin said with regards to this category.

Benchmark items in this category include:

1. Maintenance shop facilities are located in an ideal location with adequate space, lighting, and ventilation.

2. Standard tools are provided to craft employees and accounted for by a method that ensures good accountability and control.
3. An adequate number of specialty tools and equipment are available and easily checked out through a tool control procedure.
4. All personal safety equipment necessary within the operation is provided and used by maintenance employees.
5. Safety equipment for special jobs such as confined space entry and electrical system lock-out, is available and used.
6. Maintenance achieves a high level of housekeeping in its shop areas.
7. Maintenance maintains a broad awareness of new tools and equipment to improve methods, craft safety, and performance.
8. Maintenance continually upgrades tools and equipment to increase craft safety and performance.
9. An effective process to manage special tools and equipment inventory is in place.

Technical skill resources—three benchmark categories

See Table 4.2.

1. Craft skills development and *pride* in maintenance.
2. Maintenance engineering support.
3. OCE.

Craft skills development and *pride* in maintenance. Craft skill training can be a weak link in the journey toward maintenance excellence (Fig. 2.2). When people say education is expensive, try ignorance! It has been said that "What if we train them and they leave? But what are the risks and consequences if we do not train them and they stay until retirement?

"The Perfect Storm"*

I have heard many warn of the problem most companies in the near future will face with more losses of skilled labor than we have seen since the early 80s. What will our workforce look like and will a company be able to hire the right people. Bob Williamson with Strategic Work Systems calls this "The Perfect Storm."

---

*Thanks to Ricky Smith, President of Maxzor Corp for providing some important facts and solutions for our introduction to the Craft Skills Development benchmark category with "*The Perfect Storm.*"

I think America needs to wake up and look at the data or we will continue to lose our Industrial Base to foreign countries. We are a country that is losing industrial facilities at a very high rate. It is always a struggle to find out what research tells us about the future so that we can prepare for it. In looking at the data from research conducted at the U.S. Department of Labor and the U.S. Department of Education along with surveys compiled by many organizations, some things become very apparent which will help us make decisions on a skilled workforce in the future.

The perfect storn: What is the problem?

It is known that within the **next** *3–5 years most companies will lose over 50% of their skilled workforce.* It is known that across our nation high schools are shutting down or have shut down their vocational education programs. A recent survey completed by our company with over 200 maintenance organizations responding found the following:

- In the next 3–5 years they will lose over 47% of their skilled workforce
- Over 72% of these companies have no formal testing process for new hires for maintenance
- Over 98% say they have not trended the losses of their skilled workforce

Dealing with the facts

Now for some hard news, information from the results of the latest Bureau of Census Report (2004) stated the following:

- More than 51 million adults or approximately 23% of the adult population in the U.S. possess limited literacy capacity.
- In 31% of the adult population English is their second language. In five states 50% of the population stated that English is their second language.
- 25% of adults who have limited literacy are living below the poverty level in the U.S.

We must ask ourselves how far will technology go in the next 20–30 years and will a company be capable of competing in the global marketplace. By the year 2012 (6 years from now) the Department of Labor states that the demand for skilled maintenance workers will increase by 13.6% (776,000 additional jobs). The demands on industry to replace retiring workers will place an even higher demand on skilled workers.

What can you do?

1. Begin trending your workforce losses over the next 2–10 years based on a projection. This information will help you make logical decisions as to what should you do about your skill losses and how quickly should you prepare.
2. Perform a job task analysis along with a task-to-training matrix to determine the critical, most frequent, and most important tasks in your maintenance department. Most of the maintenance work does not require a top level technical person. In fact 90% of maintenance work requires a person with midlevel skills.
3. Establish a hiring process to identify the needs of your company not for just now but also the future. Most companies do not test new employees for specific abilities, skills, attributes, mechanical aptitude, and literacy levels required to perform their job to standard.
4. Train your operators to take over basic maintenance tasks thus reducing pressure on maintenance personnel and thus improving efficiency.
5. Have your current experienced maintenance personnel write PM (Preventive Maintenance) and CM (Corrective Maintenance) procedures that identify steps, specifications, tools, and material required for each task. Use your current knowledge base to prepare your department for the future.

I have confidence in our American Industrial Base and believe most companies will meet the challenges they face in the near future and will succeed.[†]

Again, special thanks to Ricky Smith, a professional person, a maintenance consultant with experience in the *plant maintenance trenches* and a current member of the U.S. Army and veteran of the Iraqi Freedom war. I too have faith in our American industrial base and the people resources, our craftspeople who make it happen despite how top leaders and even maintenance leaders perform. Maintenance excellence must begin with *pride*-in-maintenance and with craftspeople really interested in developing excellence in their chosen professional skill area.

Benchmark items in this category include:

1. The types and levels of craft skills required for an effective maintenance operation have been identified.

[†]The Perfect Storm was provided by Ricky Smith, CMRP, CPMM, ricky.smith@maxzor.com, President, *MAXZOR Corp*, www.maxzor.com.

2. Job descriptions include well-defined standards for job knowledge and skill levels required for each craft area.
3. An assessment of the current job knowledge and skill level of each craftsperson has been made to determine individual training needs.
4. The overall training needs for the maintenance staff have been developed with a plan of action and cost.
5. The organization has committed to providing the necessary resources for maintenance training and skills development.
6. A program for craft skills development has been designed to address priority training needs and is being implemented.
7. Results of training are determined by a competency-based approach, which ensures demonstrated capability to perform on newly trained craft tasks.
8. A policy to pay-for-skills gained is available or is being developed as part of the craft skills development program.
9. The benefits of developing multicraft capabilities within maintenance have been evaluated and incorporated into the craft skills training program as applicable.
10. Individual training plans for each craftsperson are being used to document compliance type training as well as training that the craftsperson and supervisor see as being needed.
11. Actual hands-on job competency is being documented (for example by the work order system) to help validate overall craft skill levels.
12. The overall craft workforce has shown the initiative for continuous craft skills development.

Maintenance engineering support. Benchmark items for this category include:

1. Engineering and maintenance work closely together during the design and specification stages to improve equipment reliability and maintainability.
2. Purchase of new equipment and modifications to existing equipment is subject to maintenance review prior to final approval.
3. Engineering provides key support to maintenance and operations for improving equipment effectiveness.
4. Engineering provides key support to maintenance during installation and start-up of new equipment to ensure that operating specifications are achieved.

5. Engineering supports maintenance as required to troubleshoot chronic equipment problems and to define/eliminate root causes.
6. Engineering and maintenance work closely to develop an effective equipment and spare parts standardization program.
7. Capital additions, building systems changes, and facility layout changes are subject to maintenance review before final approval.
8. Up-to-date prints, parts/service manuals, and other documentation for equipment and facility assets are available to maintenance.
9. Engineering coordinates material requisitioning with maintenance for project work, major overhauls, and machine building.

Overall craft effectiveness (OCE). Two important chapters: Chap. 13, "Improving Craft Productivity—An Essential Strategy for Profit and Customer Service," and Chap. 14, "Introducing OCE as a New Buzzword—Overall Craft Effectiveness," provide proven approaches to implement solutions that support improving OCE. Benchmark items in this category include:

1. Improvement of craft productivity and OCE is being established as an important element of the overall maintenance improvement process.
2. Priorities have been established with a plan of action for improving the OCE factor for craft utilization (wrench time).
3. Priorities have been established with a plan of action for improving the OCE factor for craft performance.
4. Priorities have been established with a plan of action for improving the OCE factor for craft service and quality.
5. Improvements in OCE and craft productivity are evaluated against baseline measurements to determine progress.
6. A team effort is being used that involves the craft workforce to provide a cooperative effort for improving OCE and eliminating a reactive, firefighting strategy that wastes valuable craft resources.

Physical assets and equipment resources: seven benchmark categories

See Table 4.2.

1. Production asset and facility condition evaluation program
2. Preventive maintenance and lubrication
3. Predictive maintenance and monitoring technology applications
4. Process control and instrumentation system technology

5. Energy management and control
6. Maintenance and quality control
7. Critical asset facilitation and OEE

Production asset and facility condition evaluation program. Benchmark items for this category include:

1. A process is in place to periodically evaluate the current operational status of production equipment and, in turn, define repair, overhaul, or replacement recommendations for life-cycle cost reduction opportunities.
2. A facility condition assessment process is in place to evaluate the current operational status of facility-type equipment and in turn defines repair, overhaul, or replacement recommendations for life-cycle cost reduction opportunities.
3. The PM and *predictive maintenance* (PdM) programs provide key information as to the overall state of maintenance for production asset.
4. The PM and PdM programs provide key information as to the overall state of maintenance for facility-type assets.
5. The current evaluation processes for production and facility assets supports defining the true maintenance requirements of the operation.
6. The current evaluation processes for production and facility assets provide an effective baseline for replacement planning and capital justification of new assets: both production and facilities are related.

Preventive maintenance and lubrication. Chapter 20, "Case Study: Developing an Effective Preventive Maintenance Strategy," provides case study results and a good roadmap for establishing or reinvigorating an existing PM/PdM process.

Benchmark items for this category include:

1. The scope and frequency of PM and lubrication services have been established on applicable equipment.
2. Operations staff supports and agrees with the frequency and scope of the PM and lubrication program.
3. Optimum routes for PM inspections and lubrication services are established.
4. PM checklists with clear, concise instructions have been developed for each piece of equipment. Lubrication checklists and charts are available for each asset included under the lubrication program.

5. Inspection intervals and procedures are periodically reviewed for changes/improvements and updated as required as part of well-defined PM change process.
6. Planned times are established for all PM and lubrication tasks.
7. The total craft labor requirements by craft type to accomplish the overall PM program have been established to validate staffing needs for effective PM.
8. The required level of manpower is being committed to achieve the total scope of PM services.
9. Actual craft time devoted to PM is known and evaluated as a percentage of total craft time available.
10. Goals for PM compliance are established and overall compliance and results are measured against the company benchmark.
11. All noncompliance to scheduled PM and lube services is aggressively evaluated and corrected. Lubrication services are viewed as a key part of preventive maintenance and are not neglected or overlooked.
12. Maintenance and operations work with close communication, coordination, and cooperation to schedule PM and lube services.
13. The success of PM is measured based on multiple factors: reduced emergency repairs, increased planned maintenance work, reduced downtime costs, the elimination of the root cause of problems, and improved product throughput and quality.
14. PM is a highly visible function within maintenance, and is well received as an operational strategy.
15. The PM staff are well-qualified crafts people and serves as good maintenance ambassadors and "customer service representatives."
16. A PM *master schedule* is developed to evaluate the weekly or monthly plan and to level load tasks when required.
17. Corrective repair work orders are generated and become planned work as a result of PM inspections and provide one measure of PM success.
18. PM manpower needs are adjusted to satisfy changing PM inspection requirements.

Predictive maintenance and condition monitoring technology applications. Chapter 21—Today's Predictive Maintenance Technology: Key to Continuous Reliability Improvement—provides an extensive plan of action related to PdM and condition monitoring technology applications: Also Remy Kolb, Founder of RMK Maintenance Solutions, LLC provides key points to consider for this benchmark category.

Benchmark items for this category include:

1. Equipment has been evaluated for the application of PdM technology, and the scope and frequency of PdM services has been established on all applicable equipment.
2. A plan for using current PdM technology is being developed or is now being put in action. PdM techniques used on all applicable equipment = 10; application of PdM in progress based on plan = 9; PdM plan developed, no progress = 6; no plan for PdM = 0.
3. Maintenance, engineering, and others have technical knowledge and necessary skills for using PdM techniques.
4. Optimum routes for PdM inspections are established.
5. Inspection and testing intervals and procedures are periodically reviewed for changes/improvements and updated as required as part of well-defined PdM change process.
6. Planned times are established for all PdM tasks.
7. The required level of man power is being committed to achieve the total scope of PdM services.
8. Goals for increased reliability are established and overall PM/PdM compliance and results are measured against the company benchmark.
9. All noncompliance to scheduled PdM services is aggressively evaluated and corrected.
10. Maintenance and operations work with close communication, coordination, and cooperation to schedule PdM services.
11. The success of PdM is measured based on multiple factors: reduced emergency repairs, increased planned maintenance work, reduced downtime costs, the elimination of the root cause of problems, and improved product throughput and quality, and so forth.
12 P M/PdM is a highly visible function within maintenance and is well received as an operational strategy.
13. Corrective repair work orders are generated and become planned work as a result of PdM inspections and provide one measure of PdM success.
14. Critical equipment has been evaluated for the application of continuous monitoring technology.
15. A process for evaluating and eliminating the root causes of failure is in place.
16. Positive results from PM/PdM and overall reliability improvements are being communicated throughout the entire operation.

Process Control and Instrumentation System Technology. Benchmark items for this category include:

1. List of all instrumentation systems sorted for criticality.
2. Instrument systems assigned machine unit (tracking) no. along with other vital information such as location, contract service vendor, type (process, regulatory compliance, condition monitoring, and the like).
3. Operational instructions and data, installation drawings, catalog information, maintenance documents, and parts information on file and indexed.
4. Instrument system components tagged and identified according to standards.
5. Calibration procedures and maintenance tasks written and referenced in work order system.
6. Calibration records up-to-date, filed, and indexed. Historical records dated through the previous two years or according to insurance carrier and/or legal counsel.
7. Critical system operational and emergency procedures written, up-to-date, and available to operators at control panels or other essential locations.
8. All hand-held and portable instruments used for operational checks and maintenance functions listed with tracking numbers and with service/calibration schedules entered into work order system.
9. All laboratory measurement equipment identified with tracking numbers, calibration stickers, and calibration schedules entered in work order system.

Energy management and control. Jim Ross, PE President of Resource Management Systems (RMS) Inc in Cookeville, Tennessee, provides key points for maintenance leaders to consider now and in the future as energy management is again in the news more and more as a critical global concern and an important news headline in the United States.

Why should maintenance be concerned about energy? Traditionally, energy is regarded as an uncontrollable overhead item, and less attention is given to the energy system in a facility than to the day-to-day problems that occupy most of the maintenance department's time. Energy and the systems that make it available to the facility should be of paramount importance to maintenance. Energy is a key resource that should be productively utilized, but moreover elevated to a higher level of priority because without energy and on-site utilities *nothing else runs*. Many enterprises have simply ignored the easily implemented opportunities for energy conservation and cost reduction. Of even greater concern is that the management also ignores proper maintenance of the energy systems at the risk of costly failures. Plant or facilities engineers or other managers should devote a significant amount of time to energy management and effective system maintenance.

What is the real cost of energy? We tend to take energy for granted, but continual vigilance is needed to assure that the process will run, the building will be heated and paychecks will be printed on time, all of which require energy. The real cost of energy is not just the price per gallon, MCF or KWH, but in the acquisition and maintenance of adequate supply of energy. As energy is an essential resource for almost any manufacturing or processing operation, massive effort must be expended to keep the supply of energy available in the proper quantity and quality at the minimum cost consistent with keeping the enterprise in effective operation. The real cost of energy is in its *unavailability.* When energy becomes unavailable, costs soar. Some of the costs associated with energy unavailability are:

1. Production interruptions
2. Missed schedules
3. Lost customers
4. Penalties for shutting down customer's production lines
5. Spoiled work
6. Costly cleanout of ruined product from processing equipment

Likewise, the benefits of effective energy management can be measured in terms of lost production avoided, productivity improvement, continued customer satisfaction, and profitability increases. In this vein, energy can and should not only avoid costly interruptions, but also help the organization to make a profit by driving the energy process that runs the productive processes and the business.

What is the energy process? The entire effort to obtain and maintain an adequate, dependable, and cost-effective supply of energy is more extensive than most people realize. Eleven key steps in the energy process include:

1. Planning of the process
2. Determining energy needs for process, building heat, and other uses
3. Selection of energy forms
4. Negotiation of rates
5. Optimum system installation
6. Anticipation of future needs

7. Maintenance requirements of the system (periodic, preventive, predictive maintenance)
8. Continuous improvement in energy conservation
9. Accommodation of environmental concerns
10. Use of waste heat or material for energy
11. Energy-related waste disposal

The energy process includes not only the equipment in the plant or facility, but utility systems that supply energy (gas, electric, water) to the plant. The process also includes secondary energy systems such as steam and compressed air. The distribution system, protective devices, and all equipment needed to supply energy to productive operations are part of the energy process.

How Can Maintenance optimize energy processes? Although some managers tend to take energy for granted, continual vigilance is needed to assure that the energy processes are functioning correctly. Seven key steps to make this a reality are:

1. Appoint someone in the organization to be responsible for maintaining energy systems.
2. Conduct an assessment to current energy management practices, identify best practices, and correct deficiencies.
3. Take periodic energy surveys to avoid gradual decay in the energy systems.
4. Check ownership of interface devices with utility suppliers.
5. Survey condition of all energy related equipment, and bring up to optimum condition and into code compliance.
6. Train maintenance technicians in energy system maintenance, energy conservation, and rapid response to energy problems to keep energy processes functioning at top efficiency.
7. Set up a PM program for every part of the system, and routinely check for such items as loose contacts, worn insulation, exposed wiring, missing cover plates, dust and dirt on electrical equipment, heat buildup in buss ducts and switches. Take corrective action immediately.

Economics of energy saving projects. Too often, worthwhile energy projects are killed by the one-year payback syndrome. In addition to being too strict, the one-year payback concept ignores the time value of money, a fundamental of engineering economy. In many energy and environmental projects, the payback is nowhere close to a one year undiscounted payback.

To get top management's attention, the project author must get outside of the narrow confines of labor or utility savings, and include the benefits of a dependable cost-effective energy process to justify energy-related projects. The basis for these savings is in keeping records possibly using activity-based costing. If energy problems cause lost production, customer dissatisfaction or excessive costs, cost data that quantify these problems should be presented to top management to get project approval.

Life-cycle costing/approach and expenditures justification. Although applicable to many types of operations, the life-cycle costing approach is particularly applicable to the facility equipment such as heating, air conditioning, computer controls, energy recovery equipment, environmental equipment, electrical systems, or retrofits to these systems. The technique takes into account not only the initial cost of the equipment, but also costs such as maintenance, utilities, major overhauls, operating labor, supplies, and cost of removal and disposal at the end of the life cycle. Disposal costs of residues from the process or system are also included. Benefits also are accounted for over the life of the equipment rather than for the time for cash recovery as in the discounted cash flow method.

Conclusion. The cost of poor maintenance of energy processes as cited earlier can be catastrophic to an enterprise. Energy is a prime resource of most enterprises deserving a high priority for expert maintenance. Jim Ross PE also contributed to the new energy management category for the 2005 *Scoreboard for Maintenance Excellence.*

The benchmark items in this category include:

1. Energy costs are reflected in facility budget via monthly tracking of energy usage, demand, and cost by each customer location/building. Designated individual analyzes usage, demand, and costs on a monthly basis to identify and correct problems.
2. An energy team (or designated staff position) has been formed and chartered to deal with energy management issues and is actively involved to advise on utility conservation and its recommendations and results published periodically.
3. Steam trap surveys are performed to look at operating status, inlet and outlet pressures, replacement needs, and the like, and are done on a routine basis as part of the PM/PdM program.
4. Air compressions have been properly sized and air leaks are routinely corrected.
5. Overall systems analysis of the heating and cooling systems of existing facilities is being conducted to optimize the efficiency and operation of the entire heating and cooling system.

6. Boilers in the facility are well maintained and periodically inspected and trimmed for maximum efficiency.
7. A water management strategy is in place to control process water, monitor usage, and address corrosion.
8. A comprehensive energy audit has been conducted for the facility based upon federal energy management guidelines.
9. Chronic facility breakdowns and problems impacting energy are aggressively investigated to determine root causes.
10. New technologies such as infrared analysis is introduced to check the condition of roofs, switch gears, and so forth.
11. Energy-efficient motors, lights, ballasts, and the like are used throughout the facility or are being planned as part of new facilities and major renovations.
12. Facility automatic energy control systems are in place and are being planned for use in new facilities.

Maintenance and quality control. One of the most important things that all top, maintenance, and crafts leaders must remember is that maintenance plays an important role in quality production and the delivery of quality customer services. Good examples are (a) the maintenance of clinical equipment in a hospital and (b) the sealing process for a simple oil filter where the canister is sealed to the top cap that screws into the motor. For Purolator in Fayetteville, North Carolina, this changeover process was so critical to the safety that only well-trained maintenance staff was assigned to make changeovers to this high speed oil filter sealing operation. Chapter17—Maintenance Quality and Customer Service—includes additional comment on this benchmark category. Also thanks to BSI Management Systems for the following important information as an intro to this category. BSI is the world's leading international standards: testing, registration, and certification organization. The following key steps for implementing a quality management system must consider the high cost to quality if an organization gambles without including maintenance directly in the *quality improvement process*.

Implementing a Quality Management System (QMS): Consider Maintenance's Role in Quality Registration

There are key steps that every company implementing a QMS will need to consider:

Step 1: Purchase the standard. Before you can begin preparing for your application, you will require a copy of the standard. You should read this and make yourself familiar with it.

Step 2: Review support literature and software. There are a wide range of quality publications and software tools designed to help you understand, implement, and become registered to a QMS.

Step 3: Assemble a team and agree on your strategy. You should begin the entire implementation process by preparing your organizational strategy with top management. Responsibility for a QMS lies with senior management; therefore it is vital that senior management is involved from the beginning of the process.

Step 4: Consider training. Whether you are the quality manager seeking to implement a QMS or a senior manager who would like to increase your general awareness of ISO 9001:2000, there are a range of workshops, seminars, and training courses available.

Step 5: Review consultancy options: You can receive advice from independent consultants on how best to implement your QMS. They will have the experience in implementing a QMS and can ensure you avoid costly mistakes.

Step 6: Choose a registrar. The registrar is a third party, like BSI Management Systems, who comes and assesses the effectiveness of your QMS and issues a certificate if it meets the requirements of the standard. Choosing a registrar can be a complex issue as there are so many operating in the market. Factors to consider include industry experience, geographic coverage, price, and service level offered. The key is to find the registrar who can best meet your requirements.

Step 7: Develop a quality manual. A quality manual is a high-level document that outlines your intention to operate in a quality manner. It outlines why you are in business, what your intentions are, how you are applying the standard, and how your business operates.

Step 8: Develop support documentation. This is typically a procedure manual that supports the quality manual. Quite simply, it outlines what you do to complete a task. It describes who does what, in what order, and to what standard.

Step 9: Implement your quality management system. The key to implementation is communication and training. During the implementation phase everyone operates according to the procedures and collects records that demonstrate you are doing what you say you are doing

Step 10: Consider a preassessment. A preassessment by your registrar normally takes place about six weeks into the implementation of the quality system. The purpose of the preassessment is to identify the areas where you may not be operating to the standard. This allows you to correct any areas of concern you may have before the initial assessment.

Step 11: Gain registration. You should arrange your initial assessment with your registrar. At this point the registrar will review your QMS and determine whether you should be recommended for registration.

Step 12: Continual assessment. Once you have received your registration and have been awarded your certificate, you can begin to advertise your success and promote your business. To maintain your registration, all you need to do is continue to use your quality system. This will be periodically checked by your registrar to ensure that your quality system continues to meet the requirements of the standard. Learn more about the Route to Registration with BSI, or order Standards, or learn about BSI's Training.

Thanks to the BSI Management Systems for this important information. BSI is the world's leading international standards: testing, registration, and certification organization. BSI was founded in 1901 and has issued over 35,000 registrations in over 90 countries. U.S. Corporate headquarters are in Reston, Virginia. A wide range of free guidance documentation and other information is available on their Web site www.bsiamericas.com. Maintenance operations can often play a big role to support registration services for the various international standards illustrated in Fig. 4.5.

Consider the following nine evaluation items from this category as you support quality in your organization:

1. Quality control has included maintenance processes within it span of factors impacting quality and has included maintenance factors within its baseline measurement of quality.
2. Major repairs and set-ups impacting quality have clearly defined procedures and specifications established.
3. Documentation of all equipment conditions, factors, and settings that contribute to quality performance is available.
4. Quality of maintenance repairs are evaluated and used as a key performance indicator and as a means to validate crafts skills.
5. Maintenance and quality work together with close coordination and cooperation to resolve quality issues related to maintenance processes.
6. Optimum machine speeds have been established and included in set-up procedures and operator training.

Quality Management Systems

ISO 9001:2000 – enhance the quality of product and service delivered

ISO/TS 16949 – for the automotive industry

As 9100 – for the aerospace industry

TL 9000 – for the telecoms industry

Environmental Management Systems

ISO 14001 – manage your environmental impacts

Responsible Care – RC14001/RCMS

GHGEV – greenhouse gas emissions verification

Medical Devices

ISO 13485, CMDCAS, FDA 510k review

CE Marking Services – for MDD directive, IVD directive, AIMD directive

Other Management Systems

OHSAS 18001 – occupational health and safety management systems

BS 7799/ISO 17799 – information security management systems

BS 15000 – IT service management systems

HACCP and ISO 22000 – food safety management systems

Figure 4.5 A summary of quality, environmental, medical devices, and other management systems.

7. All machine-related quality defects are aggressively evaluated and corrected.
8. Losses due to minor stoppages, idling, and minor equipment failures are addressed by operations and maintenance for corrections.
9. Chronic equipment breakdowns and problems are aggressively investigated as to cause.

Critical Asset Facilitation and Overall Equipment Effectiveness (OEE)

Critical asset facilitation is a term we first heard about when working with the Boeing Commercial Airplane Group across over 50 maintenance operations in their five manufacturing regions. The world-class metric of OEE has been around for a while as well. Both in combination can achieve significant uptime and capacity increase for critical production equipment. The most critical manufacturing operation or process that is *the true throughput constraint* in your plant is the place to start. Also more on this best practice is in Chap. 18—Case Study: Critical Asset Facilitation: Lessons Learned from Boeing Commercial Airplane Group.

Benchmark items in this category include:

1. OEE ratings have been established for major equipment assets or processes to provide a baseline measurement of availability, performance, and quality.
2. Priorities have been established with a plan of action for improving OEE.
3. The OEE factor of availability is being measured and methodology rated.
4. The OEE factor of performance is being measured and methodology is rated.
5. The OEE factor of quality is being measured and methodology is rated.
6. Equipment improvement teams have been established to focus on improving equipment effectiveness based on established priorities for critical equipment.
7. Improvements in OEE are evaluated against baseline (OEE) measurements to determine progress.
8. Optimum machine speeds have been established and included in set-up procedures and operator training.
9. All machine-related quality defects are aggressively evaluated and corrected.
10. A process for critical asset facilitation has been implemented.
11. Critical asset facilitation for condition-based maintenance (CBM) includes CBM technologies and hardware requirements, durations and/or hours for each routine, frequencies for each routine, necessary tools and equipment for each routine, and perform/record baseline of test and measures.

12. Critical asset facilitation includes developing PM plan for each craft/system, inspections/service requirements in sequence of events, schedules for lube and filter routes, duration for each routine by subsystem level and by shift, necessary parts, tools and equipment for each task, and identification of facilitated asset in CMMS.
13. Critical asset facilitation includes a complete review of required documents, manuals, and drawings required to maintain and operate the asset.
14. Critical asset facilitation includes visual management such as labeling of signal or alarm functions/operations, direction of rotation (drives, chains, motors) and replacement parts number, proper control adjustments (for e.g., pressure, temperature, speed, level, voltage), energy lockouts (for e.g., electrical, hydraulics, compressed air), electrical conductors, functions/designations of source on cabinets, panels, boxes, switches, valves, buttons, light, and the like, all fixable, adjustable, or critical fasteners, filters functions (such as hydraulic, lube, and air), and replacement part number, normal operating ranges/levels/readings and label function monitored, lube point with product number, major component functions (for e.g., coolant pump, exhaust fan), motor function being powered (for e.g., pump, axis drive, screw, conveyor), and fluid type/direction of flow/pressure on pipes/hoses/lines.
15. Critical asset facilitation includes facilitated asset documentation material with the following: list of systems, subsystems and components, CBM routines, CBM baselines, PM routines, lube and filter routes, list of prints, manuals, and so forth for PM and CBM routines, visual management digital images, job safety analysis, MSDS, lockout/tag-out documentation, craft asset specific training/certification requirements, part list and/or bill of material for PM routines, startup and shutdown procedures, final operator checklist, documented alignment, and test procedures.

Information resources—eight benchmark categories

See Table 4.2. There are many sources of information for top leaders and maintenance leaders within an organization. Hopefully the *sea of data* has been converted to *information to see and use* to manage/lead your maintenance business. Most of these benchmark categories as shown here have more details and discussion in later chapters.

1. Maintenance business operations, budget, and cost control
2. Work management and control: maintenance and repair (M/R)
3. Work management and control: major shutdowns and overhauls

4. Shop-level maintenance planning and scheduling (Chap. 23—Case Study: Planning for Maintenance Excellence at Lucent Technologies)
5. Shutdown and major maintenance planning /scheduling and project management
6. Manufacturing facilities planning and property management
7. Maintenance performance measurement (Chap. 27: Developing Your Maintenance Excellence Index to Validate Results)
8. CMMS and business system (Four chapters in Part III, "Developing CMMS as a True Business Management System")

Maintenance business operations, budget, and cost control. The benchmark items in this category include:

1. The maintenance budget is based on a realistic projection of actual needs rather than past budget levels.
2. Maintenance expenditures are charged to work centers or operating departments and budget variances monitored to highlight problem areas.
3. Deferred maintenance repairs to operating equipment and facilities-related assets are identified and presented to management during budgeting process with an evaluation as to the negative future impact of deferring maintenance.
4. Maintenance provides key input and support to long-range budget planning for new equipment, equipment overhaul and retrofit, facility expansions, rearrangements, and repairs.
5. Labor and material costs are established for all work orders accumulated to the equipment history file along with problem, causes, and action taken.
6. An equipment history file is maintained for major pieces of equipment to track life-cycle cost, types of repairs, and repair trends.
7. The equipment history file is reviewed periodically to analyze repair trends and define root causes on critical equipment as means to evaluate recurring problem and to improve reliability.
8. Labor and material costs are estimated prior to the start of major planned repair work and projects.
9. Major work order and project-related cost variances are investigated and explained to person authorizing the work.
10. Cost approval guidelines are established for large or special repair jobs as compared to normal repair.

11. The cost of downtime is known and published for major pieces of equipment or work centers and is used in determining priorities for repair.
12. Maintenance operations can operate as an internal business with current financial, budgeting, and cost accounting systems.

Work management and control: maintenance and repair (M/R). The benchmark items in this category include:

1. A work management function is established within the maintenance operation.
2. Written work management procedures, which govern work management and control per the current CMMS, is available.
3. A printed or electronic work order form is used to capture key planning, cost, performance, and job priority information.
4. A written procedure which governs the origination, authorization, and processing of all work orders is available and understood by all in maintenance and operations.
5. The responsibility for screening and processing of work orders is assigned and clearly defined.
6. Work orders are classified by type, for example, emergency, planned equipment repairs, building systems, PM, and project work.
7. Reasonable "date-required" is included on each work order with restrictions against "ASAP."
8. The originating departments are required to indicate equipment location and number, work center number, and other applicable information on the work orders.
9. A well-defined procedure for determining the priority of repair work is established based on the criticality of the work and the criticality of equipment, safety factors, cost of downtime, and the like.
10. Work orders are given a priority classification based on an established priority system.
11. Work orders provide complete description of repairs performed, type labor, and parts used and coded to track causes of failure.
12. Work management system provides means to provide information to customer: backlogs, work orders in progress, work completed, work schedules, and actual cost charge backs to customer.

Work management and control: major shutdowns and overhauls. The benchmark items in this category include:

1. Work management and control is established for major repairs, shutdowns, and overhauls, and includes work by in-house staff and contracted services.
2. Work management and control of major projects provide means for monitoring project costs, schedule compliance, and performance of both in-house and contracted resources.
3. Work orders are used to provide key planning information, labor/material costs, and performance information for major shutdown and overhaul work.
4. Equipment history is updated with information from work orders generated from major repairs, shutdown, and overhauls.
5. The responsibility for screening and processing of work orders for major repairs is assigned to one person or unit.
6. Work orders for major repairs, shutdown, and overhauls are monitored for schedule compliance, overall costs, and performance information including both in-house staff and contracted services.

Shop-level maintenance planning and scheduling. Chapter 23—Case Study: Planning for Maintenance Excellence at Lucent Technologies—includes how one organization chooses planner/scheduler implementation as one best practice need identified from a *Scoreboard for Maintenance Excellence* assessment.

The benchmark items in this category include:

1. A formal maintenance planning function has been established and staffed with qualified planners in an approximate ratio of 1 planner to 20–25 crafts people.
2. The screening of work orders, estimating of repair times, coordinating of repair parts, and planning of repair work is performed as a support service to the supervisor.
3. The planner uses the priority system in combination with parts and craft labor availability to develop a start date for each planned job.
4. A daily or weekly maintenance work schedule is available to the supervisor who schedules and assigns work to crafts personnel.
5. The maintenance planner develops estimated times for planned repair work and includes a work order for each craft.
6. A day's planned work is available for each crafts person with at least a half of a working day known in advance.
7. A master plan for all repairs is available indicating planned start date, duration, completion date, and type crafts required.

8. The master plan is reviewed and updated by maintenance, operations, and engineering as required.
9. Scheduling/progress meetings are held periodically with operations to ensure understanding, agreement, and coordination of planned work, backlogs, and problem areas.
10. Operations cooperate with and support maintenance to accomplish repair and PM schedules.
11. Set-ups and changeovers are coordinated with maintenance to allow scheduling of selected maintenance repairs, PM inspections, and lubrication services during scheduled downtime.
12. Planned repairs are completed on time and in line with completion dates promised to operations.
13. Deferred maintenance is clearly defined on the master plan, and increased costs are identified to management as to the impact of deferring critical repairs, overhauls, and so forth.
14. Maintenance planners and production planners work closely to support planned repairs, to adjust schedules, and to ensure schedule compliance in a mutual goal.
15. The planning process directly supports the supervisor and provides means for effective scheduling of work, direct assignment of crafts, and monitoring of work-in-progress by the supervisor.
16 Planners training have included formal training in planning/scheduling techniques, training on the CMMS, and on the job training to include developing realistic planning times for craft work being planned.
17 Benefits of planning/scheduling investments are being validated by various metrics that document areas such as reduced emergency work, improved craft productivity, improved schedule compliance, reduced cost, and improved customer service.
18. Planning and scheduling procedures have been established defining work management and control procedures, the planning/scheduling process, the priority system, and the like.

Shutdown and major maintenance planning and scheduling and project management. Shutdowns and turnarounds are intense periods of multitudinous simultaneous projects. Lee Peters PE, President of PETERS & company, Project Management and Engineering, Inc. provides 11 key areas for successful project planning.

1. The key to success in these intense periods is starting early.

2. First, develop a plan for how the outage will be planned, executed, and controlled. This prior planning may occur as early as a year before the major work.
3. Maintenance runs on materials—no repair parts means no repair. So it is crucial to get control of the parts first.
4. Get control of the amount of work that will be done on all equipment affected by the outage. Prepare a timeline for identifying work.
5. Do not do any work that can be done in any other environment—on the run, or unit (short-term) outage.
6. Contractors present their own set of challenges. Site access, site security, sanitation, power, canteens and food, contacts, chain of command, materials, tools, equipment, move-in, move-out, and fire safety are all examples that must be controlled.
7. Controlling work done by in-house crews is always a challenge. Do you add people from production, from the street, from contractors to these crews?
8. Knowing what is on order, what is received, where it is, who controls the storage, and how and when to issue the part are issues that remain after the materials are identified and requisitioned.
9. Maintenance departments handle similar challenges on a day-to-day basis; they do not handle the quantity that major work brings at one time.
10. Wise management sets up a project organization to manage major work. This organization may have part-time people, but each individual must know the priority for allocating their time. Project teams must have policies and procedures for managing safety, time, cost, quality, risk, and learning.
11. Now complete the project, the turn-around, and the outage.[‡]

The benchmark items for this category include:

1. The planning and scheduling function includes major repairs, overhauls, and project-type work not considered as part of day-to-day maintenance work.
2. The use of work orders, estimating of repair times, coordinating and staging of repair parts/materials, and planning/scheduling of internal

[‡]This list was provided by Lee A. Peters P.E. F.ASCE (PETERS & company, Project Management and Engineering, Inc., 70 North Main Street, Zionsville, IN 46077, 317-873-0086, Fax 317-873-0052, www.projectleader.com).

resources and contractor support is also included for major work not considered day-to-day maintenance and repair.

3. A project work schedule or formal project management system is used to manage major work.
4. Estimated labor and materials are established prior to starting the project using work orders with effective labor and material reporting to track overall cost, work progress, schedule compliance, and the like.
5. The master plan for all major repairs is available indicating planned start date, duration, completion date, and type crafts required.
6. Resources required for day-to-day maintenance work are not compromised by having to perform major repair type work, installation, modifications, and so forth, consuming in-house resources required for PMs and other day-to-day type work.
7. Scheduling/progress meetings are held periodically with operations to ensure understanding, agreement, and coordination of major work and problem areas.
8. Major work performed by contractors is preplanned, scheduled, and includes measuring performance of contracted services.
9. Planning and scheduling procedures have been established for project type work.

Manufacturing facility planning and property management. Benchmark items for this category include:

1. The equipment asset inventory system provides an accurate and complete record of asset information for both plant and facility assets.
2. New facilities planning, equipment additions, and renovations are well coordinated with both plant and facility maintenance staff.
3. Maintenance staff provides input into the engineering planning process for new facilities, major facility additions, and new production equipment additions.
4. An effective procedure for adding new facility and new equipment information to the asset inventory is used as well as the deletion of equipment being removed from the facility.
5. Maintenance requirements are clearly designated as to responsibilities for both plant maintenance and facility maintenance, and the site is effectively organized and staffed to accomplish both.
6. The overall property management function within the organization provides close coordination with the site's maintenance operation when planning new facilities, renovations, or major production equipment additions.

7. Facilities planning include adequate planning for future maintenance requirements.
8. Consideration is given to life-cycle costing of systems and/or subsystems when designing new facilities, renovations, or major production equipment additions.
9. Maximum standardization of facility systems and subsystems are planned for within new facilities, major renovations, and production equipment additions.

Maintenance performance measurement. Four chapters in Part V, "Validating Best Practice Results with the Maintenance Excellence Index," provide very detailed guides on improving your capability to measure results of your maintenance operation.

The benchmark items in this category include:

1. Maintenance performance measurement includes a wide range of performance indicators in order to evaluate the total effectiveness and impact of maintenance service throughout the operation to include craft labor, planning/scheduling, PM, asset reliability, equipment effectiveness, and cost.
2. Maintenance labor and material costs are reported monthly and reviewed against previous costs or budgeted costs to evaluate current trends.
3. Equipment downtime attributable to maintenance is monitored. The cost of downtime for major pieces of equipment or processes is known and used to measure value of increased equipment uptime.
4. Realistic labor performance standards/estimates have been developed and used for all planned work and recurring tasks.
5. Maintenance labor performance is reported monthly or weekly to evaluate actual performance against established performance standards.
6. The measurement of craft utilization is available from the labor reporting system to evaluate productive hands-on, wrench time versus nonproductive craft time.
7. Periodic reviews are done to evaluate the maintenance operation by determining overall craft utilization and the nature of delays and nonproductive time such as waiting for parts, instructions, unbalanced crew, or waiting for equipment, and the like.
8. The effectiveness of maintenance planning is evaluated by factors such as percent work orders planned versus total work orders, percent work orders completed as planned versus total planned work orders, and percent work orders with estimates versus total work orders completed.

9. Baseline performance factors and information is available to evaluate all ongoing improvements against past performance. Periodic reports to summarize and highlight the tangible benefits from continuous maintenance improvement are provided.
10. A method to measure performance of contracted services is in place.
11. The craft workforce understands the need for improved craft labor productivity and the challenge to remain competitive with contract service providers.
12. Maintenance performance measures are linked to operational performance and support total operations success and profit optimization.

Computerized maintenance management system (CMMS) and business system. Part III, "Developing CMMS as a True Business Management System," provides five chapters on this category and introduces the CMMS benchmarking system as a standard for defining and maximizing current CMMS utilization.

The benchmark items in this category include:

1. The identification of specific CMMS functional requirements has been clearly defined and a complete definition of system capabilities has been determined based on the size and type of maintenance operation.
2. Equipment (asset) history data is complete and accuracy is 95 percent or better.
3. Spare parts inventory master record accuracy is 95 percent or better.
4. Bill of materials for critical equipment includes listing of critical spare parts.
5. PM tasks/frequencies data complete for 95 percent of applicable assets.
6. Direct responsibilities for maintaining parts inventory database is assigned.
7. Direct responsibilities for maintaining equipment/asset database is assigned.
8. Initial CMMS training for all maintenance employees with ongoing CMMS training program for maintenance and storeroom employees.
9. Adequate support from supplier and consultants is budgeted to ensure a successful start-up.
10. Customization of the CMMS is planned to accommodate specific needs for part numbers, equipment numbers, work order, and management report formats.
11. Training for CMMS is a top priority and will be established as an ongoing process for new and existing users of the system.

12. System outputs have been developed into a maintenance information system that provides management reports to monitor a wide range of factors related to labor, material, equipment costs, and so forth.
13. A CMMS systems administrator (and backup) is designated and trained.
14. Inventory management module is fully utilized and integrated with work order module.
15. Reorder notification for stock items is generated and used for reorder decisions.
16. CMMS provides MTBF, MTTR, failure trends, and other reliability data.
17. Engineering changes related to equipment/asset data, drawings, and specifications are effectively implemented.
18. Maintenance standard task database is available and used for recurring planned jobs.

Parts and material resources—two benchmark categories

See Table 4.2.

1. Storeroom operations and internal MRO customer service
2. MRO materials management and procurement

Storeroom operations and internal MRO customer service. Roland Newhouse of Newhouse Associates in Athens, GA shares some important topics to introduce this benchmark category. Provided here is an excerpt from *How to Improve Your Maintenance Storeroom.*

How to Improve Your Maintenance Storeroom

The ownership of the maintenance stockroom varies from organization to organization. The ownership of the maintenance spare parts can be vested in the maintenance department itself, or operated under the auspices of the financial department, or operated under the auspices of the production or manufacturing department. There are stockrooms where operating supplies and maintenance spare parts are intermixed, departments where tooling is kept in a separate stockroom from maintenance spare parts. The integration of manufacturing supplies and/or tooling with maintenance spare parts does not present any type of organizational structure that on the surface needs to be changed. If the budgets are set up properly and the manning is done properly, so that the stockroom, regardless of the diversity of materials kept in it, can be run efficiently and service the needs of all the personnel dependent upon that stockroom operation. More importantly is the concept of ultimate responsibility and the level of input that the maintenance department will have in terms of stocking levels and items to be stocked in the stockroom.

Therefore, it is better that the stockroom, whether it contains main tenance parts exclusively or other materials, be under the direct control of the maintenance department. If it cannot be under the direct control of the maintenance department, the maintenance department must have final say on what MRO materials will be kept in the stockroom and in what quantities, subject to whatever checks or systems and procedures the organization requires.

Assume for a minute that the stockroom is 100% self contained and manned whenever the plant is operating. By 100% self-contained we mean that they issue every part, every cotter pin, and everything is controlled through perpetual inventory. Now, let us construct a list on the easel of things we can do to lighten the stockroom workload. After we construct the list, we will number them from the least impact on control to the largest risk to control.

From my nonaccounting background there appears to be one correct way to manage and control the inventory from an ownership and dollarizing point of view. Then there seems to be about 5000 other ways to do it, some blatantly illegal and some considered the slight bending of the rules and of little consequence. Most operations that are not complying 100% generally adopt an approach that if the auditors don't challenge it, it is OK. This approach leaves open the mechanism by which several different operations can manage this one area of their business in such a large number of ways.

Let us take up the "Harvard MBA" way of doing things. Under this policy the MRO inventory is "owned" by the balance sheet. That means that the MRO inventory as a whole, capitalized items and expensed items is treated as an asset. All MRO materials (except free stock) are purchased, placed into inventory, charged to the MRO inventory account on the balance sheet, and then charged to the P&L at the time of use. In a corresponding accounting entry, the MRO inventory account is relieved of the amount charged to the P&L.

Very simple as far as it goes. However, for the purposes of this analysis, there are three classifications of MRO materials:

- Capitalized spare parts (Generally over $500 with a life of greater than a year)
- Expensed spare parts (Generally less than $500 and/or a life of less than one year)

- Free stock (Commodity items placed on the floor for usage without allocation inventory control)
- Free stock is a no brainer. Buy it, place it on the floor for free access by the craftsmen, expense the purchase price as a maintenance overhead expense, and forget about it.

Expensed spare parts gets only slightly more complicated. Buy it, place it in inventory, and charge the purchase price to the MRO inventory balance sheet account. When the part is used, charge the equipment asset, and relieve the MRO inventory balance sheet account. If there is an ongoing cycle count program as part of the perpetual inventory management, then any time a correction is made as the result of cycle counting, the corresponding dollar correction must be made between maintenance overhead (or equipment asset if it is obvious) and MRO inventory balance sheet account. If there is no cycle counting program, then there must be a physical inventory count at the end of each fiscal year. Based on the physical inventory, the appropriate dollar corrections are made.

Capitalized inventory items represent the most complicated transactions. The IRS expects expenditures to be capitalized and depreciated if:

1. The asset is used in your business, or held for the production of income.
2. The asset has a useful life of more than one year and costs a material amount.
3. The asset will wear out or lose value over time.
4. If the asset is fully installed and ready for use.

Using the above rules let us analyze some real situations.

1. A spare component is purchased that costs $5000 and is placed into inventory. The spare component is a replacement in the event that the currently operating component should fail. Using the foregoing rules we find:
   a. The component is used in the business for the production of income, but is not in current use.
   b. The component has a useful life of more than one year when in use and a virtually infinite life when not in use.
   c. When the component is put into service, it will wear out and lose value.
   d. The component is not installed and ready for use.

   Under these conditions the component value of $5000 is charged to the MRO inventory balance sheet account, the value is capitalized, but no depreciation is incurred as long as the component is not in service.

2. The component in service fails and the inventoried component is installed. Using the original rules we find:
   a. The component is used in the business for the production of income and is in current use.
   b. The component has a useful life of more than one year when in use.
   c. Now that the component is in service, it will wear out and lose value.
   d. The component is installed and ready for use.

   Now the depreciation (whatever depreciation system the business is using) can begin, and will continue until the component fails or the component if fully depreciated.

3. The last example: The same component purchased in example "A" and installed in example "B" fails, however is repairable. The original component that was replaced in example "B" has been repaired and is in the storeroom as the backup spare. We need not concern ourselves with the original component, only the one purchased in example "A."
   a. The component is used in the business for the production of income, but is not in current use.
   b. The component has a useful life of more than one year when in use and a virtually infinite life when not in use.
   c. When the component is put into service, it will wear out and lose value.
   d. The component is not installed and ready for use.

The failed unit is repaired (either internally or contracted out). When the component is restored to serviceable condition it is placed back into inventory valued at the undepreciated value remaining, plus the repair cost (amount charged to the MRO balance sheet account). The depreciation of the component is suspended until is again placed into service.

Now that we have explored the theoretically correct rules for handling capitalization and expensing of MRO inventory materials, why would anybody want to deviate? First and foremost, it is easier for everybody if all the MRO inventory items were expensed at the time of purchase. The maintenance department does not have to conduct inventory chores (cycle or physical). If it is all written off at the get-go, then nobody is too excited about if any are missing (unless it creates an outage).

Plant management does not get caught flat-footed at the end of the fiscal year having to take a big write-off for MRO inventory shrinkage. Also if close attention is not paid to keeping the MRO inventory current, obsolete inventory builds up that cannot be returned or sold. In an expensed storeroom this material can be disposed of without causing a blip on the accounting scene. Under the correct accounting procedures, the plant would have to take the write-off at the time of disposal. If there is one thing that managers and accountants do not like, it is the "surprise accounting." And when one looks around in industry, this is by and large the single biggest deviation to approved accounting methods.

There are other variations on how MRO inventory is treated from an accounting basis. The good news is that as the manager of MRO inventory or the maintenance operation as a whole, you probably will not be making the decisions on how accounting treats MRO inventory. Senior management makes these decisions and generally it would not be within the scope of your responsibilities to challenge these decisions. However, it is important that you understand the correct procedures and the reasons for them.

There are various levels of security, which can be applied to maintenance spare parts stockrooms. The first and most obvious is to have a totally locked-down stockroom manned by inventory personnel reportable to the maintenance department with an issuing window and noone gets in the stockroom other than inventory personnel. If a craftsman has a need to get into the stockroom in order to match a part or find a part to accomplish a jury-rigged repair, an inventory clerk must accompany him at all times. If the plant operates 24 hours a day, 7 days a week as it does in many process industries, then the stockroom must be manned at all times while the plant is in operation.

A manned stockroom also offers many advantages other than the simple security aspect. In a manned stockroom, kitting becomes a relatively simple matter. As discussed earlier, one of our goals is to decrease the amount of time the craftsmen wait at the window of a stockroom waiting for stockroom personnel to select the materials they need and bring it to them. Also with the stockroom manned at all times, stockroom personnel can generally be assigned small hand-tool repairs that are generally kept in the stockroom. There is a single point of responsibility for the integrity of the perpetual inventory.

There are many facilities which feel that they can afford to man the stockroom on a day-shift basis when the bulk of the maintenance crew is in-plant. However, there are no economics of manning the stockroom during off-shifts when only minimized maintenance crews, or line mechanics are in-plant. Although it is preferable to man the stockroom while the plant is in operation, there are compelling economic forces that preclude the strategy. If the going rate for stockroom clerks is $8 per hour with a 40 percent fringe cost, their annual cost is $22,848 per annum ($8 × 2040 h. + 40%). To go from one shift coverage to around the clock coverage requires three additional bodies. This equates to an additional cost of $68,544 to man the stockroom around the clock.

In the average size industrial setting, inventory shrinkage of $75,000 in one year would be considered to be a sizeable loss. Yet, every survey and study we have seen says that approximately 10 percent of inventory shrinkage actually never leaves the facility. The other 90 percent are paperwork errors, constituted mainly of not filling out a requisition when removing a required part. This means that only $7500 of the $75,000 can be preserved as an out-of-pocket cost avoidance with tighter security. It could be difficult to convince the management to spend $68,544 for a gross saving of $7,500.

There can be other factors than head-count economic factors to justify around the clock storeroom coverage. If the storeroom contains a lot of expensive one of a kind spares with long lead times, inventory integrity is of paramount importance. If the operation simply cannot afford to be shut down for the lack of a part (missing because of a paperwork error), full-time coverage is justified. And there are managers who feel that good security keeps honest people honest. Full-time coverage in this case is to remove temptation from those who would not even think of stealing unless the opportunity presented itself. In reality, a maintenance department cannot protect itself from a dedicated thief. After all, this is the maintenance department and not the federal mint. If someone wants to steal from your facility bad enough, they will find a way.

There are various ways to try and exercise control over who has access to the stockroom and what is removed from the stockroom during off-shifts when the stockroom is unmanned. The most obvious way is to provide the off-shift craftsmen with a key to the stockroom. If there are two or more craftsmen with keys to the stockroom and something goes amiss, such as the stockroom left unlocked or materials are missing, there is no definitive way to ascertain who the culprit is. Also keys can be duplicated. One of the methodologies that have been used in the past for controlling stockroom egress and exit during periods of time when the stockroom is unmanned is to use the credit-card lock. Passing the craftsmen's personalized activation card through the locking device allows the craftsmen to enter the stockroom. The act of using the credit card to enter the stockroom records the entry on a computer program along with the time that entry was made and can log exit time, if that becomes important.

In a small- to medium-size maintenance operation, using bar-code technology to operate the stockroom, it is possible to run an adequately secured stockroom without having any manning in the stockroom. This requires an exceptionally dedicated workforce and controlled access rather than unlimited access.

The most effective way to get materials from the stockroom to the job site, for the craftsmen for job start-up, is to have the materials delivered to the job site by the stockroom. In this way the materials can be accumulated from the storage shelves, kitted, and delivered to the job site at the absolute minimum cost. The most expensive way to get material to the job site is to let the craftsmen stand at the window while the stockroom personnel gathered the parts and bring them to the window to the craftsmen, and then the craftsmen transport them to the job site.

Not every plant is going to have a large enough maintenance department to support a full-time person delivering stock or even a part-time person from the stockroom delivering stock to the job sites. However, if the plant is of sufficient size to support this type of operation, it then becomes economical to supply each of the craftsmen with radio communication to the central stockroom. As parts are required at job site the stockroom can deliver them to the job site on an emergency basis or on regular rounds through the facility basis.

In a very large operation, such as an oil refinery or very large plant facilities where travel time from a centralized maintenance shop becomes undesirable and uneconomical, it can become very advantageous to set up satellite stockrooms throughout the plant at strategic locations. In the satellite stockrooms there would be contained specialized materials that fit on equipment that are in the area of the satellite stockroom. Also the satellite stockroom would be stocked with a minimal amount of expendable materials such as nuts, bolts, and cotter pins. The important issue here is control. The satellite stockroom is obviously going to be unmanned and should be locked and under control. You need some methodology by which materials are drawn out of the satellite stockroom, that they get properly reported both for replacement and for charging of materials to machinery repaired. Probably the best way to insure control or at least exercise the best control possible under the circumstances would be to use bar code technology to remove materials from the stockroom. These bar code scans, as materials are removed, can be communicated back to the central stockroom through hard wiring or through radio frequency. However, if the bar code unit is a portable unit, it could be picked up on a periodic basis and downloaded batch style into the central stockroom.

Any stockroom, be it central stockroom or satellite stockroom, is not going to be effective unless a good perpetual inventory is in place. There are several ways to maintain a perpetual inventory. One is to simply inventory everything in the stockroom on a periodic basis and update the records accordingly. Another methodology is to use the cycle counting system.

Under Pareto's law, 80 percent of the total value of the parts in the stockroom is contained in the 20 percent of the most expensive parts. We assign those 20 percent of parts a designation called "A." The next 15 percent of the value are contained in the next 40 percent of the parts in the stockroom. We will assign that 40 percent segment the letter designation "B." The last 5 percent of the value of the parts in the stockroom are contained in the last 40 percent of the parts. We will assign those the letter designation "C." If we are going to be prudent managers of our materials in the stockroom, we want to pay very close attention to the "A" parts and very little attention to the "C" parts. Therefore, we make a determination that we would like to count the "A" parts four times a year. We would like to count the "B" parts at least once a year and the "C" parts we may elect to never count or we may elect to count once every three years. We now have the basis for the ABC cycle counting system. We take the total amount of "A" parts that are contained in the stockroom, multiply it times four, so that we will count them four times a year. Then we divide by 52 and that is the number of "A" parts that we have to count every week and verify bin content every week. We take the "B" parts, multiply them by 1, divide them by 52 and those are the numbers of "B" parts that we have to count every week. And if I am going to count "C" parts every three years, we divide the number of "C" parts by 3 and divide that by 52 and that's the number of "C" parts I have to count every week. When we say count we are specifically referring to going to the location where the parts are stored and verifying the quantity on hand.[§]

Benchmark items in the store room operations and internal MRO customer service category include:

1. The parts inventory system provides an accurate and complete record of information for each stock item. Parts "where used" is included in the master database along with usage, vendor information, and warranty information.
2. The "ABC" classification of stock items is known and proper storage methods and accountability is established for each.
3. "A" and "B" items have valid reorder points, and safety stock levels established.
4. "C" items (50 percent of stock items with 5 percent of total inventory value) are identified and use two-bin system or floor issue.
5. Inventory accuracy is determined by an effective cycle counting program. Cycle counting used = 10, count once per year = 7, count occasionally = 5, do no inventory counts = 0.

[§]This part of the text is provided by Roland Newhouse, Newhouse Associates, Athens, GA. E-Mail: RolandNewhouse@aol.com and Phone: 706-372-5222

6. Inventory accuracy is regularly measured and is 95 percent or above. 95 percent or above = 10; 90–95 percent = 9; 80–89 percent = 8; 70–79 percent = 7; less than 70 percent = 5.
7. An up-to-date storeroom catalogue is readily available to crafts (hard copy or electronic) and includes all stock items, storage locations, stock numbers, and so forth.
8. Parts usage history is continually reviewed to determine proper stock levels, excess inventory items, and obsolete items.
9. A critical spares listing is available for critical assets and critical spares (insurance items) are denoted in the parts inventory database.
10. Spare parts for critical instrumentation system components are listed with inventory tracking numbers and are stored in a separate area. Remotely stored items (for safety or quick access) must be listed in main inventory as to storage location.
11. Storeroom procedures are in place that define issues, receipts, inventory control, access control, parts accountability, reserving of parts, staging, quality control of parts, and the like.
12. An operations assessment has been conducted for the storeroom operations to provide overall evaluation of facilities, storage and handling equipment, staffing levels, inventory levels, systems, and procedures.

MRO materials management and procurement. Benchmark items in this category include:

1. Procedures and evaluation criteria for adding new maintenance parts and materials to stores are used.
2. Stores requisitions and issues are tied to the maintenance work order and changed directly to the repair job.
3. Maintenance planners and the storeroom personnel coordinate to reserve repair parts and material for planned work. "Kitting" and direct delivery to the job site is done whenever possible.
4. Purchasing has an effective program to evaluate vendor performance and quality.
5. Purchasing has developed partnerships with selected vendors and suppliers and is committed to purchasing based on fast delivery, quality parts, and service.
6. Maintenance storeroom staff is well-trained, customer-oriented, and provide a high level of customer service to maintenance.

7. Maintenance storeroom and MRO procurement performance indicators have been established, and are evaluated and reported on a monthly basis.
8. An effective control method is in place for emergency purchases.
9. Purchasing and maintenance work together to ensure procurement of quality parts and material and the "high cost of low bid buying is avoided."
10. Stockouts are being monitored as a performance measure of customer service from the storeroom operation and the MRO procurement process.
11. Standardization of parts and components is being pursued by maintenance and procurement staff.

Hidden Resource: The Synergy of TEAM Efforts; Two Benchmark Categories

See Table 4.2.

1. Maintenance supervision/leadership
2. *Continuous reliability improvement* (CRI)

Maintenance supervision/leadership. For this category Lee Peters PE provides some very appropriate comments and personal lessons learned as an introduction to this important benchmark category. I first met Lee in Vietnam in 1970. He was my company commander for a short period with those 72 Yellowbird dump trucks. Lee Peters is not a relative but a great friend and a true leader. I personally agree when he says that "Maintenance supervisors are the heroes of every maintenance shop."

Maintenance supervisors are the heroes of every maintenance shop. As the conductor of the maintenance orchestra, supervisors make wonderful music by playing their mechanics. Great supervisors know their people, their skill levels, their attitudes, and their desires. Individual mechanics skills grow and develop by thoughtful work assignment. Productive work teams are welded together by the supervisor mixing and matching work crews.

Supervisors dramatically impact the total crew productivity by scheduling mechanics by name to work a weekout. Besides scheduling, the maintenance supervisor teaches new skills, troubleshoots reliability problems, expedites materials, plans tomorrow's jobs, oversees work completion, coaches improved performance, coordinates with production, and assigns work to mechanics. Every day of every week each maintenance supervisor works wonders.

Quality work is obtained daily because the supervisor is attentive, skilled, and conscientious. The physical touches to the maintenance process sets example and expectations. Being of the floor stamps out poor practice, assures safety, and enforces cleanliness. They vividly demonstrate their technical knowledge, incredible supervisory skill, drive for completion, and relentless discipline by all they deliver. They earn wizard status daily.

Tragically, well-run maintenance organizations are transparent to production management. No one knows they are there because nothing fails. We must remember and recognize that good maintenance supervisors are indeed hidden assets that never show on a balance sheet.

The benchmark evaluation items for this category include:

1. Nonsupervisory work is minimized as a result of supervisors having adequate clerical support, storeroom support, planner support, and the like.
2. Supervisors perform primarily direct supervision of maintenance to include scheduling work assignments, verifying quality of completed work, evaluating performance, and so on.
3. Supervisors actively support good housekeeping and the safety program by conducting/attending meetings, providing ideas, and having an attitude that creates greater safety awareness.
4. An effective supervisory development program is available to increase supervisory leadership and technical skills.
5. Maintenance supervisors are team players and are able to gain cooperation and support from operations management staff for the improvement of overall customer service.
6. Supervisors actively support continuous maintenance improvement with ideas and suggestions.
7. Supervisors promote *pride* in maintenance and encourage ideas from craft employees.
8. Supervisors have the technical background to identify training needs of their craft workforce and create positive support to craft skills development.
9. Supervisors are managing and leading their work groups as if "they owned the company."**

**The introduction to this benchmark category was provided by Lee A. Peters P.E. F.ASCE (PETERS & company, Project Management and Engineering, Inc.70 North Main Street, Zionsville, IN 46077, 317-873-0086, fax 317-873-0052, or at www.projectleader.com).

Continuous reliability improvement (CRI). CRI is a process developed and used by staff from The Maintenance Excellence Institute that goes beyond current *reliability-centered maintenance* (RCM) and TPM approaches focused only upon parts of the total maintenance business process. CRI combines continuous improvement, methods of improvement, and the principles of Alan Mogenson's Work Simplification to outline a continuous, integrated process for *Working Smarter—Not Harder.* CRI improves total reliability of the following resources in all types of maintenance environments:

- Equipment/facility resources (asset care/management and maximum uptime via RCM techniques)
- Craft and operator resources (recognizing the most important resource: crafts people and equipment/process operators)
- MRO Resources (establishing effective materials management processes)
- Maintenance information resources (effective information technology applications for maintenance)
- Maintenance of technical knowledge/craft skills base (closing the technical knowledge resource gap)
- Synergistic team processes (topping the value-added resource of effective leadership-driven teams to support total operations success)

CRI brings about a positive pride-in-Maintenance culture and proactive attitudes/actions for the synergy of team efforts to improve each maintenance resources area. All this is focused on contributing to gained value of all resources to support total operations success of all type of public and private sector organizations.

Benchmark items in this category include:

1. CRI is recognized as an important strategy as evidenced by the current status of maintenance and the ongoing improvement activities.
2. The current level of commitment to CRI is based on results of overall assessment.
3. Maintenance improvement opportunities have been identified with potential costs and savings established.
4. Improvement priorities have been established based on projected benefits and valid economic justifications.
5. Top management has reviewed, modified, and/or approved maintenance improvement priorities and has made a commitment to action.

6. Sufficient resources (time, dollars, and staff) have been established to address priority areas.
7. Implementation plans and leaders for each priority area are established.
8. A team-based approach is used to identify and implement practical solutions to maintenance improvement opportunities.
9. CRI for the maintenance resource of physical asset and equipment resource is rated.
10. CRI for the maintenance resource of craft labor resources is rated.
11. CRI for the maintenance resource of MRO material resources is rated.
12. CRI for the maintenance resource of information resources is rated.
13. CRI for the maintenance resource of technical skills and knowledge is rated.
14. Maintenance employees participate on functional teams within maintenance and on cross-functional teams with other department employees to develop maintenance improvements.
15. Written charters are established for each team to outline reasons for the team, process to be used, resources available, constraints, expectations, and results expected.

Going beyond RCM and TPM

We understand the need for the RCM and TPM types of improvement processes and Six Sigma. On the shop floor we see today's trend toward forgetting about the basics of "blocking and tackling" while going for the long touchdown pass with some new analysis paralysis scheme. The approach presented is built upon the basics and then goes well beyond the traditional RCM/TPM approaches with CRI of all maintenance resources. It is about improving the total maintenance process and all maintenance resources as shown in Fig. 4.6.

We strongly believe in basic maintenance best practices as the foundation for maintenance excellence. Our improvement process includes all maintenance resources, equipment, and facility assets as well as the crafts people and equipment operators. It also includes MRO materials management assets, maintenance informational assets, and the added value resource of synergistic team-based processes. CRI improves the total maintenance operation.

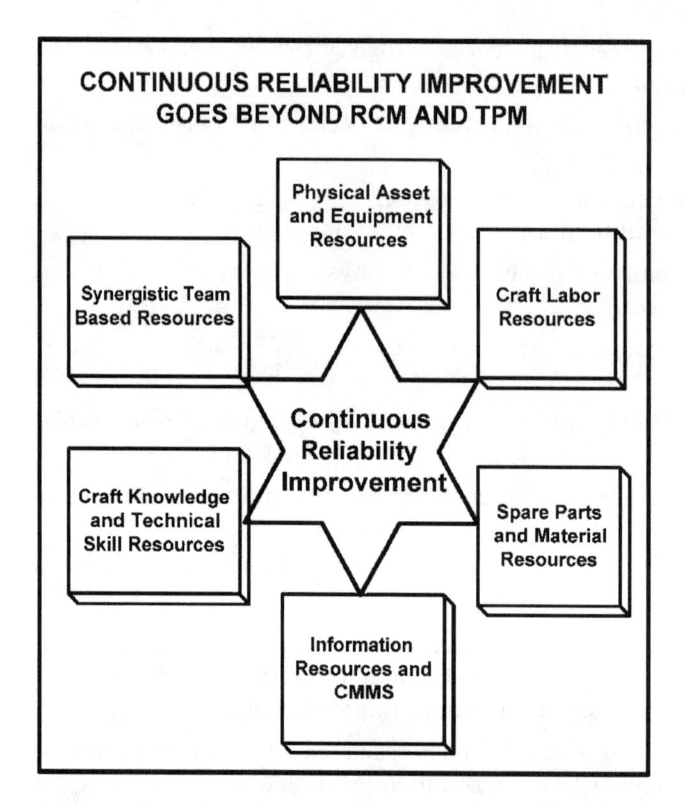

Figure 4.6 Improve maintenance in total and go beyond RCM and TPM.

The Scoreboard for Excellence

The Scoreboard for Excellence concept and the various versions have been used to perform over 200 maintenance evaluations and over 5000 organizations have requested and received copies of them for their internal use. *The Scoreboard for Maintenance Excellence* was used by plant maintenance operations for example at Honda of America after making slight modifications and then used it extensively as a self-assessment to help direct their maintenance strategy. It was then translated into Japanese for presentation to key Japanese executives visiting Honda plants in the United States. Another excellent example is where the Boeing Commercial Airplane combined elements from this same scoreboard with their companywide maintenance goals to develop *The Boeing Scoreboard for Maintenance Excellence.* Over 50 facilities maintenance work units, at region, group and team levels were then evaluated with structured on-site visits. The use of this comprehensive best practice guideline specifically tailored to maintenance of aircraft manufacturing equipment (and the associated manufacturing and test facilities complexes) across the United States is still one of the largest internal benchmarking efforts ever undertaken.

The Scoreboard for Maintenance Excellence, as shown previously in Table 4.2 The Scoreboard for Maintenance Excellence—Summary of Benchmark Categories includes 27 evaluation categories (maintenance best practices areas). It evaluates the total maintenance operation within the scope of coverage for a manufacturing organization. But there can be also be well-defined focus areas when an evaluation is conducted such as on CMMS needs and current system gaps, on planning/scheduling, on MRO materials management, the application of CRI and the need for predictive maintenance technologies.

A complete scoreboard assessment is recommended and so *Just Do It!*

For example, MRO materials management, storeroom operation, and procurement may be an area needing special attention. Shop-level planning and scheduling is often a typical need and can be a primary focus area. Regardless of the different areas creating the obvious concerns and "organizational pain," a short-term, piecemeal approach to an evaluation is not recommended. A complete benchmark evaluation of the total maintenance operation is highly recommended. There are 300 specific evaluation items that are evaluated through direct shop floor interviews, close observations, and review of information or procedures. Each one is important; some provide more value than others. But each of the 300 items on *The Scoreboard for Maintenance Excellence* is part of establishing a solid foundation for profit and customer-centered maintenance. Long-term CRI is also a very important connecting link that ensures we consider all maintenance resources in our improvement process. Later in Chap. 7—Case Study: The Scoreboard Self-Assessment: Just Do It!—we encourage you to do a self-assessment with or without external resources like The Maintenance Excellence Institute. The key is to *never, never, never give up* on the fact that maintenance and reliability improvement must be continuous. A continuous commitment and faith that you, as the maintenance leader demonstrate, will make a positive difference wherever you might practice the art and science of maintenance. The journey is not to *maintenance excellence* but rather a journey that *continuously moves toward maintenance excellence.* Another key point is all about what this book encourages you to do with all the topics in this book. That key point borrowed from the Nike commercial says to . . . "JUST DO IT."

Chapter

5

Guidelines for Conducting a Scoreboard for Maintenance Excellence Assessment

A number of organizations like The Maintenance Excellence Institute (TMEI) stand ready to support your mission-essential plant-maintenance operation with an assessment performed by well-qualified staff. While a self-assessment has many benefits, we believe an assessment conducted by an outside resource provides a greater sense of the "big picture" in terms of objectivity and completeness. Regardless of your situation it is important that you do something to determine "where you are." Should you want to begin with an internal self-assessment of maintenance, here are some guidelines to consider when using "The Scoreboard for Maintenance Excellence." Chapter 7—"Case Study—The Scoreboard Self-Assessment: Just Do It!"—also reinforces that you can do a self-assessment at anytime with the overall tools in this book. That has been our stated goal from the Preface. Figure 5.1 provides the visual road map.

Obtain Leadership Buy-in

a. Establish a firm commitment from the organization's top leadership for conducting a total maintenance operation assessment. Figure 5.2 illustrates this important key to your success. Not only must the commitment be from top leadership it must begin with the maintenance leader.

b. Establish a firm commitment from the organization's top leadership to take action based on your current needs. Every organization is different and may have a need to focus on one or more specific best practice categories. In this case the self-assessment can focus on

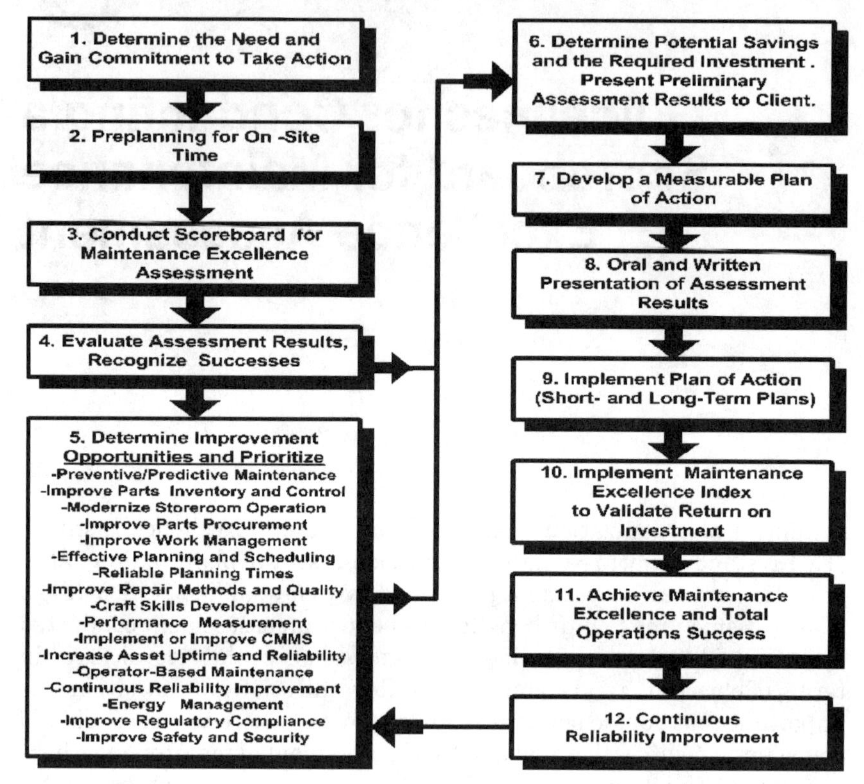

Figure 5.1 The Scoreboard for Maintenance Excellence assessment.

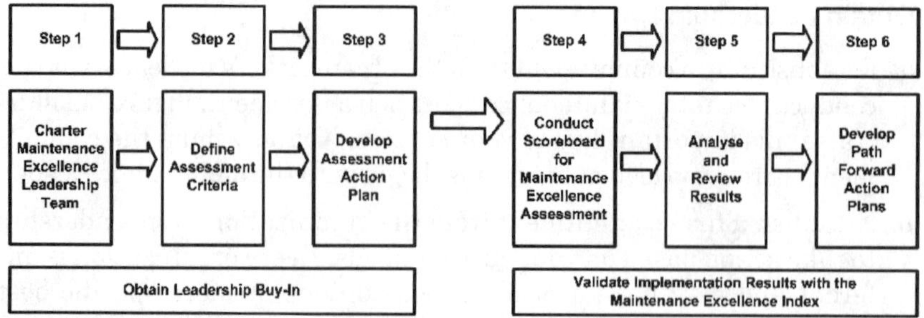

Figure 5.2 Key steps for your self-assessment.

areas such as improving storeroom operations or planning and scheduling processes. But the assessment must take a complete look at all scoreboard categories and get a complete "where we are now" picture of the total maintenance process.

c. Maintenance leaders must be brave enough and prepared to share both good news and bad news based upon results of the evaluation to top leaders. When we perform an assessment, we strive to make it a very positive process for all involved. There will always be successes to highlight and no real bad news, but we focus on opportunities.

Step 1: Charter maintenance excellence strategy team

a. Establish a maintenance excellence strategy team to guide and promote improved maintenance practices within your organization, whether single or multiple sites.

b. Utilize a team-based approach with a cross-functional evaluation team specifically chartered for conducting and preparing the results of your evaluation.

c. Have at least one team member with solid background and knowledge in each of the 27 evaluation categories.

d. Consider bringing on board third-party outside support to be a part of your team and the assessment process.

Step 2: Understand the evaluation categories and evaluation criteria

a. Gain complete understanding of each evaluation of 27 categories.

b. Gain complete understanding of the 300 evaluation items.

c. Modify existing evaluation criteria as required for your organization.

d. Define importance and weighted value of evaluation categories.

e. Add or delete evaluation criteria as required for your unique operation.

f. Ensure that all team members understand the scoring process for each evaluation item.

g. Ensure that consistency in scoring each evaluation category is applied using standard guidelines.

Step 3: Develop assessment action plan

1. Determine baseline information requirements, persons to interview, and observations needed prior to start of evaluation. A preassessment checklist follows.

Preassessment checklist for baseline information. The following checklist is not all inclusive of the information required for an assessment. It does represent very important areas that we try to get prior to performing an assessment. Typically all of the data/information listed in the following checklist may not be readily available. As much information as possible should be assembled and reviewed prior to the start date of the assessment. When we perform an assessment this step allows us to gain a better understanding of the client's operation prior to the on-site visit and helps save a lot of time when on site.

Organization charts/job descriptions.

- Mission statement/quality statement for your overall organization
- Mission statement for your maintenance operation
- Staff directory of personnel and contact information/e-mail, and the like
- Organization chart for your plant/facility
- Maintenance organization charts by each craft area
- Head counts in each craft area
- Position descriptions of key maintenance leaders—managers, supervisors, foremen, maintenance engineers, planners, team leaders, storeroom supervisor, and so forth
- Craft job descriptions—sample from each craft area

Craft skills development.

- Information on craft training completed in past several years, that is, in-house programs, vendor seminars, or ongoing apprenticeship programs
- Results of past craft skills assessments or employee surveys on training needs, and the like

Craft labor rates/overtime history

- Average hourly rate by craft area
- Average fringe benefit percentage factor for your organization
- Overtime rates
- Past 3 to 5 years history of overtime
- Overtime hours (by craft area, if available)
- Overtime payroll costs (by craft area, if available)

Maintenance budget and cost accounting.

- Total maintenance budget (3 years)
- Total craft labor cost (3 to 5 years)
- Total parts/materials (3 to 5 years)
- Contract maintenance costs (3 to 5 years) (by type of service, if available)

- Copy of current maintenance budget
- Copy of previous years' maintenance budget
- Procedures for charging craft labor and parts/materials to equipment history or to maintenance budget accounts
- Procedures for monitoring equipment/process uptime

Preventive, predictive, reliability centered maintenance (RCM) and total productive maintenance (TPM) processes.

- Sample lubrication services checklists or charts
- Sample *preventive maintenance* (PM) checklist/instructions
- List of equipment having *predictive maintenance* (PdM) services for vibration analysis, oil analysis, infrared, or other technologies
- Any summary results of major success with PM/PdM and reliability improvement at the plant/facility site
- Summary of experiences with RCM type processes
- Summary of experiences with TPM type processes

Maintenance storeroom and maintenance repair operations (MRO) purchasing operations.

- Total dollar value of current maintenance-related inventory (ABC classification, If available)
- Inventory dollar value of critical spares/insurance-type items
- Inventory dollar value of all other items
- Estimated value of items not on inventory system
- Total number of stocked items (stock-keeping units)
- Total number of critical spares/insurance items
- Total stock items—all other parts not classified as critical spares
- Copy of storeroom procedures for:
  - Purchase requisitions/purchase orders
  - Additions to stock/establishing stock levels
  - Issuing/receiving
  - Receiving requirements for incoming quality validation
  - Obsolete parts
  - Parts/material inventory classification
- Results of most recent physical inventory or cycle count results
- Accuracy level, write-offs or adjustments
- Copy of storeroom catalog (if on-line will review on-site)
- Information on vendor stocking plans and vendor partnerships currently in place

Work orders and work control.

- Copy of work order currently being used and priority system description
- Copy of work order/work control procedures

- Current backlog by craft area (if available) planning and scheduling procedures
- Work management procedures
- Time-keeping methods (operations personnel and maintenance personnel)

Computerized maintenance management system (CMMS)

- System name and date initially installed
- CMMS system administrator's name and contact information
- Names of other key staff with responsibilities for data integrity:
  - Parts information
  - Equipment/asset information
  - PM/PdM procedures
  - Maintenance budgeting and services charge backs
  - Shop-level planning and scheduling
  - Project type planning and scheduling
- Your primary CMMS vendor support person and contact information

Maintenance performance reports.

- Copy of any reports or current information that is being used to evaluate maintenance performance
- Summary of *key performance indicators* (KPIs) that you feel are needed
- Operations performance report
- Copy of any reports or information used to evaluate operations performance that uses maintenance-related data, that is, equipment uptime (availability)

2. Develop schedule and implementation plan for the assessment.
3. Develop and implement a communication plan within the organization to inform all about the process.

Step 4: Conduct assessment of total maintenance operation

a. Assign team members to specific evaluation categories (ideally, in two-person teams for each category).

b. Conduct kickoff meeting, firm up specific interview and observation schedules.

c. Conduct the assessment, record observations, and assign scores to each evaluation item.

d. Ensure CMMS is an effective business management tool for maintenance.

Step 5: Analyze, review, and present results

a. Review all scoring for consistency.

b. Develop final results of the assessment and document in a written report.

c. Determine strengths/weaknesses and priorities for action.

d. Define potential benefits either direct or indirect savings or gained value from existing resources.

e. Gain internal team consensus on methodology for determining benefits and the value and type of savings.

f. Present results to top leaders with specific benefits and improvement potential clearly defined.

g. Refine results based on feedback from top leaders.

h. Gain commitment from top leaders for investments to implement recommendations.

Step 6: Develop path forward for maintenance excellence

1. Develop a strategic plan of action for implementing best practices.
2. Define tactical plans and operational plan of actions.
3. Define key performance measures, especially those that will validate projected benefits.
4. Implement methodology to measure performance and results.
5. Measure benefits and validate *return on investment* (ROI).
6. Maintain a *continuous reliability improvement* (CRI) process (i.e. repeat assessment process).

   Follow up initial use of "Scoreboard for Maintenance Excellence" with periodic assessments every six to nine months.
   Follow up initial use of CMMS benchmarking system.
   Continuously validate results with maintenance excellence index.

Invest in External Resources

It is extremely important to know where your organization stands on physical asset management and maintenance issues and challenges so that it can quickly identify areas for improvement. Every delay along the way delays receiving the potential benefits and added value. Self-assessments and very good starting points are recommended when nothing else is available for using external support. But a more comprehensive, objective assessment performed by external consulting resources (or possibly qualified corporate-level staff (with decades of maintenance-focused expertise is highly recommended). In the long run, external resources will provide additional justification and measurable results. Total operations' success depends upon effective maintenance of physical assets. So take a look at using external resources to support this essential first step after your organization makes the initial commitment. Chapter 6 provides, "A Strategy for Developing a Corporate-Wide Scoreboard," something that your *chief maintenance officer* (CMO) can coordinate and lead as a first job when taking over this new position.

Chapter

6

Strategy for Developing a Corporate-Wide Scoreboard

The objective of this chapter is to provide guidelines for developing a "Scoreboard for Maintenance Excellence" that looks to standardize maintenance best practices across a multiple-site operation. Then use this organizational specific benchmarking guide to conduct a pilot assessment, to provide a recommended path forward, and to implement results with a measurement process that will validate *return on investment* (ROI). The assessment scope is the total maintenance operation at the pilot site, including assessment of the current *computerized maintenance management systems/enterprise asset management* (CMMS/EAM) installation, the maintenance storeroom operations, *maintenance repair operation* (MRO) procurement, and the overall *continuous reliability improvement* (CRI) initiatives. The recommended path forward strategy that we use is outlined here.

Recommended Path Forward Strategy

1. *Define impact on capacity, quality, and internal customer satisfaction.* Determine specific areas where the total maintenance process can have or has had an impact on throughput, quality, and customer satisfaction. The team determines key assessment areas related to capacity constraints due to maintenance, negative impacts on quality, internal customer service, or other functional areas that should be incorporated into the assessment process. A cooperative link to maintenance leaders and engineering staff at other sites should be initiated. Also there may be a need to focus the assessment on specific best practice areas.

- Typically, at this point, we recommend that a maintenance excellence strategy team be chartered to help guide the resulting improvement process across the multisite operation. (See App. F: Maintenance Excellence strategy team sample charter.)
- For a large operation, it is also highly recommended that another new C position be established; the *chief maintenance officer* (CMO) to support profit optimization via maximizing physical asset management and maintenance of production assets and facilities.
- A recommended maintenance excellence strategy team charter and a recommended position description for a CMO would be developed at this point.

2. *Establish scoreboard for maintenance excellence.* At this point, enhanced version of the "Scoreboard for Maintenance Excellence" incorporates evaluation criteria specific to the organization into the baseline assessment. The client and *The Maintenance Excellence Institute* (TMEI) then develop a "Scoreboard for Maintenance Excellence" that defines maintenance best practices and standard procedures that the client wants to adopt as its corporate best practice baseline.

The Boeing Commercial Airplane example used in Part II combined elements from this same scoreboard with their companywide maintenance goals to develop "The Boeing Scoreboard for Maintenance Excellence." All plant and facilities maintenance workunits, at region, group, and team levels were then evaluated which encompassed a crafts workforce of over 5000. This effort is still one of the largest internal benchmarking efforts ever undertaken. To evaluate maintenance information technology resources, the pilot assessment will also utilize *The CMMS Benchmarking System* (discussed later in Part III) to define current CMMS utilization plus detect any functional limitations of the current system.

3. *Conduct pilot assessment at a client site.* The benchmark assessment at the pilot site is then conducted using the client-specific "Scoreboard for Maintenance Excellence" developed for the client. The approach for interviews, observations, on-site time, and preassessment time is to:
   - Outline data and information requirements prior to start date of pilot assessment.
   - Preview data prior to assessment start date.
   - Establish tentative interview schedules for on-site time.
   - Conduct short kick-off meeting to confirm schedules and make introductions.
   - Conduct staff-level interviews, observe operations, and interview shop floor staff.
   - Conduct customer and operations staff interviews.

- Evaluate the total maintenance and facilities operation (both in-house and contract operations) based upon the "Scoreboard for Maintenance Excellence" best practice categories.
- Perform baseline assessment of the current CMMS installation using *CMMS Benchmarking System.*
- Interview maintenance leaders, selected customers, procurement and storeroom personnel, and craftspeople.
- Complete the assessment.
- Prepare the results of the assessment off-site.

4. *Define successes and opportunities.* Specific observations are noted, baseline ratings established, and recommendations made for each evaluation category based on the results of the assessment. We then prepare the initial "draft" results from the assessment, and the client and the MEI team develops a consensus on final recommendations. We discuss observations in the report and define the "state of maintenance processes" at the pilot site. It is also very important to highlight client successes that have been noted during the assessment and ensure that the maintenance leader and others get recognition for their past successes.

5. *Define potential benefits, gained value and ROI.* At this point we also determine potential benefits and ROI possibilities for the pilot site. This step involves obtaining baseline information in a number of key areas that will form the foundation for projected savings and total operations benefits. The client and the TMEI team develop a consensus on the final scope of projected savings and gained value of productivity improvements. If these benefits are representative of the potential at other sites, then this is denoted to project the relative potential for the entire organizations. From this information, a consensus on projected benefits is established for:

   a. Value of asset/equipment uptime providing increased capacity and throughput
   b. Value of increased quality and service levels due to maintenance
   c. Value of facility availability or cost avoidance from being non-available
   d. Value of increased direct labor utilization (production operations)
   e. Gained value from increased craft labor utilization/effectiveness via gains in wrench time
   f. Gained value from increased craft labor performance/efficiency
   g. Gained value of clerical time for supervisors, planners, engineering, and admin staff
   h. Value of MRO materials and parts inventory reduction
   i. Value of overall MRO materials management improvement

j. Value of overall maintenance costs reductions with equal or greater service levels
k. Value of increased facility and equipment life and net life-cycle cost reduction
l. Other manufacturing and maintenance operational benefits, improved reliability, and other reduced cost

The basic foundation of our consulting approach (as outlined in this book) is that for all projections of benefits and gained value we make, there will be a recommended method to validate results from successful implementation of our recommendations. This helps validate our services, but most importantly it provides return on investment for maintenance improvements. It is also important for the client team to do the same thing if they are not using external resources. Make sure you do two things to sell getting your investments for maintenance improvement.

a. Clearly define all potential benefits and gained value you expect to receive.
b. Put in place methods to validate results to all concerned.

6. *Document results of pilot assessment and plan best practice implementation.* TMEI then prepares the assessment report with client reviews and defines specific improvement opportunities. A recommended path forward for implementation is then developed. The client and TMEI team plan and present results of the assessment to key leaders in the organization. The strategic-level plan with both tactical and operational items that evolves from this step requires a close team effort between the client, TMEI, and leadership at the pilot plant facility. Additional internal and external resources that are needed for implementation are also identified. TMEI provides proposals as appropriate for support to implementing recommendation to achieve implementation of potential results. The client and TMEI team ensure that the strategic plan of action becomes fully integrated with the business plan for the operation.
7. *Measure results of implementation.* To measure the results we provide a recommended maintenance excellence index (MEI) (discussed later in Part V) that will measure and validate results of CRI and maintenance process improvements. This key deliverable utilizes MEI's proven methodology for managing, improving, and measuring maintenance as a profit center. It involves developing a consensus on 10 to 15 metrics that will measure overall performance of the maintenance operation.

   This process of "internal benchmarking" includes high-priority metrics in the specific areas where benefits and savings were projected in any of the 12 areas from Step 5 mentioned earlier. Typically this step requires the maintenance excellence strategy team's consensus on the metrics to be included in the MEI or a leadership decision is made on the metrics to be used. This step also requires establishing baseline values and performance goals for each metric selected and continuous measurement of improvement during implementation to validate projected results defined in Step 5.

8. *Refine scoreboard for maintenance excellence.* There may be a need to revise and update the client-specific "Scoreboard for Maintenance Excellence" based on actual results of the pilot assessment. Baseline evaluation criteria are adjusted to fit the unique maintenance and organizational culture requirements of each client along with incorporating lessons learned from the pilot assessment.
9. *Plan assessments at other sites.* MEI team with the client develops the plan of action for the assessment at the other sites. At this point a CMO should be firmly established to support the other assessments and the maintenance improvement processes that continuously evolve at each site. Two important recommendations:
   - TMEI highly recommends the establishment of a top level maintenance champion (CMO) responsive to the needs of multiple sites within the organization. This internal person will be essential for providing technical leadership to future best practice implementations, standard practice development, and the measurement process that will be common to all sites.
   - TMEI also highly recommends that future assessments continue with the independent expertise from TMEI for objectivity and consistency of initial baseline ratings at all sites. The client's maintenance champion would then conduct all follow-up assessments, support best practice implementation, support the measurement process, and follow up on implementation progress across the total organization.
10. *Present assessment results and recommendations to top leaders.* The assessment report with specific recommendations and the recommended path forward for implementation is presented to key leaders by the client and TMEI team. Included herein is a summary of key deliverables that can be expected from successful execution of Step 1 to Step 10.

Summary of deliverables to achieve results

a. A recommended maintenance excellence strategy team charter and functioning team
b. "The Scoreboard for Maintenance Excellence" developed for the organization and revised if required after the pilot assessment
c. In-depth assessment of current maintenance practices at a pilot site

d. Recommendations in all 27 maintenance evaluation categories
e. Benchmark assessment of the current CMMS/EAM installation based on the *CMMS Benchmarking System* criteria
f. Recommendations to improve utilization of the current CMMS/EAM installation
g. Recommended path forward for a strategic-level plan with tactical/operational items
h. Definition of improvement opportunities
i. Summary of potential benefits with an estimate of savings and gained value for each
j. Documented assessment of results in a written report
k. Recommended performance measurement process ready for immediate implementation
l. An MEI that validates overall performance improvement
m. Recommended metrics, data sources, and documentation in a standard operating procedure guide for the client-specific MEI
n. Plan of action for future assessments at other sites
o. An organization maintenance champion (CMO) established
p. Oral presentation of results to top leaders
q. A proposal that defines where TMEI can best support the recommended path forward.

The project benefits and the measurement process, now ready to implement, helps to ensure commitment to action by the client's top leaders. TMEI provides proposals as appropriate to help implement maintenance best practice recommendation to support early achievement of potential benefits that we have projected.

11. *Conduct assessments at other sites and begin maintenance excellence process.* This step begins the scheduled assessments at other sites, defines additional opportunities, plans and implements recommendations, and validates results. CRI and refinements to Steps 3 through 7 are made as the maintenance excellence process continues.

Typical Project Plan of Action

The recommended path forward offers an excellent opportunity for immediate results at the pilot site, plus the time to learn from this assessment before conducting future assessments. MEI highly recommends having a maintenance excellence steering team in place to provide overall leadership, support, and direction. The measurement of results ensures that initial projections of benefits are achieved and that the ROI for this pilot effort exceeds expectations. Table 6.1 shows a typical project schedule where the pilot assessment includes two sites.

TABLE 6.1 Typical Project Plan of Action: Recommended Next Steps after the Scoreboard for Maintenance Excellence Assessment

Path forward action items	Weeks after project initiation					
	Prework	Week 1	Week 2	Week 3	Week 4	Week 5
Step 1: Define impact on quality, capacity and customer service						
Step 2: Establish client scoreboard						
Step 3: Conduct pilot assessments						
• Site one						
• Site two						
Step 4: Define specific opportunities						
Step 5: Define potential benefits						
Step 6: Plan for pilot implementation						
Step 7: Establish measure pilot results						
Step 8: Refine client scoreboard						
Step 9: Plan additional assessments						
Step 10: Present pilot assessment results						
Step 11: Begin other assessments					After week 5	

Document assessment results

After "The Scoreboard for Maintenance Excellence" assessment has been completed, a written and oral report to top leaders will document the results with a presentation of recommendations and a plan of action. Key areas of the report presentation will help you to:

- Determine strengths/weaknesses and priorities for action
- Benchmark your CMMS installation
- Maximize benefits of CMMS
- Develop maintenance as a profit center
- Define potential savings
- Develop recommended plan of action (and implement)
- Develop method to measure and validate results
- Initiate an MEI

Determine strengths/weaknesses and priorities for action

After an objective assessment is completed, it is very easy to identify strengths and weaknesses, which then leads to defining the priorities for action. In some operations it is very often back to the basics such as:

- Preventive maintenance has been neglected, no time to do it.
- Understanding of predictive technologies is limited.

- Application of CRI never initiated.
- The parts storeroom was never been given the proper attention it needed.
- Accountability for craft time is not being done.
- Charge back to customer not done or incomplete.
- A reactive, firefighting repair strategy is in place.
- Valuable craft time is wasted, chasing parts/materials, waiting, unplanned work.
- Never trying to do the job right the first time.
- Asset uptime uncertain and manufacturing operation not reliable.
- Quality inconsistent due to maintenance processes.
- Never time for craft training.
- The CMMS was purchased as "the solution" not "the tool."
- The existing CMMS functionality is not being fully used.

Very often the CMMS takes the hit as the cause of all the weaknesses. CMMS is blamed for not being able to do this and that, and causing all types of problems and extra work. This attitude will generally always be the case when the CMMS was purchased as "the solution" and not "the tool." Bottom line here is that most systems are underutilized and when fully used with all their intended functionality will serve their primary IT purpose.

So, just as we can benchmark a total maintenance operation and its best practice application with the 27 category "Scoreboard for Maintenance Excellence" in Part II, we can also benchmark the CMMS that is in place. Here we need to evaluate the CMMS and its current application as to its effectiveness in supporting best practices. Is the CMMS enhancing current and future best practices or not? Are we getting maximum value from this IT investment? Is our CMMS truly a maintenance business management system? How can we improve current use of the system? In Part III, we will now take a look at how to get answers and take action on these key questions such as CMMS for better maintenance business management.

But first let's take a look at the following case study which can help both the single site or a total operation with multiple sites to define it own unique scoreboard. All and much more for less than one-week investment in training for maintenance excellence. You can easily do a self-assessment when we make the choice to just do it.

Chapter

7

Case Study: The Scoreboard Self-Assessment: Just Do It!

You Can "Just Do It"

A "Scoreboard for Maintenance Excellence" assessment can be done without outside consulting services. This book will help you to do a self-assessment and to use all four benchmarking tools included within it immediately. That has significant value, so as Nike says, *"Just Do It."* It will have value for you and it will pay back the cost of this book investment 100 fold. As I stated before the ratings you give for each of the 300 criteria are "your scores" and your personal baseline of "where you are now" with well-accepted best practices. The methodology and baseline point assignments (Table 7.1) we use for rating based on our comparative baseline from assessment of over 300 maintenance shops. But for your personal rating the same relative values can still be used as your personal baseline.

Your personal assignment of baseline points for each benchmark criterion can be completed right now

Let us say you have our Excel version of the scoreboard and do a self-assessment, or each member of your team does an independent assessment. This can be done really very quickly with the Excel version. What you get will be a number of personal baselines; rating of all benchmark items. Your team can come together, use the same basic Delphi technique process as described for the ACE Team process in Chap. 15 "The Ace Team Benchmarking Process" and get a consensus on the rating values. The rating values and the total score again are not absolute. You could

TABLE 7.1 Guide to Baseline Scoreboard Point Assignments

Scoreboard for maintenance excellence		
Total benchmark point ranges	Overall scoreboard rating summary The baseline total score	Baseline points per criteria
90– 100% of total points	Excellent: 2700– 3000 points	10
80– 89% of total points	Very good: 2400–2699 points	9
70–79% of total points	Good: 2100–2399 points	8
60–69% of total points	Average: 1800–2099	7
50–59% of total points	Below average: 1500–less than 17699 points	6
49% or less of total points	Poor: 1499 points or less	5

say that they are "precisely inaccurate" and maybe like some other analysis paralyse data manipulations. The first overall rating done by you personally or by your team is the initial baseline or your "line in the sand." The next time, say six months later, you use your personal judgment (or the team's) for assigning the numbers again for a hopefully higher overall rating. The key is doing periodic self-assessments concurrent with successful implementation taking place for your priority plans of action. So that is why I say: *Go ahead and just do it*! We will in fact send you the Excel version of this book's current "Scoreboard for Maintenance Excellence" to use by e-mailing Pete @ PRIDE-in-Maintenance.com. That is if you send *The Maintenance Excellence Institute* (TMEI) a copy of your first self-assessment—your line in the sand—on *where you are with your baseline scores*. But, there is another option that does not require a consultant, only some time for what we call "training for maintenance excellence."

Once again, outside help is highly recommended, but?

However, most people/consultants in maintenance agree that outside help for this process has great value when well-qualified support is used with a professional and objective vision of seeing where you are. And if your consultant makes it a very positive proactive process and not a top leader scheme like our previous example of a true story tomato catsup manufacturing plant. But the other option that I will share with you as part of this case study will let you complete a self-assessment via a true workshop and training for maintenance excellence where the training is not over when it's over, somewhat like Yogi Berra used to say.

How This Training for Maintenance Excellence Event was Conceived

The reason that this workshop was developed was due to two things:

1. McGraw-Hill called and asked me in March 2005 to do this book.
2. Our work with the world's best airline—the U.S. Air Force and the challenges we had with developing their unique scoreboard in October/November 2004.

So here is the case study on how a new workshop was conceived.

Case Study: How a Workshop the Maintenance Benchmarking and Best Practice Workshop: *Developing Your Scoreboard for Maintenance Excellence* Was Conceived

Background. During the period October 2004 to June 2005, we provided benchmark assessments of selected air-base maintenance operations for the largest Air Forces Command in the U.S. Air Force. This was for the *Civil Engineering Operations* (CEO) and infrastructure division of the *Air Combat Command* (ACC). With *headquarters* (HQ) at Langley Air Force Base, Virginia. ACC is a major air force command headed by a four-star general, The ACC Civil Engineer; over all of ACC's civil engineering responsibilities is a one-star general. ACC was created on June 1, 1992 by combining its predecessor's Strategic Air Command and Tactical Air Command. ACC is the primary provider of air-combat forces responding in seconds to air borne aggressor force. This organization of over 100,000 military and civilians is the primary force provider of combat airpower to America's war-fighting commands. To support global implementation of national security strategy, ACC operates fighter, bomber, reconnaissance, battle-management, and electronic-combat aircraft. It also provides command, control, communications and intelligence systems, and conducts global information operations.

The ACC mission. ACC organizes, trains, equips, and maintains combat-ready forces for rapid deployment and employment while ensuring strategic air-defense forces are ready to meet the challenges of peacetime air sovereignty and wartime air defense. ACC numbered air forces provide air componency to U.S. Central and Southern Commands with Headquarters ACC serving as the air component to U.S. Northern and Joint Forces Commands. ACC also augments forces to U.S. European, Pacific, and Strategic Commands.

Scope of ACC base maintenance. Our work involved benchmark assessment for CEO Squadrons that provide maintenance of an air base, which is basically a city with an airport. The total scope of ACC's base maintenance responsibility represents the air force's largest overall basing infrastructure (Table 7.2) and includes the following:

TABLE 7.2 Scope of Base Maintenance for the Air Combat Command

Facilities category	Quantity	% of air force	Air force standing
Major bases	15	18	1st
Plant replacement value	$28.7B	15	2nd
Buildings(Sq Ft)	100 M Sq Ft	15	2nd
Airfield pavement	32M Sq Yds	16	2nd
Housing	24,324 units	18	1st
Dormitory rooms	13,324	20	1st
Annual facility budget	2.4M	25	1st

The air force's largest overall basing infrastructure (As of September 2005)

We saw many successes and very positive factors including:

1. People assets: Capable military and civilian craft resources and effective maintenance leadership/supervision, especially the ACC project team of Major Paul Martin, Waltrina Davis, Tim Yuen and Rene Peniza.
2. Current organizational structure with functional support from HQ ACC civil engineering operational staff and the *Air Force Civil Engineering Support Agency* (AFCESA).
3. "Standardized" maintenance management procedures.
4. The *Civil Engineering Maintenance Assistance Team* (CEMAT) process where functional experts visited each base with "inspection quides" to help improve each functional area: craft training, supply operations, HVAC, electrical, utilities, power generation, energy management, pavements, and the like.
5. Capability for concurrent mission training for deployment, actual war-time deployment from the actual craft work and then a very challenging base-level maintenance mission. Something not seen within the private sector.
6. Many positive maintenance improvement initiatives being addressed at HQ ACC and at base level.
7. Craftspeople and supervisors (military and civilian) who were demonstrating PRIDE in maintenance as *people really interested in developing excellence in maintenance.*
8. An existing base construction program (extensive new maintenance requirement to handle with an air force growth policy of zero new positions (challenge No. 4 in Chap. 3).
9. A number of current ACC maintenance management practices and philosophies that the private sector should immediately adopt.

Getting started. The first step was developing a "Scoreboard for Maintenance Excellence" specific to the USAF's air-base maintenance operations, the civil engineer Squadron's, and their respective CEO flights doing the actual maintenance work. This became a very important lesson learned for the TMEI.

Lesson learned. However, as we began with the ACC team to merge our scoreboard with their unique maintenance best practice criteria, two things became obvious:

1. It was hard to get the time with the ACC team members to review our criteria, for them to suggest ACC criteria, and for them to make final scoreboard revisions/comments.
2. A broad base of overall maintenance management experience and current best practice knowledge was missing from members of the team. Having a better understanding of the broad scope of best practices by this team would have helped tremendously. The functional experts on the team were outstanding with many key maintenance improvement projects under their responsibility such as improving supply operations. But their expertise needed better understanding in other functional areas.

Therefore, the effort to develop the ACC scoreboard took much longer than I felt it should have taken. And also selected final recommendations did not have the impact due to a lack of a thorough understanding of selected best practice by some on the ACC team. We also may not have made a strong enough case for some, but there appeared to be a reluctance to rock the boat because they felt all was well. Also during this time, deployment for the war in Iraq was taking place as well as the base realignment initiative being performed by the DOD “BRAC” Commission. So our issue was not at a premium from the ACC CEO team's standpoint, nor at base level in each CEO squadron/flight. And at the time of writing this, the case is still out on their total commitment to high priority recommendations of over 100. Basically, 100 items that if I owned the ACC business of base maintenance would implement by prioritizing:

1. Strategic plans
2. Tactical plans
3. Operational plans
4. Do it now actions (which ACC did almost immediately!)

Budget for this organization (and the military with a war in progress) became very tight in June 2005, so funds for implementation has had to be put on hold. So the case is still out on many recommendations, which will add gained value, potentially craft resources/ productivity of well over $10,000,000. This is a number I would bet on and take a percentage of documented savings if it were legal from a consulting ethics standpoint. But it is not legal, unless you are a personal litigation lawyer or an an accounting firm like Accenture that shared in phantom shadows of projected E-procurement savings here in North Carolina State purchasing and procurement operations. Time will tell what true return on investment ACC gets from their benchmark assessment process investment.

Workshop evolution So after several ACC-base assessments and during the final result reviews, I started reflecting back on these two questions:

1. How could we have improved the ACC scoreboard developmental process? One thing we should have definitely performed is a pilot-base assessment, like we did for the Boeing Commercial Airplane Group at their Portland, Oregon complex back in the late 1990s.
2. How can we help ensure a self-assessment has good value and just doing it, which was something this chapter advocates for the buyers of this book.

So, what about a workshop where we reviewed and discussed each of the 27 best practices categories and each of the 300 benchmark criteria? What if we had a team of two to four people coming from each organization and they did some preworkshop work and a great deal of actual workshop work. And what if we had it at a location that was somewhat like a minivacation spot where they could also take their spouses for fun. The workshop would allow each organization attendee to first learn about important maintenance best practices and then to:

1. Actively participate via a workshop setting to develop their very own very unique "Scoreboard for Maintenance Excellence."
2. Perform a self-assessment with their own unique scoreboard and get a complete copy in easy-to-use Excel format.
3. Develop a strategic plan of action for their top five priorities.
4. Attack just one key item for improvement when they returned to the trenches.
5. Identify their key metrics for *their* maintenance excellence index.

So being a very methodical Methodist, I said let's "just do it" and see how it works out. And that is what we did for our very first public "workshop" at Oak Island, North Carolina, from May 17–20, 2005. This four-day professional development, training for maintenance excellence event was called, you guessed it: *Maintenance Benchmarking and Best Practices*. It is tailored for four types of maintenance operations: plant, fleet, facilities, and healthcare operations. Figure 7.1 shows the charter attendees for the first event.

Another lesson learned. A clear understanding of today's best practices is very important to their successful application. We have helped several very large organizations develop a client-specific "Scoreboard for Maintenance Excellence." From the ACC experience, we have found that concurrent best practice training is necessary when a client-specific scoreboard is being developed. Due to our lessons learned from successful projects, TMEI decided to offer this workshop to the public. Interested readers can learn specifics about the workshop (materials provided, topics covered, plant tours, costs, discounts, dates, and location) by contacting TMEI via our web site at PRIDE-in-Maintenance.com by email at Info @ PRIDE-in-Maintenance.com.

Why Training for Maintenance Excellence?

Our training for maintenance excellence service is a very essential element of our approach to providing both maintenance excellence services and operational services. We recognize the importance of maintenance to an organization's successful pursuit of world-class status. The training for maintenance excellence suite of offerings focuses on creating organizational awareness and the internal understanding that maintenance must be managed and led as profit or service-centered. Successful implementation of today's best practices requires changes in philosophies, attitudes, and the application of technical knowledge. Training for

Figure 7.1 The charter workshop attendees for maintenance benchmarking and best practices: developing your "Scoreboard for Maintenance Excellence."

maintenance excellence can provide a measurable return on investment to justify your training dollars.

Our training capabilities

For helping one achieve total operations success, we provide extensive training capabilities when needed as part of a maintenance excellence services project. Our sessions have been presented worldwide through our alliance team member group on site, by selected conference producing organizations in the USA, Canada, UAE, Argentina, Brazil, Peru, Puerto Rica and Ireland and by various Major universities. A complete course catalog with "workshop" descriptions for 2006 and 2007 is available at www.PRIDE-in-Maintenance.com.

Understanding the true value of maintenance!

Regardless of the type of operation, top leaders must understand the true value of maintenance. Maintenance leaders must develop and nurture an organizational culture that clearly supports long-term continuous maintenance improvement. Training for maintenance excellence supports our belief in the basics and building upon basic best practices as the foundation for maintenance excellence.

Customized training for your operation

It is essential that excellent attitudes toward maintenance exist within each client operation whether it is a manufacturing site, a large university facilities complex, a healthcare facility, or a fleet operation. An in-house presentation tailored specifically to your organization will normally be the most cost-effective approach. Remember that you can "just do it" yourself, using The Scoreboard for Maintenance Excellence, The Scoreboard for Facilities Management Excellence and all other available scoreboards.

Part

3

Developing Your CMMS as a True Maintenance Business Management System

Chapter

8

Maximizing the Value of CMMS for Profit-Centered Maintenance

A fully utilized *computerized maintenance management system* (CMMS) to support the business of maintenance is an essential information technology tool. Information that we can see and use is what we are after, not a sea of data. CMMS to support effective physical asset management is about good maintenance data into CMMS and good maintenance information coming out (Fig. 8.1) to help make the right decisions for the business of maintenance.

Today's information technology for CMMS/*enterprise asset management* (EAM) offers the maintenance leader an exceptional tool for managing the overall maintenance operation and maintenance processes as an internal business and "profit center." However, maintenance surveys and benchmark evaluations conducted by *The Maintenance Excellence Institute* (TMEI) and others validate that poor utilization of existing CMMS systems is a major improvement opportunity. What are some of the typical benefits of an improved CMMS that could be missing from your operation?

- *Improved work control.* Better work management with improved control of work requests by craft, monitoring of backlogs, determining priorities, and scheduling decisions for overtime effectively. Full accountability of craft time/labor cost to work orders, which accrues to asset history and ensures charge backs to customers/tenants.
- *Improved planning and scheduling.* The systems and procedures to establish a more effective day-to-day maintenance planning and scheduling process contribute to improved craft labor utilization and customer service. Better planning and scheduling with our customers

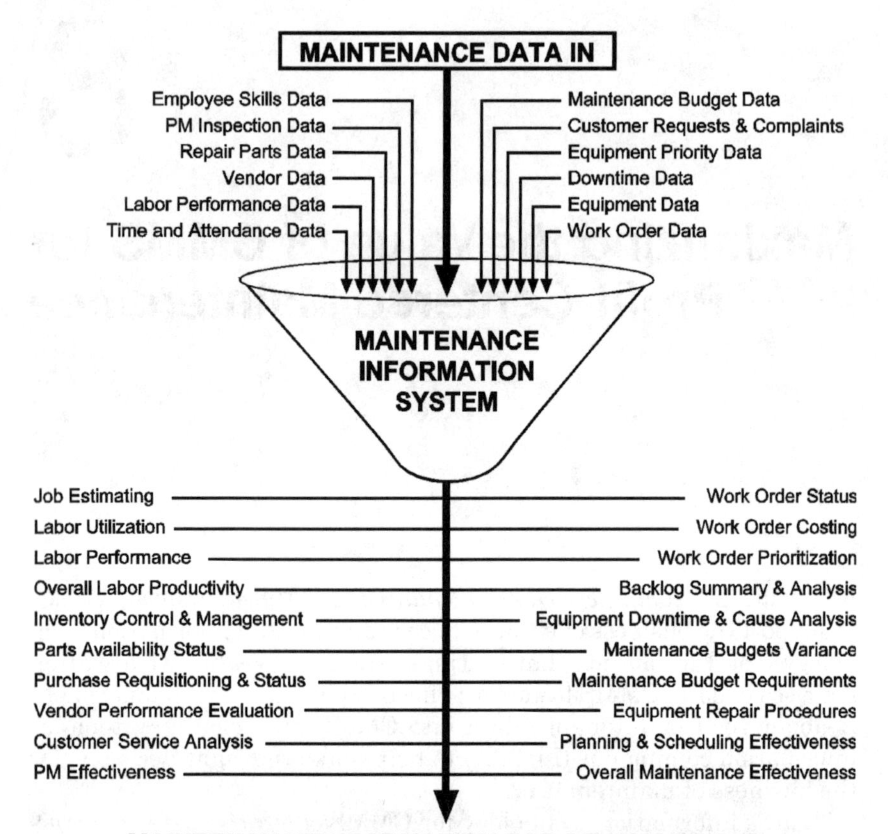

Figure 8.1 Maintenance data in and effective CMMS for good information out.

is an important benefit. We must plan for maintenance excellence because it does not occur naturally.

- *Enhanced preventive and predictive maintenance* (PM/PdM). Automatic scheduling of repetitive PM activities with PM tasks and inspection frequencies documented on the PM module and printed as part of the PM work order. CMMS enhances PM by providing a method to monitor failure trends and to highlight major causes of equipment breakdowns and unscheduled repairs.
- *Improved parts and materials availability.* Well-organized stockrooms with accurate inventory records, a stock locator system, accurate stock levels, and a storeroom catalog can significantly improve the overall maintenance operation. Having the right part at the right time is the key to effective maintenance planning, increased maintenance customer service, and reduced downtime.

- *Improved MRO materials management.* The means for more effective management and control of maintenance parts and material inventories. Information for decisions on inventory reduction is readily available to identify parts usage, excess inventory levels, and obsolete parts.
- *Improved reliability analysis.* The means to track work order and equipment history data related to types of repairs, frequencies, and causes for failure. It allows maintenance to have key information on failure trends that lead to eliminating the root causes of failures and to improving overall equipment reliability.
- *Increased budget accountability.* Provides for greater accountability for craft labor and parts/materials through the work order and storeroom inventory modules. Increased level of control, greater visibility and accountability of the overall maintenance budget by individual piece of equipment and by using department, or work order. Replacement and renovation decisions for facilities and other building systems can be supported by cost information from the CMMS.
- *Increased capability to measure performance and service.* A vast source of maintenance information to allow more effective measurement of maintenance performance and service to provide measurement of improvements in areas such as craft labor productivity, PM compliance, downtime, store inventory control, backlog, level of maintenance, service, and reliability.
- *Increased level of maintenance information.* A major benefit of CMMS comes from developing the historical database that becomes readily available as critical maintenance information. An effective CMMS helps turn data into information you can use to manage maintenance as an internal business.

Your starting point for developing a "Class A" CMMS installation begins taking shape many months and often years prior to a "go live" date. There will be decisions on functionality requirements, the evaluation and selection process, the CMMS implementation plan and team, and the original projection of benefits. But most importantly, it requires decisions on maintenance best practices that change the way you do business—your internal maintenance process business. Hopefully the results from the "Scoreboard for Maintenance Excellence" assessment were the right ones and provided a clear path forward for best practices needs.

CMMS Justification Is Much More Than Economic Factors

The process to justify an effective CMMS must be based upon the foundation that it is an investment that provides measurable benefits to maintenance and the total operation. The strategy to justify a CMMS must start with determining the true need for a CMMS and understanding that there are benefits to support economic justification of the investment. While the economic benefits provide us with the bottom line results for the CMMS investment, the overall justification strategy must include other key elements.

These other elements provide a total justification strategy to support the overall management decision. We need to show management that this is not just another investment in software, it is a well-planned process to change the way we do business. A successful strategy to support the CMMS economic justification process will include a well-defined plan to address each of the following seven key elements.

A strategy to gain maximum value from CMMS

1. Determining the true need for CMMS
2. Determining maintenance best practices needed
3. The CMMS evaluation and selection process
4. Determining CMMS functional requirements
5. The maintenance best practice implementation process
6. The CMMS implementation process
7. The CMMS benchmarking process

Determining the true need for CMMS. The true need for CMMS can best be determined by conducting a total benchmark evaluation of the maintenance operation as per the strategy from this book. We highly recommend this as the very important first step because it is during this process that we determine the maintenance best practices needed in your operation. Also the CMMS functional requirement becomes very obvious when this is done per our assessments at Marathon, Siderar, Boeing, BigLots, Wyeth Medica Ireland, Atomic Energy Canada Limited, the Air Combat Command, and so forth. We recommend that it be performed by an objective, third-party consultant when at all possible. If your company has a large corporate engineering with knowledgeable maintenance engineers this may be a possibility for using internal staff. Whatever your situation, it is really important to evaluate your total operation for best practices that will enhance your CMMS investment. "The Scoreboard for Maintenance Excellence" included in Chap. 4 and our approach in Chaps. 5, 6, and 7 can become your benchmark guides if your only option to it is with internal resources.

During the total maintenance evaluation is where, for example, we look closely at the storeroom operation and the current status of our parts database to support improved parts inventory management and lower inventory. Does the storeroom facility and the current layout support good utilization of space, access control, and overall control of the inventory? Is storeroom owned by maintenance or the finance people such as plant purchasing?

During this process we will evaluate how we use (or do not use) a work-order system to account for labor, parts, and repairs performed to determine cost and asset history information. Are we able to fully account for all craft labor and able to charge back maintenance costs to our customer? It is here that we evaluate our preventive maintenance program for accurate scheduling and completion of tasks. We review our maintenance planning and scheduling process to determine if we can use this best practice to improve customer service and create a proactive, planned process that also improves craft productivity. Can we justify the investment in planner positions based upon improving craft productivity through planned work?

When we take the time to complete a total evaluation of the maintenance operation as the first step in our CMMS justification process, we make an important statement to management. This statement is we know that the CMMS software alone is not the total solution. We recognize that we must change the way we do business. It lets management know that we must adopt improved maintenance practices that will really provide the key to potential savings opportunities with a new CMMS.

Determining maintenance best practices needed. The results of the total maintenance evaluation will highlight key best practice needs that must be addressed within your organization. If the results of your "Scoreboard for Maintenance Excellence" looks like the case study summarized as follows, you can be confident that CMMS software alone will produce little results. This operation required that almost a complete change to its current maintenance practices take place.

Case Study

We will now take a brief look at some of the changes that had to occur in this company before a CMMS could ever begin to effectively support the maintenance operation. This was conducted with the 1993 version of the "Scoreboard for Maintenance Excellence" (Summary shown in Table 8.1) with 18 benchmark categories and 200 evaluation criteria. This represents our lowest scoreboard rating value from 1981 to the present.

Case study comments—some good news. The following are some comments about this manufacturing operation and the maintenance best practices needed before an investment in CMMS would even begin to provide benefits:

- Previous holding companies had neglected maintenance as well as the total operation to the point that serious issues related to safety, production, employee morale and deferred maintenance needs were present.

TABLE 8.1 Case Study: Bad News Now and Good News Later

CAT	Scoreboard for Maintenance Excellence Benchmark Category	Benchmark Items	Maximum Points by Category	Actual Benchmark Points by Category
A	Maintenance and organization culture	10	100	17
B	Organization and administration	12	120	48
C	Work authorization and work control	10	100	8
D	Budget and cost control	11	110	9
E	Maintenance planning and scheduling	12	120	3
F	Maintenance storeroom	16	160	7
G	Preventive and predictive maintenance	22	220	35
H	Lubrication program	11	110	25
I	Overall equipment effectiveness (OEE)	9	90	2
J	Operator-based maintenance	8	80	15
K	Engineering support	9	90	43
L	Safety, housekeeping and regulatory compliance	12	120	62
M	Craft skills development	9	90	2
N	Maintenance performance measurement	9	90	10
O	Maintenance supervision/ leadership	6	60	16
P	Computerized maintenance management systems (CMMS)	13	130	10
Q	Maintenance facilities, equipment, and tools	7	70	59
R	Continuous maintenance improvement	14	140	32
			2000	403
	Total baseline benchmark points This is bad news and good news as you will see		21 % of total possible points	

- The company, however, became employee owned after a long period of decline, almost to the point of having to close the doors. New leadership realized there were many opportunities for improvement and one of the biggest was in the area of maintenance.
- There was actually a DOS-based CMMS in place at one point in time. It had been abandoned, but there was still access to a PM database with previously used PM tasks and frequencies. The maintenance manager wanted to put in a new CMMS, but knew that it would not really solve all of his current problems.
- The company decided based on advice from the maintenance manager that it needed a complete evaluation of its maintenance operation. Operational reviews were also in progress to improve production operations.

Here are a few key points from the "Scoreboard for Maintenance Excellence" benchmark evaluation shown in Table 8.1. This became the maintenance best practice baseline for this company after key best practices did get implemented for both maintenance and manufacturing operations.

a. *Maintenance and organization culture.* The low rating here reflected past management practices. New leaders were now taking action to put maintenance back as one of their top priorities and providing the resources to put maintenance back into good shape.
b. *Work authorization and work control.* There was virtually no accountability for work accomplished. A manual work-order system was in place but it was rarely being used. Maintenance was basically reacting to verbal work requests without any way to establish priorities or to know what level of backlog existed. Work control was almost nonexistent except for an Excel spreadsheet that the maintenance manager had developed to try and establish some means to monitor work backlog. Equipment history was nonexistent.
c. *Budget and cost control.* Without an effective work-order, system costs accrued to major budget accounts and not at the level to support maintenance cost control. Deferred maintenance was extensive and getting worse without an effective PM program being in place.
d. *Maintenance planning and scheduling.* A pure reactive, fire-fighting strategy was in place that wasted valuable craft time by having primarily emergency repair with minimal parts support.
e. *Maintenance storeroom.* Approximately $500,000 of parts inventory was totally uncontrolled without an inventory management system. Crafts would come to the stockroom and search for parts stocked in locations by machine type.
f. *Preventive and predictive maintenance.* PM tasks and frequencies had been developed years ago, but the PM program had been discontinued for about two years due to not having enough craft time to do it because of trying to keep up to day-to-day emergency work. Equipment had changed and almost all of the PM procedures needed review and updating.
g. *Operator-based maintenance.* Certain maintenance-related tasks for start-up and shutdown of equipment had previously been included in equipment operator's job descriptions. Due to an incentive plan being put in place, operators were not doing these tasks and equipment uptime was being severely affected. Needed repairs often went unreported during a shift due to operators wanting to stay on direct labor work for incentive payment.
h. *Engineering support.* Equipment documentation and engineering support was minimal.
i. *Maintenance performance measurement.* Nothing in place to measure any areas of maintenance performance.
j. *Computerized maintenance management systems (CMMS).* They were beginning the process to evaluate application of CMMS and made a wise choice to conduct a "Scoreboard for Maintenance Excellence" benchmark assessment.
k. *Continuous maintenance improvement.* The overall company was beginning to establish a team-based continuous improvement process and maintenance was to be one of the key areas for teams.

CMMS was not their solution. In this actual case study example, CMMS software was very obviously only part of the total solution. A new

CMMS plus a new way of doing business was needed. As a result of the evaluation, the company redirected its efforts and developed a strategic maintenance plan that eventually improved the key best practices listed earlier. Their CMMS justification strategy ensured a commitment to maintenance best practice implementation along with their new CMMS.

The CMMS evaluation and selection process. A justification strategy for CMMS now progresses to a very important area of evaluation and selection. Very closely linked to this element is the definition of CMMS functional requirements which are included within the next chapter. With clearly defined functional requirements we can now begin to make valid comparisons between vendors, determine the recommended vendor and our final cost for CMMS software.

It is important to understand that successful CMMS implementation impacts the entire organization, not just the maintenance function. To provide full benefits, effective CMMS implementation must cross departmental lines and boundaries. Therefore, the evaluation and selection decision and the implementation process do not fall on one department or functional workgroup; it must cover a wide range of functional areas.

Establish the CMMS team. A CMMS Team should be formally chartered and established using a cross-functional approach. For multiplant sites there should be representatives from all sites. We should include representatives from maintenance, operations, engineering, accounting, MRO materials management and purchasing, and maintenance storeroom along with either plant level or corporate information systems staff. The CMMS vendor will also become a vital member of the CMMS team after the final decision is made. Each functional area will have vested interests in the system. Their acceptance and buy-in is important during evaluation and selection and critical during the actual implementation process.

Get outside help when needed. The CMMS team must also consider maintenance best practices that are needed concurrent with implementation of a new CMMS. In many organizations, internal resources are not available to support maintenance best practice implementation and systems integration that is needed. Value adding, engineering consulting services should be considered at this point. CMMS vendors are in the business to develop and install software and often lack the shop-level maintenance experience needed to get full value from their systems over the long term. Depending on the size of your organization and scope of the CMMS project, you should consider adding outside resources as support to your CMMS team. Consider engineering consulting firms that:

- Understand maintenance and the importance of total operations solutions and has had management of productions operations.
- Have a focus on successful implementation of CMMS and best practices and good past references.
- Have sound maintenance experience at the grass roots level as well as experience in operations management.
- Understand total enterprise systems integration and how other parts of your operation are impacted by CMMS.
- Understand how to work as part of a team process or have the capability to establish leadership-driven team processes.
- Provide well-defined deliverables at a fixed cost.
- Uses proven project management techniques in their work to ensure meeting deliverable timelines.
- Can become CMMS team member in your operation very quickly and provide measurable results that can be validated.

The next step, "determining CMMS functional requirements," is a key part of overall CMMS evaluation before we begin the final selection. It is during this step that each functional area—maintenance operations, the storeroom and MRO material management, purchasing, accounting, and information systems—determines what functional capabilities are needed in the new CMMS.

Chapter

9

CMMS Functional Requirements That Support the Business of Maintenance

Functionality—The Important CMMS Features

The functionality capabilities of *computerized maintenance management system* (CMMS) are the features that provide the benefits from information technology for maintenance operations. Effective CMMS enhances and is an important enabler for almost all best practices. The important first step is deciding, "Where you are" with today's best practices and current CMMS functionality. Next to "define where you want to go" a plan of action for best practices such as an effective storeroom, formal planning and scheduling, more effective *preventive maintenance* (PM) and *predictive maintenance* (PdM) is developed. A new or upgraded CMMS may in fact be a major part of the plan. Now we can define the functional requirements needed for a new CMMS or gaps within an existing CMMS. This is also about benchmarking—your functionality needs versus what the CMMS vendors say they can provide to meet your needs.

This chapter will help to define CMMS functional requirements so that an evaluation and selection process can address specific needs and then support the right choice for your specific maintenance operation. We will present extensive definitions of CMMS functional capabilities and provide a convenient rating method to use for making a detailed comparison of CMMS functional capabilities of multiple vendors. Appendix C provides a comprehensive CMMS functionality rating benchmarking process (checklist).

Making the Right CMMS Choice

Before progressing to the evaluation and selection phase of CMMS, we must understand our specific needs. It is important to benchmark and formally evaluate current practices of your maintenance operation (previous chapters). This identifies best practices that are needed. A strong commitment to implement best practices brings with it the need for effective information technology in the form of CMMS. To make the right CMMS choice and gain maximum value from our investment we must define CMMS functional requirements. Most CMMS systems include the basic functional capabilities associated with the core modules for:

a. Work order/work management

b. Equipment/asset management

c. Inventory management/maintenance repair operations (MRO) materials management

d. Planning and scheduling

e. Preventive maintenance and predictive maintenance

f. Purchasing

g. Other CMMS functional capabilities

Advanced levels of functional capabilities may include the need for purchasing capabilities or greater capabilities for budget and cost control, project management, or being able to use CMMS with links to prints, assembly drawings, parts catalogs, or other documents. The following CMMS functional capabilities are defined according to the CMMS module where they are typically included. We will cover functional capabilities of the core modules first and then cover advanced functional capabilities.

Work order /work management

1. *Work- order control.* Manages and controls approval of work orders and provides easy access to the entire work order database for backlog and planning analysis.
2. *Work request.* Creates a work request as a preliminary step before creating a work order to execute the work. Provides query capability for the person requesting work and notifies requesters that an open work order exists for the equipment listed on the request.
3. *Work-request review.* Provides a review and approval process for work requests. This online functionality can be tailored to specific levels of detail and approval authority required prior to submitting the work request to maintenance for creation of a work order. In

some cases, engineering approval or engineering support could be required before the request is entered into the work order database.

4. *Warranty status*. Notifies requester that equipment listed on the work request is under warranty. [*Note*: For many operations this alone can save enough real cash to support *return on investment* (ROI) on CMMS plus best practices, for example, the air combat command with $1,200,000,000 and the State of North Carolina University System with $3,000,000,000 billions in new construction. All require effective commissioning creating added maintenance requirements as our challenge four in Chap. 3.]
5. *Work-request status*. Provides requester the capability to query the status of their work requests. This could be a number of self-defined work order status codes such as *work in progress* (WIP), *held awaiting parts* (HAP), *engineering approval required* (EAR), *held awaiting equipment* (HAE), or a number of other status codes related to specific needs for tracking status of maintenance work. This inquiry can also give other information to include expected or planned completion date and/or time.
6. *Consolidate work requests*. Consolidates a number of work requests into a single work order.
7. *Work request/work order priority*. Provides the means to prioritize and rank work requests and work orders. May include capability to assign equipment criticality index (10 most critical down to 1 least critical) within the equipment database and a criticality/priority code for the type of work (i.e., safety, PM, project work, environmental, emergency, contractor, and the like) and criticality codes for the work types. With this information, each work order can have a CMMS system calculated RIME Index to determine priority based upon combining equipment/asset criticality and the criticality of the work order type. This process for prioritization was developed by the staff at Albert Ramond & Associates of Chicago in the early 1980s.
8. *Work request/work order entry screen*. Designed to guide data entry so that information to diagnose the problem and the cause of the failure is provided via scroll down tables that prompt failure code entries.
9. *Establish and view job plan steps*. Establishes and provides view of an existing work-order job plan by individual job steps.
10. *Work-order job steps*. Associates a work-order job step with specific equipment, craft supervisor, estimated time and/or standard time, list of tasks, tools, and special equipment. Also associates work-order job step with resources, parts requisition or purchase order, work-order bill of materials, routes and checklists, safety documents, material costing, measurements, and labor costing.

11. *Work- order job plan entry / update.* Enters a new job plan or updates an existing job plan from job plan templates. Once a job plan is established as a template, it can remain in the database as a text file for easy update for the same job or a similar job.
12. *Work-order status.* Posts actual hours against work order or work-order steps. Progresses work orders that have been started via direct entry of craft labor hours to the work order, via time and attendance system or other work order progress methods. Can show percent craft hours (by craft types) completed and compared to budgeted time (or planned time) to actual time worked. *Note*: In some CMMS systems, this information is in "real time," being updated continuously via a craft shop floor time reporting system.
13. *Contractor work-order inquiry.* Inquires on work orders and work-order status for a given vendor/contractor.
14. *Job plan.* Creates the work-order job plan and include resources: crafts, material, equipment, or contract services. Provides capability to copy materials requirements directly off the bill of materials for equipment. Matches craft job skills to the job requirements to ensure qualifications are met.
15. *Job plan requisitions.* Automatically creates pick lists for stocked items or purchase requisitions for nonstock items or orders for contract services needed for the job plan.
16. *Job plan documents.* Attaches drawings, specifications, procedures, permits, and other documents to the job plan and specifies which documents should print out with the job packet.
17. *Job plan print.* Generates a work-order job plan to a local printer along with job plan documents needed for the job packet.
18. *Work-order scheduling.* Work orders created, planned, approved, scheduled, worked, progressed, completed, and transferred to history. This function turns the work request into a "statused" work order that may be ready for execution or put on hold for a number of specified reasons.
19. *Quick entry work orders.* Provides for expedient entry of an approved work order of short duration such as emergencies and small-unplanned jobs. This feature bypasses the entry of a work request and is used to account for miscellaneous work performed and accounted for and entered to the system from a simple, manual crafts log for such type of jobs.
20. *Just-in-time scheduling.* Blanket work orders can be scheduled for a specific time during the day or as needed.
21. *Work-order planning notes.* Creates a preclosure work-order screen to allow remarks and narrative to be input before the work is completed.

In some operations, the use of internal e-mail may be linked to this CMMS function.

22. *Standard maintenance practice library.* Provides easy access to standard maintenance tasks and repair practices library, preplanned work scope, job plan templates, and historical averages of previous repair times. Repetitive work can be captured, cataloged, and reused. Standard job plans or job plan templates can be copied to a work order. May include ability to identify procedures, drawings, and other documents related to a standard task to include standard repair times.
23. *Work-order master update.* Adds, changes, or deletes work-order master records and data.
24. *Initiate project or shutdown work orders.* Selects and initiates project or shutdown types of work orders.
25. *Process project type work orders.* Checks by work-order job step for material availability. Commits and prints MRO pick lists by job step when material is available for the work order.
26. *Work-order inquiry by equipment/unit/parent/project ID.* Provides a display of all open-in-maintenance work orders for a particular equipment/unit/parent/project ID/subcomponent.
27. *Work-order inquiry.* Inquires against work-order master information, job steps, materials, and costs.
28. *Subcomponent inquiry.* Inquires on work orders for a given subcomponent of the equipment/asset.
29. *Close work orders in maintenance.* Closes a work order in maintenance, updates the appropriate equipment history and cost, and provides capability to link the total maintenance repair cost to the financial system.
30. *Reopen work orders.* Reopens a closed-in-maintenance work order to adjust time and materials charged or to add, delete, or change key information.
31. *Time and attendance interface.* The time and attendance system interfaces with the CMMS to provide immediate and real time actual craft labor updates to the work order file.
32. *Work-order material status inquiry.* Obtains the material status for a specific work order or job step.
33. *Closed work-order history inquiry.* Inquires the history for a closed work order
34. *Work-order transaction history inquiry.* Inquires the transaction history for a work order for tasks performed along with labor and material charges accrued to the job.

35. *Work- order transaction accounting history.* Inquires the accounting transaction history for a work order.
36. *Work-order text file.* Narrative text files and/or attachments can be viewed and/or modified while editing the work order. Allows for text documents, such as job plans, checklists, and procedures to be attached to a work order.
37. *Online, paperless work orders.* Provides for work requests and work orders to be created, approved, and closed online through a paperless system. Dispatches, completes and reassigns work orders to maintenance supervisors, individual craftspeople, technicians, operations, and contractors.
38. *Material issues to work order.* Updates stock and nonstock issues to work orders and related equipment hierarchies.

Equipment/asset management

1. *Equipment hierarchies.* Equipment can be identified and structured into parent/child hierarchies at higher level (cost center, maintainable unit, and department) or at lower level (component, subcomponent, or spare part). Defines equipment classes and assigns each equipment to a class to include function, location, vendor, or criticality.
2. *Equipment identification.* Provides equipment master descriptive information for a piece of equipment (parent/child), manufacturer, serial number, spare parts/bill of materials listing, equipment specs, repairs performed, failure causes, costs, and so forth. May include capability to record the installation and removal of components within process type operations that require tracking by serial number and compliance to process safety management requirements. Performs full text searches using key words in the description.
3. *Equipment inquiry.* Online top-down view of equipment structure with a minimum of five levels and online view of equipment components/parts. May include graphical capabilities to show site layout with drill down to department/work center, specific asset, and subcomponents included as part of the equipment parent/child relationship.
4. *Regulatory and special requirements.* Provides data supporting the definition and maintenance of equipment and component regulatory information, permit requirements, and special requirements such as fire protection.
5. *Equipment spares inquiry.* Obtain spare parts/bill of material list from equipment master data that are component parts of an equipment asset. Add, change, or delete items from spares list or copy spares list

to another asset. Also, copies all or part of spares list to a work-order job plan and creates parts requisition and pick list to storeroom.

6. *Equipment spares update.* Adds, changes, or deletes equipment spares information via mass updating capability.
7. *Parts interchangeability inquiry.* Obtains a list of all the equipment assets that use a stock item based upon MRO information. Provides cross-reference link to specifications data for the MRO item.
8. *Warranty information.* Tracks warranties on newly installed or repaired equipment/components and provides information to maintenance when a work order is created for item still under warranty. Enters warranty and shelf-life information into the system when item is actually received.
9. *Equipment maintenance history.* Provides complete repair history for equipment/components with online access.
10. *Equipment failure information.* Tracks failure information at various levels: equipment, component, subcomponent, or spare part. Provides for recording of failure cause and corrective actions taken.
11. *Equipment parent/child relation.* Attaches/detaches/reassigns equipment parent/child hierarchies. Adds, changes, deletes, or reactivates equipment asset information and asset specifications.
12. *Equipment operations data.* Provides for equipment usage or measurement data being entered directly or tied to process control or equipment monitoring systems as means to trigger actions such as inspections. Data on operating factors and their high and low trigger points, which indicate abnormal operating conditions are established and maintained.
13. *Equipment service.* Provides data on operational status, outage schedules and out-of-service history based upon downtime captured from work order or manually entered. Data may be used to support "availability element" calculation for establishing the *overall equipment effectiveness* (OEE) factor.
14. *Equipment inquiry by parent.* Obtains all subsidiary ("child") equipment information for a specified equipment "parent."
15. *Equipment cost history inquiry.* Obtains equipment historical actual costs for equipment "parent or child" or rolled-up costs for an equipment hierarchy. Updates historical maintenance costs for a piece of equipment (parent/child) via work-order closing.
16. *Equipment support documentation.* Information in the equipment master data that provides reference to drawings, specifications, calibrations data, technical reference manuals, performance standards, and other asset support documentation that needs to be viewed or

printed for repair or troubleshooting activities related to the equipment. Direct links to *document management systems* (DMS), *computer-aided drafting/design systems* (CAD), regulatory compliance data, and other systems to support asset documentation may be established.

17. *Location history.* Supports the entry and review of location and installation history data for serial numbered items and regulated components that must be specifically tracked.

Inventory management/MRO materials management

1. *Materials master catalog.* Includes all descriptive reference information for spare parts, tools, and consumables stocked in support of work activities, special purpose items available through direct purchase and materials referenced on bill of materials parts lists. Adds, changes, deletes, reinstates, or inquiries on master data for a stock item. Provides common stock catalog, standard stock numbers, and common database. May include nonstock items not warehoused by the organization, but where item master information is maintained and available for creating purchase requisition from nonstock item data. Provides method to classify items based on characteristics, use, value, criticality, turnover as well as other factors.
2. *Inventory management.* Provides state-of-the-art MRO inventory management procedures and control to include tracking inventory cost by last-in-first-out(LIFO), first-in-first-out (FIFO), or by moving average cost accounting. Provides capability for ABC analysis and EOQ analysis and establishes min/max inventory levels for reorder. Provides automatic reorder notification based on reorder point and capability for cycle counting or wall-to-wall physical inventory. Tracks materials stored in single or multiple warehouses and designated controlled storage areas.
3. *Stock item inquiry.* Provides descriptive information, current purchase order status, work order, and statistical information about a stock item. Inquires against stock items based upon stock noun, manufacturer, manufacturer's part number, and other key word attributes. May include capability to provide list of material requests by work order, project, or other parameter.
4. *Stock item availability.* Provides inquiry on stock item availability and usage in a single warehouse, a group of warehouses, and a controlled storage area or rolled up to the enterprise level.
5. *Stock item availability for schedule.* Provides capability to run a trial schedule to determine material availability and automatically create parts back order when available quantity is insufficient.

6. *Material request.* Provides for the request of material (stocked, nonstocked, and consignment) either as a stand-alone or referenced to a work order or project. Approval automatically allocates material to the request and creates purchase requisition for replenishment if material is not available in warehouse. Reports available may summarize material requests for analysis by various criteria: quantity status, by work order, or by project.
7. *Reorders.* Reorders stock items manually, emergency only, or automatically based on reorder parameters.
8. *Automatic reorders.* Produce automatic reorder requisitions for a preferred vendor or for direct purchase orders against previously established blanket orders.
9. *Reorder information.* Reports to determine whether the reorder policy should be changed based on usage rates, stockouts, and safety stock levels.
10. *Qualified vendor.* Provides list of specifically qualified or disqualified vendors for the stock item and respective vendor profiles.
11. *Update part/vendor reference data.* Automatically updates of part/vendor reference data from inventory processing.
12. *Nonstock item status.* Maintain online list of nonstock items that have been received but not issued to a work order.
13. *Cycle count schedule.* Provides for scheduling of cycle counts based on established schedule per ABC code or cycle count code. Provides for adding or deleting stock items into the cycle count and adjusting schedule based on labor resources.
14. *Inventory count verification.* Inputs the results of a physical inventory or cycle count and make an inventory/accounting adjustment.
15. *Issue material via pick list.* Issues material for a particular work order or a work-order job step as the result of a pick list. Prints pick list at user-defined interval before the material is due at the job site.
16. *Automatic pick list.* Initiates a pick list for materials that are received, which were deficient on work orders prior to their receipt. May include capabilities for cross docking of back-ordered items directly to staging areas or the job site without going into storage. May provide charging of parts costs directly to a work order from receiving activities.
17. *Direct issues (no pick list).* Provides for direct issue of stock items to a work order, work-order job step, project, and person or cost center without a pick list or without going through the requisitioning process.

18. *Pick scheduling.* For larger, high-volume operations, this feature maintains backlog of all pick requirements and provides schedule to create the most efficient order picking operation.
19. *Rebuilt items.* Establishes inventory costs of rebuilt, serialized items at replacement (new) value, or reconditioned value calculated as a percentage of replacement value.
20. *Transaction inquiry by rebuilt stock item and condition.* Provides review of stock item rebuild condition code transactions and the resulting status of rebuilt items.
21. *Adjust deficient quantity.* Adjusts the deficient quantity for a stock item on a work order.
22. *Spare history (deactivated items).* Allows stock warehouse records to be retained for historical purposes when deactivating a stock item.
23. *Receive materials.* Receives materials against an existing document such as a purchase order or material transfer document. Provides means to register problems or exceptions with the order such as damaged goods or short shipment. May include capability to enter warranty, shelf-life information, or quality inspection requirements during receiving process, or to denote hazardous material or a hot item for expediting.
24. *Receive material from other warehouses.* Enables material movement between multiple warehouses and the receipt of material into one warehouse from another warehouse in the overall system.
25. *Put away control.* For larger, high-volume operations, this feature provides computer-assisted selection of put-away locations, schedules put away, and assigns crew to perform put away. May use portable data terminals to capture actual put-away locations, which are automatically uploaded to the host system via *automated data collection* (ADC) equipment.
26. *Pick scheduling.* Provides backlog of all planned pick requests to evaluate the picking backlog and supports scheduling of the parts to be picked, issued/transferred, and shipped in a timely manner.
27. *Stock history inquiry.* Provides inquiry on the recent transaction history for a stock item. Inquiry by stock transaction history, by group identification number or commodity code.
28. *Stock usage analysis.* Provides database search to determine issues over designated time periods and helps identify surplus or obsolete items. Maintains status of surplus and obsolete review recommendations.
29. *Stockout analysis.* Evaluates stockouts that have occurred over a specific period for a stock item based on the item not being available at the requested pick date.

30. *Materials usage and forecast.* Evaluates the historical usage patterns and provides a materials forecast.
31. *Transaction inquiry by document.* Provides inquiry on stock transaction history by document number.
32. *Material quarantine, inspection, and testing.* Transfers material to and from rebuild, quarantine, inspection, and testing activities; and tracks the status of inspection and testing either from internal or external sources. Tracks the inspection/testing location, person responsible, and the inspection/test procedure. May provide summary of all traceable stock items and shelf-life expiration data in the warehouse.
33. *Material inspection/testing procedures.* Identifies and codes the inspection procedure directly to the stock item, requisition, purchase order(PO) line, and receiving record and maintains complete history of inspection results.
34. *Materials management procedures.* Provides a standards and procedures library to define how material is to be processed at a specific warehouse to include required quality control inspections, special handling procedures, receiving procedures, or storage instructions.
35. *Stock owner warehouse record update.* Adds, changes, deletes, or reactivates stockowner warehouse records including consignment vendors. Includes bin location, cycle count, and recorder data. Reconciles consignment information for invoicing.
36. *Warehouse location data.* Provides inquiry on the physical quantity and quantity type of a particular bin location. May define bin locations such as replenishment bins, job box, and temporary storage. May include capability to store multiple stock items in one bin location or store material in multiple bin locations. Quantity types may be used to define various stages that the quantity passes through while in the warehouse such as received/returned/accepted, rejected, on hold for *quality check* (QC), staged for specific work order, ready for surplus, ready for scrap, and so forth.
37. *Location utilization.* Provides storage utilization information with description of all storage locations and summarizes current utilization. Used primarily for computer-assisted put-away.
38. *Material returns.* Provides capability for processing issues, receipts, transfers, and returns of stock to include credit to previously closed work orders.
39. *Material reservation.* Provides for reserving stock items for planned jobs and denoting in the inventory record that an item has been committed and is not available for issue to other material requests.

May include automatic commitment of material when work order is in scheduled status. May include diverting reserved quantities from one work order to another.

40. *Staging location control.* Provides control of parts going to staging/kiting areas for short-term or long-term periods for outages and major projects.
41. *Establish new stock item.* Provides controlled process for adding new stock items to master catalog. Initial order requirements are entered and the request to add new stock item is routed on line for necessary approvals.
42. *Tool and special equipment tracking.* Provides for tracking of tools and special equipment issued to maintenance personnel or to a material request.
43. *Electronic data interchange (EDI) and electronic commerce.* Capability for EDI, fax transmissions, and paperless internal/external requisitioning approvals, and purchase order creation and capability for electronic commerce technology use via the Internet.
44. *Bar coding capability.* Provides interface with industry standard bar coding systems. Provides print capability for bin locations, work order, asset tags, vehicle tags, and the like.
45. *Document management system (DMS) interface.* Direct interface with document management systems such as DEMS (software solutions), PRISM (ATR, Inc.) or Machine Maintenance II (Applied Image Technology), and others.
46. *Vendor electronic catalog interface.* Provides capability to interface directly with selected vendor part information systems and to input data directly to either the inventory or purchasing functions from electronic catalogs.
47. *Materials management performance reporting.* Provides for initiating manual physical counts or automatic cycle counting requirements and measure inventory accuracy. May be designed for reporting of stock activity, stock-outs, stock status, and inventory accuracy, receipts/issues, delivery service (internal or external), vendor service, and quality.
48. *Change stock accounting class.* Changes accounting class for a particular stock item.
49. *Sell material or return to vendor.* Records the sale of materials or the return of material to a vendor.
50. *Material transfer between work orders.* Transfers material from one work order to another.

51. *Work order adjustment.* Adjusts costs, initiates/updates accounting fields, adjusts actual material costs, and opens or closes a particular work order that has been closed and is in accounting history.
52. *Material transfer.* Makes material transfers from warehouse to and from surplus and obsolete accounts.
53. *Stock value and quantity adjustments.* Adjusts the value (price) and quantity of a particular stock item and update accounting system history.

Planning and scheduling

1. *Planning.* Provides for detailed planning of the craft labor, materials, special equipment, and other resources needed to complete a work order. Permits developing specific job steps with type of craft labor and estimated times. Allows for scheduling of the work order when all planned resources are available. Planning functions may be included as a separate module or be part of the work-order module.
2. *Work-order search.* Provides search of work-order file based on specific criteria such as priority code or "request date" to support building a schedule. Selects prioritized work-order tasks for scheduling.
3. *Craft scheduling.* Supports ability to assign estimated labor hours by craft for work-order tasks assigned to a specific crew and shift or specific individual.
4. *Schedule review.* Provides review of schedule to check status of tasks and others.
5. *Individual assignment review.* Provides review of work that has been assigned to individuals. Personal notes may be added by individuals to specific tasks.
6. *Shift turnover log.* Provides for making notes to support turnover of scheduled work from one shift to the next shift supervisor or craftsperson.
7. *Job plan.* Creates work-order job plan and includes resources: crafts, material, equipment, or contract services. Copies materials requirements directly off the spare parts list/bill of materials for equipment. May include matching of craft job skills to the job requirements to ensure craft qualifications are being met for the job.
8. *Job plan requisitions.* Automatically creates purchase requisitions or orders for contract services included on the job plan.
9. *Job plan documents.* Attaches drawings, specifications, procedures, permits, and other documents to the job plan and specify which documents should print out with the job packet.

10. *Job plan print.* Generates a work-order job plan to a local printer.
11. *Work-order scheduling.* Work orders created, planned, approved, scheduled, worked, progressed, completed, and transferred to history.
12. *Establish job plan steps.* Establishes and provides on-line view of an existing work-order job plan by individual job steps.
13. *Work-order steps.* Associates a work-order step with specific equipment, craft supervisor, estimated time and standard time, list of tasks, tools, and special equipment. In addition, associates work-order step with resources, parts requisition, work-order bill of materials, routes and checklists, safety documents, material costing, measurements, and labor costing.
14. *Work-order job plan entry / update.* Enters a new job plan or updates an existing job plan from job plan templates.
15. *Craft resource availability.* Determines craft hours and type craft availability (internal and contract) as compared to craft hours required for a specific planning period. Provides quick reference to availability of other resources required for the plan.
16. *Craft resource leveling.* Provides visibility of current backlog that is ready to be scheduled and the capability for leveling of craft labor over a specific planning period or handling of exceptions to the plan.
17. *Material status.* Provides status of material requests for planned work orders or work-order steps. May include material forecasting feature and generation of future demand and automatic requisitioning of necessary parts/materials.
18. *Craft skills and qualifications.* Provides matching of planned tasks with the skills and/or qualifications required for the job. This would typically be linked to a craftspersonnel module as part of CMMS or to an enterprise system human resources database or a training management system.
19. *Daily maintenance schedule.* Produces a daily maintenance schedule by craft area, lines, and the like including PMs, lube services, and PdM data collection that can be performed at regular predetermined intervals by craftspeople. It may also include planned tasks performed by operators.
20. *Craft capacity leveling.* Schedules work orders against known labor availability to ensure fully effective craft redeployment, labor utilization, and craft leveling to include contractors.
21. *Schedule compliance.* Provides method to monitor to what degree the schedule is being adhered to, based on work planned/scheduled versus actual execution of planned work.

Preventive maintenance (PM) and predictive maintenance (PdM)

1. *PM procedures.* Establishes PM tasks, frequencies, and activity descriptions by equipment/component. Creates standard PM task library and uses to build PM jobs.
2. *Multiequipment PM tasks.* Common PM tasks for more than one piece of equipment can be established with different due dates or with the same due dates. Equipment with the same PM due date can be combined onto the same PM work order.
3. *In-service PM tasks.* Provides information that PM is to be performed only when the equipment is "in service or operational."
4. *Initiate/deactivate PM work orders.* Initiates/deactivates PM work orders based on last scheduled, last completed, or on a demand basis when conditions require PM action.
5. *Automatic PM work-order generation.* Provides automatic generation of PMs based on time interval, condition based, or on meter reading of hours, miles, units produced, and the like. Time interval is in days, weeks, months, quarterly, or semiannual. Condition based may also include number of units produced.
6. *PM work-order closing.* Allows for a partial closure of PM work orders (a work order with multiple PMs to do) that would update PMs accomplished and allow those not completed to be generated again for scheduling later.
7. *PM work order suspension update.* Suspends/unsuspends PM work orders scheduling by equipment/component.
8. *PM forecasting and leveling.* Creates forecast for PMs over a specified period and provides for level loading of scheduled tasks. Creates simulation (what if) of PM workload per equipment/component.
9. *PM routes.* Associates PMs to specific equipment locations and provides capability to establish the most effective routes for PM tasks, lubrication services, instrumentation checks, and the like. In addition, associates routed PMs and work orders to multiple equipment.
10. *Tracking required certifications.* Tracks PM and other activities requiring certification or recertification to support ISO/QS 9000 requirements.
11. *PM/PdM compliance.* Provides ability to measure PM/PdM compliance continuously based on PMs generated to schedule versus PMs actually completed (within a specified window of time). *Note*: PMs to include all lube services, calibrations, and scheduled quality inspections.

12. *PM/PdM results.* Provides capability to track and document work orders initiated due to results of PM/PdM inspections that have eliminated catastrophic failure.
13. *PM/PdM costs.* Provides capability to track PM/PdM cost to include inspections and PM/PdM related work, to document PM results and to provide justification support for PM/PdM continuation and expansion as a top priority.

Purchasing and MRO procurement

1. *Create purchase requisitions.* Initiates material requests on-line to the purchasing system with all costing, vendor, and quantity information when requisitioning. Include add/delete items from requisition. May include capability to create purchase requisitions directly from the work planning process for nonstock, direct-buy items, and contracted services. May allow an online approval process to be used prior to creating the purchase order and user-defined business rules for automatic approval or rejection of requisition.
2. *Requisition status.* Provides online status of requisition as to approved/nonapproved, complete, incomplete, or closed. May include triggers for automatic requisition follow-up.
3. *Open requisition backlog.* Provides view of current backlog of requisitions to ensure priorities are considered.
4. *Creating purchase orders.* Automatically creates/prints purchase order with or without a requisition using standard purchase order format or blanket/contract purchase order. Includes add/delete items from purchase order. May include capability to combine all requisitions from the same vendor into a single purchase order. May include printing line items as they appear on requisition or adjusting line item order.
5. *Purchase order status.* Provides online viewing and edit for status of purchase order as to a number of status types: ready, open/placed, incomplete/partial receipts, and all items received, completed, or closed. May include triggers for automatic purchase order follow-up.
6. *Receiving report.* Provides a receiving report generated for receiving an item to cost center, stock location, work order, or employee and provides update to purchase order, requisition, purchase item, receive data, and inventory. May include quick receive capability with receiving directly from packing slip without accessing each individual purchase order. May include multiple deliveries per item or order line and return to vendor options.
7. *Delivery date update.* Provides status and update to the promised delivery date for a particular work-order event.

8. *Quotations.* Provides request for quotation with key information on vendor and items and vendor quotation file with purge logic.
9. *Purchase order history.* Provides complete review of purchase order history and for the entering of adjustments to the order.
10. *Expedited purchase orders.* Provides for tracking of expedited orders using special purchase order lines with promised dates, special expediting letters, and review of all expediting requests.
11. *Multiple languages and currencies.* Supports multiple languages and multiple currencies.
12. *Vendor Information.* Provides for online inquiry of items supplied by vendor, multiple vendor addresses, vendor part numbers and descriptions, and automatic or manual lead-time calculation.
13. *Vendor selection.* Provides automatic vendor selection based on multiple factors: price, lead time, preferred vendor, and so forth.
14. *Vendor performance evaluation.* Provides means to evaluate vendor performance by a number of performance criteria.
15. *Electronic purchase orders.* Provides for purchase orders in FAX or EDI formats.

Additional CMMS functional capabilities

1. *Management of change.* Provides central database for maintenance, engineering change control, mechanical integrity data, and equipment/process specification data. Records all changes in asset/equipment history and statistical process control information and who/when/why/how changes occurred. Provides integration with engineering database to ensure maintenance input on maintainability issues.
2. *Craft training database.* Records and tracks craft training history, training requirements certification, licensing, and/or interface to enterprise databases for training and human resources. May provide capability to maintain individual craft training plans, lesson plans, and the like, as required for craft and operator personnel. Also, may include contractor craft skills information and certifications.
3. *Contract management.* Provides system to manage maintenance contract specifications, work requirements, evaluating contractor response/performance, full contract administration, and progress monitoring. Provides capability to manage performance-based contracts.
4. *Compliance regulation library.* Provides capability to interface with compliance regulations or procedures library for safety, health, and environment. May include cut/paste of text from current regulations

or procedures directly to work order. May include print capability for MSDS sheets and specific work permits as required for safety.

5. *Interfaces with radio frequency (RF) technology.* Provides interface for use of RF technology for online condition monitoring, real time data transmission of work-order information, real time shop floor reporting, MRO warehouse information, and the like.
6. *Reliability performance indicators (RPI).* Provides effective database on reliability with ability to calculate MTBF (*mean between failures*) and other key reliability indicators as mean time to repair (MTTR) and mean time before repair (MTBR).
7. *Continuous reliability improvement database.* Provide access to a centralized reliability database that is maintained locally, but integrated/shared globally throughout a multiplant operation as needed.
8. *Rebuild/reconditioning costing.* Tracks in-house cost and quality for rebuilt items to compare in-house rebuild cost versus vendor cost for items such as relief valves, motors, gearboxes, and special spares.
9. *Interface with online condition monitoring.* Capabilities for direct interface with real time continuous monitoring systems via RF technologies or hard-line links.
10. *Reliability and FMECA information.* Provides user-defined, time-stamped condition fields to monitor equipment conditions each time a repair task is performed. Defines and tracks process variables, temperature, vibration and amps/RPM, and so forth, and analyzes and establishes trend data. Provides *failure modes, effects, and criticality analysis* (FMECA) information to identify and evaluate what items are expected to fail and the resulting consequences.
11. *Equipment usage.* Updates the usage factor for an equipment asset (either fixed or mobile type) for scheduling inspections based on usage not fixed intervals. May include tons, linear meters, and the like.
12. *Client report writer.* Ability to produce tabular and graphical reports effectively at the client level. Specific reports and metrics that are needed should be provided to the CMMS vendor to see how their specific system can provide custom reports as compared to their standard reports.
13. *CMMS-project management interface.* Provides interface with various project management systems that are currently available.
14. *Speech recognition technology interface.* Provides interface with speech recognition technology for work requests, time reporting, work-order closing, work order and project status updates, and so on. Includes all next generation data-input technologies.

Chapter

10

Case Study: Quantitative and Qualitative Factors for CMMS Selection

The overall strategy for *computerized maintenance management system* (CMMS) justification must consider the specific CMMS functional requirements that your operation requires per Chap. 9. By clearly defining CMMS functional requirements with input from a cross-functional CMMS team, we can progress to the final selection process and develop the economic justification for the capital expenditure request to management. This case study will provide a way for you to integrate both qualitative and quantitative factors in your final decision-making process. A more in-depth case study focused on defining quantitative factors is included in Chap. 11 that evaluated SAP R/3 against MAXIMO; 1999 versions of both world-class systems.

In some cases, we may need advanced functionality to support shop-level planning and scheduling, project management capabilities for major shutdowns, links to electronic document management systems, financial systems, and corporate purchasing systems. Advanced functionality and integration to other systems comes at a cost so we must ensure that these requirements are valid. We must also ensure that the maintenance best practice areas they are to support are in fact implemented.

Again, the CMMS team must determine the total scope of maintenance best practices needed and the CMMS functionality to support each one. The maintenance operation with planning and scheduling of both major shutdown work and day-to-day work for a large operation of 100 plus craft people plus contractors will need a much more vigorous planning capability and overall project management and cost-accountability system. This operation may need detailed work order information to ensure full utilization of contractor labor and contractor billing.

In the area of purchasing, a key question might be "Can we purchase directly with the purchasing module from CMMS or do we need to link our CMMS parts requisitions to the corporate enterprise system for purchasing?" Defining functional requirements that support all maintenance best practice areas and all functional areas of the operation now establishes our true CMMS needs for the final selection.

The CMMS Selection Process

The scope of functional requirements will quickly narrow down the list of possible CMMS vendors. With functional requirements clearly defined, the CMMS team can now begin the final selection process. The final selection may be pursued by starting with a formal *request for proposal* (RFP). Or it may start through an invitation of selected vendors to provide structured, on-site demos of their software. We recommend that the CMMS team narrow the vendors down to a "short list" and proceed without an RFP unless one is required by your company. We recommend that the top 3 to 5 vendors be selected before scheduling actual on-site vendor demos.

Prepare the vendors

To prepare for vendor demos, we recommend that the complete list (or at least the high-priority functional requirements) of functional requirements be provided to each vendor. Each vendor should clearly understand beforehand that these functional areas should be specifically shown during the demos. This process clearly defines upfront, what your company requires in terms of standard requirements and for more advanced functionality requirements. This allows the vendors to prepare their demo accordingly and to address your specific needs more thoroughly. The vendor should also be provided information about the size of the company, scope of the CMMS application, current operating systems in place (or required for the CMMS), the number of equivalent CMMS users, and any other relevant information to support the initial vendor cost proposal.

We recommend that the CMMS team members independently rate each functional area using the CMMS vendor evaluation methodology we have included in App. C. This will provide independent evaluations to be made and then the opportunity for the team to determine a consensus on their overall ratings. After defining functional requirements, each CMMS team member should understand what is needed and be able to rate each vendor consistently. Normally, at least a full 4 hours should be allocated for demos with sufficient time scheduled at the end to review any functional requirements that there might have been questions about. Site visits to see a vendor's Class A CMMS installation are a must. Ask the vendor to give you three references at their top three Class A installations (per Chap. 11—Maximizing Your IT Investment with The CMMS Benchmarking System).

Case study: CMMS selection criteria

Functional capabilities of competing CMMS vendors to meet your requirements are a major factor. However, functionality should not be the only factor to use for the final selection. We recommend that the final selection be made considering five major areas with recommended weighed values for each area. The values shown below are your CMMS team's choice, a consensus across all functional areas represented:

1. Functional requirements (40 percent)—How well will the system do what you need it to do?
2. Technical requirements (15 percent)—How well the system fits your operating system, computer platform, and meets your overall computer system technical requirements?
3. CMMS software costs (25 percent)—How much are the total software costs considering license continuation, updates, and new releases?
4. CMMS implementation and continuation cost (10 percent)—How much are the full implementation costs and the continuation costs for vendor support, license continuation fees, and new releases?
5. Vendor qualitative evaluation (10 percent)—What can we expect from vendor in terms of service, reputation, software enhancements, customization, interface capabilities, and the like?

Case Study CMMS Vendor Evaluation Methodology

Introduction

The vendor evaluation process for this case study was focused on an evaluation of five major categories with weighted importance as follows along with possible evaluation points by category.

Functional requirements	40%/2000 points
Technical requirements	15%/750 points
Software costs excluding license continuation/ updates/new releases	10%/500 points
Implementation/support including training by vendor	25%/1250 points
Qualitative evaluation	10%/500 points
Total maximum points	5000

TABLE 10.1 CMMS Qualitative Selection Criteria

CMMS qualitative selection criteria	Weighted value (%)
1. Ease of implementing and learning the system	10
2. Reliability of support and service	10
3. Quality of documentation	4
4. Vendor reputation	6
5. Vendor ability to customize system	2
6. System's ability to generate custom reports	2
7. On-site training and assistance	10
8. Full integration among modules	2
9. Prepurchase applications analysis and consulting	6
10. Adaptability to existing hardware	2
11. Advanced system features	10
12. Ease of adding modules	2
13. Networking ability	4
14. Availability of additional modules	3
15. Compatibility with existing systems	3
16. Availability of supplier's database	4
17. Service bureau type of support	10
18. Other qualitative criteria defined by CMMS team	10
Total	100

Qualitative Evaluation (Methodology Example)

1. This section with 10 percent of the total evaluation points (500) included 18 qualitative evaluation criteria provided to over 25 vendors in the initial RFP. Each criterion in turn was given a weighting factor as to its relative importance to the other criteria. Total weighting factor sums to 100. See Table 10.1.
2. Rating levels of 5 (highest) to 1 (lowest) were assigned for each vendor depending on the respective evaluation for the criteria being considered.
3. Therefore if all rating levels were 5 (highest), the maximum points for the qualitative part of the overall evaluation would be 500 points or 10 percent of 5000 total maximum points. (See Table 10.2.)

CMMS Qualitative Rating Factors and Weighted Values

Final Evaluation/Selection from Top Five Vendors (A Steelmaking Operation—8 sites)

1. This methodology allows for final judgment of the "CMMS team" to be integrated into the final selection process. Weighted values of each major evaluation category can be reconsidered by the CMMS steering team, software costs can be refined by the top vendors, actual demos can be factored in as well as many other factors not considered in the initial evaluation process. See Table 10.3
2. This methodology also allows for the next evaluation to begin with a methodology that strives to consider all factors related to fitting specific (COMPANY) CMMS needs.

The CMMS team may adjust weighted values of each factor somewhat but we think it is a good practice to determine the final selection based on these five factors listed in Table 10.4.

TABLE 10.2 CMMS Qualitative Evaluation; Points and Rating Levels by Five Major Selection Categories

Ranges of points and rating level by five major categories					
Rating Level	5	4	3	2	1
1. Functional requirements = 40%					
Points	2000	1600	1200	800	400
2. Technical requirements = 15%					
Points	750	600	450	300	150
3. Software costs = 10%					
Points	500	400	300	200	100
4. Implementation/support = 25%					
Points	1250	1000	750	500	250
5. Qualitative factors = 10%					
Points	500	400	300	200	100
Total Points	5000	4000	3000	2000	1000

The Maintenance Best Practice Implementation Process

The results of the maintenance benchmark evaluation with your scoreboard will determine the scope of work for this element of the CMMS justification strategy. Best practices should then be developed into an overall strategic maintenance plan that also incorporates within it the CMMS implementation plan. The strategic maintenance plan provides the road map for the maintenance improvement process with timelines, resources required, the associated costs, and the projected benefits.

Many of the costs/benefits related to maintenance best practices implementation would be captured as part of the CMMS economic justification process. To add increased credibility to the overall maintenance improvement process and to validate projected benefits, we must include measurement of overall maintenance performance. As we will discuss later under economic justification, we must include in our measurement process the key elements we use for economic justification. The major elements for economic justification that are detailed in Chap. 13 will typically come from four areas:

- Craft productivity improvement
- MRO inventory reduction
- Value of increased uptime/capacity
- Major projects completed sooner

We recommend that a high priority be established very early for the best practices that impact developing or upgrading three key CMMS databases: parts inventory data, equipment master data, and *preventive maintenance/ predictive maintenance* (PM/PdM) procedures. Establishing these three databases either initially into a new CMMS or preparing to convert existing data to a new system is an important first step for successful CMMS implementation. The parts inventory master file and the equipment asset master file

TABLE 10.3 CMMS Qualitative Selection Criteria by Five Major Selection Categories

		CMMS qualitative evaluation: top 5 CMMS options									
		CMMS A		CMMS B		CMMS C		CMMS D		CMMS E	
CMMS qualitative evaluation criteria	Weighting factor	RT	SC	RT	SC	RT	SC	RT	SC	RT	SC
1. Ease of implementing and learning the system	10	5	50	5	50	5	50	5	50	5	50
2. Reliability of support and service	10	5	50	5	50	5	50	4	40	5	50
3. Quality of documentation	4	5	20	5	20	5	20	4	16	5	20
4. Vendor reputation	6	5	30	5	30	5	30	5	30	5	30
5. Vendor ability to customize system (if required)	2	5	10	5	10	5	10	5	10	5	10
6. System's ability to generate custom reports	2	5	10	5	10	5	10	4	8	5	10
7. On-site training and assistance	10	5	50	5	50	5	50	5	50	5	50
8. Full integration among modules	2	5	10	4	8	4	8	4	8	4	8
9. Prepurchase applications analysis	6	5	30	5	30	5	30	4	24	5	30
10. Adaptability to existing hardware	2	5	10	5	10	5	10	5	10	5	10
11. Advanced system features	10	5	50	4	40	5	50	4	40	4	40
12. Ease of adding modules	2	5	10	5	10	5	10	5	10	5	10
13. Networking ability/connectivity	4	5	20	5	20	5	20	5	20	5	20
14. Availability of additional modules	3	5	15	5	15	5	15	5	15	5	15
15. Compatibility with existing systems	3	5	15	5	15	5	15	5	15	5	15
16. Avoidance of capital investment	4	4	16	5	20	5	20	4	16	3	12
17. Vendor strategy for next generation CMMS	10	5	50	5	50	5	50	5	50	5	50
18. Quality of refinery type demonstration site(s)	10	5	50	5	50	5	50	5	50	5	50
Total qualitative scores		496		488		498		462		480	
Top 5 CMMS options		CMMS A		CMMS B		CMMS C		CMMS D		CMMS E	
Overall qualitative ranking		2		3		1		5		4	

RT = rating SC = score where the SC = rating (RT) times the weigh ting factor

TABLE 10.4 Final CMMS Ranking with Qualitative and Quantitative Factors Considered

	Final CMMS evaluation summary and ranking				
Final evaluation category	CMMS A	CMMS B	CMMS C	CMMS D	CMMS E
1. Functional requirements	5	4	4.5	4.75	2
	2000	1600	1800	1900	1900
2. Technical requirements	5	4	5	5	5
3. Software costs	3	4	5	4	3
	300	400	500	400	300
	$1,423,700	$630,000	$388,220	$699,000	$1,800,450
4. Implementation/support costs	NE	$140,000	$127,149	NE	$2,883,000
a. Data conversion	$185,000	$60,000	$104,000	$38,400	$594,000
b. Interface	NE	$186,000	$83,200	NE	NE
c. Customization	NE	NE	$62,400	NE	NE
d. Training	NE	$60,000	$106,667	NE	$520,000
e. License/Ongoing support	$301,500	$113,400	$69,880	$155,767	$324,000
	$373,000				
Total implementation CMMS vendor	$859,500	$559,400	$553,296	$194,167	$431,000
Implementation support rating	4	5	5	4	3
	1000	1250	1250	1000	750
5. Qualitative evaluation rating	496	488	498	462	480
Overall rating points	4546	4338	4798	4512	4180
	CMMS A	CMMS B	CMMS C	CMMS D	CMMS E
Rank by CMMS	No. 2	No. 4	No. 1	No. 3	No. 5

provide the foundation for the work-order system, costing, inventory management, and overall work-control and planning. Updates to these two databases before start-up is essential. Do not underestimate the time and effort required getting these two databases established. Take the time to do this part right the first time. Once the job is complete and the new CMMS is up and running, designate specific responsibility for maintaining these two databases to someone in your operation.

The PM procedures with task/frequency data are essential to a proactive, planned maintenance operation that is able to avert major failures through regularly scheduled inspections, lubrication services, and overhauls. Application of predictive maintenance technologies such as vibration analysis, infrared testing, ultrasonic testing, and oil annuluses likewise can be used for condition-based monitoring to go beyond the limitations of a PM program. Most operations will have at least something in place that can be converted to the new CMMS if a complete PM program review and update is not possible before actual start-up. It is best to try and upgrade PM/PdM procedures before starting the new CMMS, but it is not always possible. The key here is to get the commitment to complete the review and update even if the old procedures must be converted to the new system.

Work in each of these areas is ideal for using the team process as subteams under the CMMS team. Also within multiplant operations, we recommend using a *data standards team* to ensure that the data structure for the parts master file, the equipment master file, and PM/PdM procedures are consistent across all sites. An example of consistent multisite data is from catapiller, one of our clients. As support to CMMS implementation, the case study example includes recommended charters for an *equipment master team*, a *PM review team*, and a *parts review team*.

The CMMS Implementation Process

The actual CMMS and best practice implementation presents by far the greatest challenges and the toughest part of the total process. The reason is that the real work is just starting. Typically, only about 10 percent of the real work is during the up-front steps devoted to planning, evaluation, and selection. Ninety percent or more of the real work comes during implementation. This is the step where the predefined CMMS functional requirements will be compared to actual system performance, capabilities, and results. During implementation it is where other maintenance best practices will be coming together to establish the new way of doing business, that is, a total process for improving the maintenance function and achieving full utilization of CMMS capabilities. Organizations that have used a cross-functional CMMS team during evaluation/selection should continue their approach during actual implementation. There are a number of important considerations for a successful CMMS implementation which include:

- Do not underestimate the time and effort required for getting the parts database and the equipment master file established, particularly if there is a need to develop this database from scratch. The full use of the basic CMMS module requires that these two databases be in place.

- For organizations still using a manual system, this data may be incomplete or not available, and will require a major project to identify the inventory parts and to identify and number equipment. Resources to collect and input data for all features of a CMMS must be established very early as part of the initial planning process.
- Reducing CMMS start-up costs by reducing vendor or outside consultant help can result in long-term loss of system performance as well as failure to receive planned benefits. A "do-it-yourself approach" will often cost much more over the long term than getting resources up-front when they are needed the most.
- Do not use CMMS to automate existing procedures within a reactive maintenance environment. Most organizations fail to achieve the full benefits of CMMS because they did not take the time to completely revaluate and refine current practices first.
- Full integration of the basic CMMS modules is the key to achieving maximum benefits. Many CMMS fail because the easy tasks get completed first and the harder tasks get put off or never get completed. Without effective use of the work-order and equipment history modules, for example, we can very easily have just a computerized version of the manual "fat file" where work orders are stored in manila folders by equipment number. If the storeroom inventory module never gets developed, the equipment history module loses its capability for providing total repair costs or parts listing by equipment asset number. Likewise, if a PM program is never developed or if maintenance planning is never formalized, full benefits from an integrated CMMS never occur.
- The CMMS team must work closely with the CMMS vendor and all internal resources and external consultant to define a CMMS implementation plan. The plan should include key activities, time lines, and designated responsibilities. Key implementation activities may include tasks such as the following discussion.

Making the right choice for a new CMMS requires clearly understanding CMMS functional requirement (Chap. 9) and defining key qualitative and quantitative factors for justification as defined in this chapter, and all essential benchmarking tools as well for the business of maintenance.

Chapter 11

Maximizing Your IT Investment with the CMMS Benchmarking System

This chapter introduces the *computerized maintenance management system* (CMMS) benchmarking system* as the second benchmarking tool and the improvement process for gaining better use of CMMS and your current information technology for maintenance. This benchmarking tool is a means to evaluate the effectiveness of your current CMMS, to define functional gaps and enhance current use or to help upgrade functional gaps. It is also a methodology to help develop and justify a replacement strategy. The CMMS benchmarking system which has 9 benchmark categories and 50 benchmark items is easily adaptable and can be specifically tailored to all CMMS systems and to their intended application. It has been used for Maximo, SAP R/3, 3 Datastream versions, PHC 2000, JD Edwards One World CMMS module and others.

It is designed as a methodology for developing a benchmark rating of your CMMS (Class A, B, C, or D) to determine how well this tool is supporting best practices and the total maintenance process. It is not designed to evaluate the functionality of various CMMS systems nor is it intended to compare vendors. It can also be used as a method to measure the future success and progress of a CMMS system implementation that is now being installed. Maintenance information is one of the key maintenance resources and must be a part of your approach to *continuous reliability improvement* (CRI).

---

*The concept for CMMS benchmarking system was developed as a result of using Oliver Wight's benchmarking of an installed MRP II system in 1987 for manufacturing shop floor control and a production and inventory management system (For more information, google "Oliver Wight", or go to http.//oliverwight.com)

TABLE 11.1 Summary—The CMMS Benchmarking System

The CMMS benchmarking system	
CMMS benchmark categories	Benchmark items
1. CMMS data integrity	6
2. CMMS education and training	4
3. Workcontrol	5
4. Budget and cost control	5
5. Planning and scheduling	7
6. MRO materials management	7
7. Preventive and predictive maintenance	6
8. Maintenance performance measurement	4
9. Other uses of CMMS	6
Total CMMS benchmark items	50

A summary of the CMMS benchmarking system is shown in Table 11.1 with the 9 assessment categories that include a total of 50 benchmark items for benchmarking your CMMS installation. The CMMS benchmarking system rating scale is shown in Table 11.2.

Conducting the CMMS Benchmark Evaluation

The CMMS benchmark evaluation can be conducted internally by the maintenance leader or via an internal team effort of knowledgeable maintenance people. Other options include using support from an independent resource to provide an objective maintenance benchmarking resource. "The Scoreboard for Maintenance Excellence" process in combination with the CMMS benchmarking system provides powerful tools to help achieve greater value from all types of maintenance operations.

The CMMS benchmarking system provides a means to evaluate and classify your current installation as either "Class A, B, C, or D." A total of 9 major categories are included along with 50 specific benchmark items. Each benchmark item that is rated as being accomplished, satisfactorily receives a maximum score of 4 points. If an area is currently being "worked on," a score of 1, 2, or 3 points can be assigned based on the level of progress achieved. For example, if spare parts inventory accuracy is at 92 percent compared to the target of 95 percent, a score of 3 points is given. A maximum of 200 points is possible. A benchmark

TABLE 11.2 The CMMS Benchmarking System Rating Scale

CMMS benchmarking system rating scale	
Class A	180–200 points (90% +)
Class B	140–179 points (70–89%)
Class C	100–139 points (50–69%)
Class D	0–99 points (up to 49%)

TABLE 11.3 Sample The CMMS Benchmarking System

CMMS benchmarking categories and item descriptions	Yes (4 points)	No (0 points)	Working on it (1, 2, or 3 points)
A. CMMS date integrity			
1. Equipment (asset) history data complete and accuracy 95% or better			
2. Spare parts inventory master record accuracy 95% or better			

rating of "Class A" is within the 180 to 200-point range. The complete CMMS benchmarking system is included as App. B and a sample is included in Table 11.3.

Developing a future "Class A" CMMS installation requires that each organization starts early in the implementation phase with establishing how they will determine the overall success of their installation. The CMMS benchmarking system provides the framework for internal benchmarking of the CMMS installation as it matures. It is recommended that a team process be used for the CMMS benchmarking evaluation and that it be included as part of the CMMS evaluation team's initial work.

Establishing a "Class A" CMMS requires that a number of key databases be established and that a number of maintenance best practices be in place. Data integrity, accuracy, and continuous maintenance of the key databases provide the foundation for a "Class A" CMMS installation. There are a number of other factors related to the CMMS and to maintenance best practices that in combination produce a future "Class A" installation.

We will now review each of the 9 major categories from the CMMS benchmarking system and provide key recommendations on each of the 50 benchmark items for getting your CMMS implementation started on the right track from day one.

A. CMMS data integrity
1. Equipment (asset) history data complete and accuracy 95 percent or better.
2. Spare parts inventory master record accuracy 95 percent or better.
3. Bill of materials for critical equipment includes listing of critical spare parts.
4. *Preventive maintenance* (PM) tasks/frequencies data complete for 95 percent of applicable assets.
5. Direct responsibilities for maintaining parts inventory database is assigned.
6. Direct responsibilities for maintaining equipment/asset database is assigned.

1. *Accuracy of equipment history database.* The equipment database represents one of the essential databases that must be developed or updated as part of implementing a new CMMS. It requires that a complete review of all equipment be made to include all parent/child systems and subsystems that will be tracked for costs, repairs performed, and so forth. The work to develop or update this database should begin as soon as possible after the data structure of the equipment master file for the new CMMS is known.

   The equipment master information for a piece of equipment (parent/child), manufacturer, serial number, equipment specs, and location will all need to be established. If the installation and removal of components within certain process type operations requires tracking by serial number and compliance to process safety management requirements, these equipment items will have to be designated in the equipment database.

   If an equipment database exists as part of an old CMMS, now is the time to review the accuracy of the old equipment database prior to conversion to the new system. Conversion of the new equipment master database into the new system should be done only after a thorough and complete update of the old database has occurred. Once the new equipment master database has been converted to the new CMMS, a process to maintain it at an accuracy level of 95 percent or above should be established.

2. *Accuracy of spare parts database.* The spare parts database represents another key database that must be developed or updated as part of implementing a new CMMS. For operations not having a parts inventory system this will require doing a complete physical inventory of spare parts and materials. All inventory master record data for each item will need to be developed based on the inventory master record structure for the new CMMS and loaded directly to the inventory module.

   Operations that have an existing spare parts database should take the time to do a complete review of it prior to conversion. Typically, this will allow for purging the database of obsolete parts and doing a complete review of the inventory master record data. This can be a very time-consuming process, but it allows the operation an excellent chance to finally take the time to revise part descriptions, review safety stock levels, reorder points and vendor data, and start the new CMMS with an accurate parts inventory database.

3. *Bill of materials.* One key functional capability of CMMS is to provide a spare parts listing (bill of materials) within the equipment module. This requires researching where spare parts are used and

linking inventory records with equipment master records that are component parts of an equipment asset. This function would also add, change, or delete items from an established spares list or copy a spares list to another equipment master record. Also this feature would copy all or part of a spares list to a work-order job plan and create a parts requisition or pick list to the storeroom.

The process of establishing a spares list is time consuming and would involve only major spares that are currently carried in stock. Most CMMS systems have the capability to build the spare parts list as items are issued to or purchased for a piece of equipment. It is recommended that equipment bill of materials be established, but the conversion of equipment master data can take place without this information being available. Because bill of materials for spare parts is so beneficial for planning purposes, it is recommended that the process to identify key spares in the equipment master be a priority area.

4. *PM tasks/frequencies.* The PM and *predictive maintenance* (PdM) database is another key database necessary for establishing a "Class A" installation. If a current PM/PdM database is present, it is recommended that the existing procedures be reviewed and updated prior to conversion to the new system. If the existing PM/PdM database has been updated continuously on the old system, conversion can probably occur directly from the old to the new PM/PdM database; this however, will depend on the PM/PdM database structure of the new system.

   It is recommended that in the very early stages of a new CMMS benchmark/selection process that the status of the current PM/PdM program be evaluated. If a process for the review/update of PM/PdM procedures has not been in place then it is very important to get something started as soon as possible. This provides an excellent opportunity to establish a team of experienced craft people, engineers, and maintenance supervisors to work on PM/PdM procedures to review and update task descriptions, frequencies, and making sure that all equipment is covered by proper procedures.

5. *Maintaining parts database.* After a new CMMS is installed, it is highly recommended that one person be assigned direct responsibility for maintaining the parts database. This person would have responsibility for making all additions and deletions to inventory master records, changing stock levels, reorder points and safety stock levels, and changing any data contained in the inventory master records. This person could also be designated responsibility for coordinating the development of the spares list if this information is not available. This person would be responsible for recommending obsolete items based on monitoring of usage rates

or due to equipment being removed from the operation. The practice of having one primary person assigned direct responsibility for the inventory master records can help ensure that parts database accuracy is 95 percent or greater.

6. *Maintaining equipment database.* It is also highly recommended that one person be assigned direct responsibility for maintaining the equipment database. This person would be responsible for making all changes to equipment master records. Information on new equipment would come to this person for setting up parent/child relationships of components in the equipment master records. Information on equipment being removed from the operation would also come to this person to delete equipment master records. Coordination between this person and the person responsible for the parts database would be required to ensure that obsolete parts were identified and/or removed from the inventory system due to removal of equipment.

B. CMMS education and training
7. Initial CMMS orientation training for all maintenance employees. 8. An ongoing CMMS training program for maintenance and storeroom employees. 9. Initial CMMS orientation training for operations employees. 10. CMMS systems administrator (and backup) designated and trained.

7. *Initial CMMS training.* One of biggest roadblocks to an effective CMMS installation is the lack of initial training on the system. Many organizations never take the time up-front to properly train their people on the system. Shop-level people must gain confidence in using the system for reporting work-order information and knowing how to look up parts information. The CMMS implementation plan should include an adequate level of actual hands-on training on the system for all maintenance employees prior to the "go live" date. It is important to invest the time and expense to "train the trainers" who in turn can assist with the training back in the shop. Many organizations set up "conference room pilots" where the CMMS software is set up and training occurs with actual data using CMMS vendor trainers or in-house trainers. It is highly recommended that competency-based training be conducted so that each person trained can demonstrate competency in each function that they must perform on the system.
8. *Ongoing CMMS training.* The CMMS implementation plan must consider having an ongoing training program for maintenance and storeroom personnel. After the initial training there must be someone

in the organization with the responsibility for ongoing training. If a good "trainer" has been developed within the organization prior to the "go live" date, this person can be the key to future internal training on the new system. Ongoing training can include one on one support that helps to follow up on the initial training.

9. *Initial CMMS training for operations personnel.* The customers of maintenance must gain basic understanding of the system and know how to request work, check status of work requested, and understand the priority system. During implementation, operations personnel need to get an overview of how the total system will work and the specific things they will need to do to request work. If the organization has a formal planning and scheduling process, they will also need to know the internal procedures on how this will work.

10. *CMMS systems administrator/backup trained.* It is important that each site have one person trained and dedicated as the systems administrator with a backup trained whenever possible. Typically, this person will be from information services and have a complete knowledge of system software, hardware, database structures, interfaces with other systems, and report-writing capabilities. The systems administrator will also have responsibility of direct contact with the CMMS vendor for debugging software problems and for coordinating software upgrades.

C. Work control
11. A work-control function is established or a well-defined documented process is being used. 12. Online work request (or manual system) used to request work based on priorities. 13. Work-order system used to account for 100 percent of all craft hours available. 14. Backlog reports are prepared by type of work to include estimated hours required. 15. Well-defined priority system is established based on criticality of equipment, safety factors, cost of downtime, and so forth.

11. *Work-control function established.* A well-defined process for requesting work, planning, scheduling, assigning work, and closing work orders should be established. The work-control function will depend on the size of the maintenance operation. Work control may involve calls coming directly to a dispatcher who creates the work-order entry and forwards the work order to a supervisor for assignment. The work request could also be forwarded directly to an available craftsperson by the dispatcher for execution of true emergency work.

Work control can also be where work requests are forwarded manually or electronically to a planner who goes through a formal planning process for determining scope of work, craft requirements, and parts requirements to develop a schedule. PM/PdM work would be generated and integrated into the scheduling process. The status of the work order would be monitored which might be in progress, awaiting parts, awaiting equipment, awaiting craft assignment, or awaiting engineering support. A work-order backlog would also be maintained to provide a clear picture of work-order status.

Effective work control provides systematic control of all incoming work through to the actual closing of the work order. The work-control process should be documented with clearly defined written procedures unique to each maintenance operation.

12. *Online work request based on priorities.* Requesting work on line represents an advanced CMMS functional capability where the customer enters the work request directly into the system on a local area network or via e-mail. Online work requests would include basic information about work required, equipment location, date work to be completed by, name of requestor, and priority of the work. This information would go to the work-control function where the jobs would be planned, scheduled, and assigned based on the overall workload. The requestor would have the capability to track the status of their jobs on line and even give final approval that the work was completed satisfactorily.

13. *Work-order system accounts for 100 percent of craft hours.* All craft work should be charged to a work order of some type. Accountability of labor resources is an important part of managing maintenance as an internal business. Quick reporting to standing work orders can be established for jobs of short duration within a department or for the reporting of noncraft time such as meetings, delays in getting the equipment to work on, training, and chasing parts.

14. *Backlog reports.* Maintaining good control of the work to be done is essential to the maintenance process. Having the capability to visually see the backlog helps to effectively plan and schedule craft resources.

    The CMMS reporting system should provide the capability to show backlog of work in a number of ways:

    - By type of work
    - By craft
    - By department
    - By overdue work orders
    - By parts status
    - By priority

15. *Priority system.* A "Class A" CMMS installation will have in place a priority system that allows for the most critical repairs to get done first. An effective priority system adds professionalism to the maintenance operation and directly supports effective planning and scheduling. There are two basic systems for establishing priorities:
    - *Straight numeric.* Priority 1, 2, 3, 4, 5, and so on where each priority level is defined by a definition such as priority 1: A true emergency repair that affects safety, health, or environmental issues.
    - *Ranking index of maintenance expenditures (RIME) system.* This system (originated by Albert Ramond Associates in Chicago) combines the criticality index of the equipment (10 highest to 1 lowest) with criticality of the work type (10 highest to 1 lowest) to compute the RIME priority number. The RIME priority number equals the equipment criticality index multiplied by the criticality number of the work type.

D. Budget and cost control
16. Craft labor, parts, and vendor support costs are charged to work order and are accounted for in equipment/asset history file. 17. Budget status on maintenance expenditures by operating departments is available. 18. Cost improvements due to CMMS and best practice implementation have been documented. 19. Deferred maintenance and repairs identified to management during budgeting process. 20. Life-cycle costing is supported by monitoring of repair costs to replacement value.

16. *Craft labor, parts, and vendor support costs.* The equipment history file should provide the source of all costs charged to the asset. Here it is important to ensure that all labor is charged to the work orders for each asset and that parts are charged to the respective work orders.

17. *Budget status-operating departments.* Operating departments should be held accountable for their respective maintenance budgets. With an effective work-order system in place for charging of all maintenance costs, the accounting process should allow for monitoring the status of departmental budgets. One recommended practice is for maintenance to be established as a zero-based budget operation and that all labor and parts be charged back to the internal customer. This practice helps ensure accountability for all craft time, parts, and materials to work orders.

18. *Cost improvements due to CMMS.* The impact of a successful CMMS installation should be reduced costs and achieving gained value in terms of greater output from existing resources. The CMMS team should be held accountable for documenting the savings that are achieved from the new CMMS and the maintenance best practices that evolve. The areas that were used to justify the CMMS capital investment such as reduced parts inventory, increased uptime, and increased craft productivity should all be documented to show that improvements did occur.

19. *Deferred maintenance identified.* It is important that maintenance provides management with a clear picture of total maintenance requirements that require funding for the annual budget. Deferred maintenance on critical assets can lead to excessive total costs and unexpected failures. Benefits from CMMS will provide improved capability to document deferred maintenance that must be given priority during the budgeting process each year.

20. *Life-cycle costing supported.* Complete equipment repair history provides the base for making better replacement decisions. Many organizations often fail to have access to accurate equipment repair costs to support effective replacement decisions and continue to operate and maintain equipment beyond its economically useful life. As a result, the capital justification process then lacks the necessary life-cycle costing information to support replacement decisions.

E. Planning and scheduling
21. A documented process for planning and scheduling has been established. 22. The level of proactive, planned work is monitored and documented improvements have occurred. 23. Craft utilization (true wrench time) is measured and documented improvements have occurred. 24. Daily or weekly work schedules are available for planned work. 25. Status of parts on order is available for support to maintenance planning process. 26. Scheduling coordination between maintenance and operations has increased. 27. Emergency repairs, hours, and costs tracked and analyzed for reduction.

21. *Planning and scheduling.* This maintenance best practice area is essential to better customer service to operations and greater utilization of craft resources. For most maintenance operations with

20 to 30 craftspeople, a fulltime planner can be justified. The CMMS system functionality must support the planning process for control of work orders, backlog reporting, status of work orders, parts status, craft labor availability, and the like. The planning and scheduling function supports changing from a "run to failure strategy" to one for proactive, planned maintenance.

22. *Planned work increasing.* The bottom line results for the planning process is to actually increase the level of planned work. Percent planned work should be monitored and included as one of the overall maintenance performance metrics. In some organizations with effective preventive and predictive maintenance programs the level of planned work can be in the 90 percent range or more.

23. *Craft utilization measured and improving.* Effective planning and scheduling is essential to increasing the level of actual hands-on wrench time of the craft workforce. Improving craft utilization allows more work to get done with current staff by eliminating non-craft activities such as waiting for equipment, searching for parts, and scheduling the right sequence for different crafts on the job

24. *Work schedules available.* One key responsibility of the planning process is to establish realistic work schedules for bringing together the right type of craft resources, the parts required, the equipment to be repaired or serviced along with having the time available to complete the job right the first time. The actual schedule may only start with a one-day schedule and gradually work up to scheduling longer periods of time. Work schedules provide a very important customer service link with operations that help to improve overall coordination between maintenance and operations.

25. *Spare parts status is available.* One of the most essential areas to support effective planning is the maintenance storeroom and the accuracy of the parts inventory management system. Jobs should not be put on the current schedule without parts being on hand. The planner must have complete visibility of inventory on hand balances, parts on order, and the capability to reserve parts for planned work.

26. *Scheduling coordination with operations.* As the planning function develops there will be improved coordination with operations to develop and agree upon work schedules. This may involve coordination meetings near the end of each week to plan weekend work or to schedule major jobs for the upcoming week. Direct coordination with operations allows maintenance to review PM/PdM schedules or to review jobs where parts are available to allow the job to schedules based on operations scheduling equipment availability.

27. *True emergency repairs tracked.* Many organizations like the Marathon Refinery in Robinson Il focus on reducing true emergency repairs, which create uncertainty for operations scheduling and contribute to significantly higher total repair costs than planned work. Improved reporting capabilities of an effective CMMS will allow for better tracking of emergency repairs, document causes for failures, and assist in the elimination of the root causes for failures.

F. *Maintenance repair operations* (MRO) materials management
28. Inventory management module fully utilized and integrated with work-order module.
29. Inventory cycle counting based on defined criteria is used and inventory accuracy is 95% or better.
30. Parts kiting is available and used for planned jobs.
31. Electronic requisitioning capability available and used.
32. Critical and/or capital spares are designated in parts inventory master record database.
33. Reorder notification for stock items is generated and used for reorder decisions.
34. Warranty information and status is available.

MRO materials management. The overall area of MRO parts and materials procurement, storage, inventory management, and issues represents another best practice area that often needs major work when implementing a CMMS and developing a "Class A" installation. Many organizations never take the time to set up a well-planned and controlled storeroom operation, and often find out that their parts database is a weak link needing major updates before CMMS can truly be used effectively.

28. *Inventory management module.* The work-order module must be fully integrated with an accurate parts inventory management module to charge parts back to work orders, to check parts availability status for planned work, to reserve parts, and to check status of direct purchases. A "Class A" CMMS installation will develop, maintain, and fully utilize the inventory module and ensure that it is fully integrated with the work-order module.

29. *Inventory cycle counting established.* Inventory accuracy should be one of the key metrics for MRO materials management and it can best be accomplished by cycle counting rather than annual physical inventories. Most CMMS systems will allow for your own criteria to be developed such as doing an ABC analysis of inventory items (based on either usage value or frequency of issue) and then

scheduling of periodic counts for each classification of inventory item that you want to cycle count. For example, *A* items would be counted more frequently than *B* and *C* items. The real value of cycle counting is that it is a continuous process that creates a high level of discipline and allows for inventory problems and adjustment to be made throughout the year rather than once following the annual inventory.

30. *Parts kiting.* This best practice area is key to the planning process and can evolve over time as the planning process matures to the point of being able to give the storeroom prior notification on the parts required for planned jobs. Controlled staging areas are set for parts that are either pulled from stock or received from direct purchases.
31. *Electronic parts requisitioning.* This functional capability can provide paperless workflow for requisitioning of parts directly from maintenance to the storeroom for creation of a pick list for the item or go to purchasing to create a purchase order for a stock item or direct purchase. In some cases electronic requisitioning might go directly to the vendor using e-commerce capability.
32. *Critical spares identified.* Critical spares (or insurance items) that may be one of a kind, high-cost spare are often part of the parts inventory system. It is recommended that these items be classified and identified in the item master record as such. This practice will help to separate these items from the regular inventory management process and identify them as a separate part of the total inventory value that has been fixed. Critical spares should also be identified in the spares list for the equipment they have been purchased for.
33. *Reorder notification process.* The capability to determine when and what to reorder based on a review of stock level reorder points is an important feature for a "Class A" installation. A recommended reorder report should be generated periodically and reviewed for validity as well as for any future needs that may not be reflected in current on hand balances. Based on final review of the recommended reorder report, electronic requisitioning then could occur directly to purchasing.
34. *Warranty information.* Many organizations fail to have a process in place to track warranty information and in turn incur added costs by not being able to get proper credit for items under warranty. Tracking specific high-value parts or components and specific equipment under warranty should be a CMMS functionality of the equipment master or the inventory item master database. The system should provide a quick reference and alert to the fact that the item is still under warranty and that a follow-up claim to the vendor is needed.

G. Preventive /predictive maintenance (PM/PdM)
35. PM/PdM change process is in place for continuous review/update of tasks/frequencies. 36. PM/PdM compliance is measured and overall compliance is 98 percent or better. 37. The long-range PM/PdM schedule is available and leveled loaded as needed with CMMS. 38. Lube service specifications, tasks, and frequencies included in CMMS database. 39. CMMS provides MTBF, MTTR, failure trends, and other reliability data. 40. PM/PdM task descriptions contain enough information for new craftsperson to perform task.

35. *PM/PdM change process.* This best practice area simply ensures that PM/PdM procedures are subject to a continuous review process and that all changes to the program are made in a timely manner. The CMMS system should provide an easy method to update task descriptions and task frequencies and allow for mass updating when the procedure applies to more than one piece of equipment.
36. *PM/PdM compliance is measured.:* One key measure of overall maintenance performance should be how well the PM/PdM program is being executed based on the schedule. Measuring PM/PdM compliance ensures accountability from maintenance and from operations. Normally, a scheduling window of a week will be established to determine compliance. A goal of 98 percent or better for PM/PdM compliance should be expected.
37. Long-range PM/PdM scheduling. As a PM/PdM schedule is loaded to the system, peaks and valleys may occur for the actual scheduling due to frequencies of tasks coming due at the same time period. The CMMS system should provide the capability to level load the actual PM/PdM schedule and to view upcoming PM/PdM workloads to assist in the overall planning process.
38. *Lubrication services.* Ideally lubrication services, tasks, frequencies, and specifications should be included as part of the PM/PdM module. A continuous change process for this area should also be put in place as well as an audit process established to ensure all lube and PM/PdM tasks are being performed as scheduled.
39. *CMMS captures reliability data.* The elimination of root causes of problems is the goal rather than just more PM/PDM. One important feature of a "Class A" installation is being able to capture failure information that can in turn be used for reliability improvement.

This requires that a good coding system for defining causes for failures be developed and that this information be accurately entered as the work order is closed.

40. *Complete PM/PDM task descriptions.* PM/PdM task descriptions often provide vague terminology to check, adjust, inspect, and so forth; and do not provide clear direction for specifically what is to be done. Task descriptions should be reviewed periodically and details added to the level that a new craftsperson would understand exactly what is to be done and be able to adequately perform the stated task.

H. Maintenance performance measurement
41. Downtime (equipment/asset availability) due to maintenance is measured and documented improvements have occurred. 42. Craft performance against estimated repair times is measured and documented improvements have occurred. 43. Maintenance customer service levels are measured and documented improvements have occurred. 44. The maintenance performance process is well established and based on multiple indicators compared to baseline performance values.

41. *Equipment downtime reduction.* Another key metric for measuring overall maintenance performance is increased equipment uptime. The improvement in this metric is a combination of many of the previously mentioned best practices all coming together for improved reliability. Downtime due to maintenance should be tracked and positive improvement trends should be occurring within a “Class A” installation.

42. *Craft performance.* Three key areas affecting overall craft productivity are craft utilization (wrench time), craft performance, and craft service quality. Measurement of craft performance requires that realistic planning times be established for repair work and PM tasks. A standard job plan database can be developed for defining job scope, sequence of tasks, special tools listing, and estimated times. The goal is measurement of the overall craft workforce and not individual performance. Planning times are also an essential part of the planning process for developing a more accurate picture of workload and to support scheduling of overtime and staff additions.

43. *Maintenance customer service.* The results of improved maintenance planning must be improved customer service. The overall measurement process should include metrics such as compliance to meeting established schedules and jobs actually completed on schedule.

44. Maintenance performance measurement process. In this area it is important to have a performance measurement process that includes a number of key metrics in each of the following major categories:
    - Budget and cost
    - Craft productivity
    - Equipment uptime
    - Planning and scheduling
    - Customer service
    - MRO materials management
    - Preventive and predictive maintenance

    The overall maintenance performance process should be established so that it clearly validates the benefits being received from the CMMS and maintenance best practice implementation.

I. Other uses of CMMS
45. Maintenance leaders use CMMS to manage maintenance as internal business.
46. Operations staff understands CMMS and uses it for better maintenance service.
47. Engineering changes related to equipment/asset data, drawings, and specifications are effectively implemented.
48. Hierarchies of systems/subsystems used for equipment/asset numbering in CMMS database.
49. Failure and repair codes used to track trends for reliability improvement.
50. Maintenance standard task database available and used for recurring planned jobs.

45. *Maintenance managed as a business.* One true indicator for a successful CMMS installation is that it has changed the way that maintenance views it role in the organization. It should progress to the point that maintenance is viewed and managed as an internal business. This view requires greater accountability for labor and parts costs, greater concern for customer service, better planning, and greater attention to reliability improvement and increased concern for the maintenance contribution to the bottom line.

46. *Operations understand benefits of CMMS.* There is direct evidence that operations understand that an improved CMMS is a contributor to improve customer service. The scheduling process is continuously improving through better coordination and cooperation between maintenance and operations within a "Class A" installation.

47. *Engineering changes.* Accurate engineering drawings are essential to maintenance planning and to actually making the repairs. Asset documentation must be kept up-to-date based on a formal engineering change process. Feedback to engineering must be made on all changes as they occur on the shop floor. Engineering must in turn ensure that master drawings are updated and that current revisions made available to maintenance.
48. *Equipment database structure.* To provide equipment history information in a logical parent/child relationship, the equipment database structure has been developed using an identification of systems and subsystems. Accessing the equipment database should allow for drilldown from a parent-level to lower-level child locations that are significant enough for equipment master information to be maintained.
49. *Failure and repair codes.* The reporting capability of the CMMS should provide good failure trending and support analysis of the failure information that is entered from completed work orders. Improving reliability requires good information that helps to pinpoint root causes of failure.
50. *Maintenance standard task database.* Developing the maintenance standard task database (or standard repair procedures) for recurring jobs is an important part of a planner's job function. This allows for determining scope of work, special tools and equipment, and for estimating repair times. Once a standard repair procedure is established it can then be used as a template for other similar jobs resulting in less time for developing additional repair procedures.

Summary

Developing a "Class A" CMMS installation requires the combination of a good system functionality and improved maintenance practices. The CMMS team should begin very early during implementation with how it will measure the success of the installation. The recommendations provided here for using the CMMS benchmarking system can help your organization achieve maximum return on its CMMS investment.

Understand the Power of CMMS/EAM to Support Potential Savings

The evaluation of your CMMS using *the CMMS benchmarking system* will identify improvement opportunities that translate into direct savings. It is important that these areas be highlighted and that the future process for performance measurement is focused upon these specific areas which may have been used initially for CMMS/enterprise asset management (EAM) capital project justification. The opportunities to realize both quantifiable and qualifiable benefits are numerous. Maintenance must be given the best practice tools, the people resources, and capital investments to address the improvement opportunities and in turn are held accountable for results. As summarized in Part II, there are 12 key areas where direct savings, cost avoidances, and gained value can be established, and documented. Effective CMMS/EAM will contribute to all of them and help to increase:

1. Value of asset/equipment uptime providing increased capacity and throughput
2. Value of increased quality and service levels due to maintenance
3. Value of facility availability or cost avoidance from being unavailable
4. Value of increased direct labor utilization (production operations)
5. Gained value from increased craft labor utilization/effectiveness via gains in wrench time
6. Gained value from increased craft labor performance/efficiency
7. Gained value of clerical time for supervisors, planners, engineering, and administrative staff
8. Value of MRO materials and parts inventory reduction
9. Value of overall MRO materials management improvement
10. Value of overall maintenance costs reductions with equal or greater service levels
11. Value of increased facility and equipment life and net life-cycle cost reduction
12. Other manufacturing and maintenance operational benefits, improved reliability, and other reduced costs

Use CMMS to Develop Your Maintenance Operation as a Profit Center

A fully utilized CMMS is your business management system to support the business of maintenance. It is a mission-essential information technology tool; and effective physical asset management and maintenance is also mission-essential and a core requirement for success. Often we see the CMMS being purchased as "the solution," never really integrated with the business system or the necessary basic best practices initiated to really make the IT investment work. Often maintenance is only viewed as a "necessary evil" and not as a valid "profit center" and internal business. Many times the maintenance leaders cannot sell management on doing maintenance the right way or to convince them that the right thing to do is to shut down for preventive maintenance. Conversely, when maintenance is viewed as a "profit center" the opportunities to realize quantifiable benefits are numerous. In turn, maintenance support to the profit-optimization process continues when CMMS is used effectively to develop your maintenance operation as a profit center.

Chapter

12

Case Study: Effective CMMS Plus Best Practices: A Powerful Combination for Profit at Argentina's Largest Steel Maker

This case study includes three parts. Part I first provides a client overview that involves the challenges, the approach, and a summary of the overall results. Part II provides the details of the justification process used for *computerized maintenance management system* (CMMS) *and best practice implementation* and the step-by-step calculations that were used for determining *return on investment* (ROI) possibilities. Part III outlines the cost evaluation process for the two top choices: SAP R/3 and MAXIMO. This work was performed by the TMEI/Biasca Alliance Team. Translations of final report and presentation in Spanish was done by Rodolpho Biasca and Pablo Domenquez both Argentine citizens and long-time business and maintenance consulting professionals.

Part I: Client Overview

Siderar is the largest steel company in Argentina with a production level of over 2-million tons per year. Siderar operates seven plants within the province of Buenos Aires. It is a fully integrated producer, which uses iron ore and coal as raw materials to produce coke, pig iron, and steel, in order to manufacture hot and cold rolled sheet and coated products.

The challenge

Siderar wanted to upgrade their in-house developed legacy CMMS and to implement an improved CMMS system throughout its seven plants. They also wanted to improve corporatewide maintenance best practices, have systems in place to measure results, and to ensure that the functionality of a new CMMS would fully support their maintenance and total steel making process improvement efforts. The current Siderar organization represented the privatization results over the previous operation which was an Argentine government owned organization. Extensive before and after documentation revealed extraordinary improvement by the current Siderar staff's profit and customer-centered leadership strategy.

Siderar's strategic plan required many things. As a result they wisely requested a complete benchmark assessment of their total maintenance operation at all seven plants. They needed to define current success, recommended best practice improvements and Siderar's functional requirements for a world-class CMMS. The objective was to provide a recommended CMMS solution that included evaluating the SAP R/3 plant maintenance module as one option. Siderar requested specifically that SAP R/3 (recently purchased by Siderar for financial and purchasing)) be closely evaluated as to its capability to provide the required functionality needed in maintenance operations as well.

The approach

The approach included a "Scoreboard for Maintenance Excellence" assessment at the seven Siderar plants, specific recommendations for improvements at all plants, the definition of functional requirements for a future CMMS, (based upon Appendix C- CMMS Functionality Checklist of this book) and a recommended performance measurement process (a Siderar Maintenance Excellence Index). The approach included an important team effort with Siderar staff at both corporate engineering and plant levels to provide a true Siderar solution. Results over the course of the project included:

- Development of functional requirements for a corporatewide CMMS
  - Established method to evaluate vendors
  - Developed CMMS vendor short list
  - Evaluated SAP R/3 plant maintenance and procurement module
- Established a recommended Siderar CMMS strategy
  - Recommended CMMS vendor
  - CMMS implementation plan established
  - Established a Siderar *CMMS benchmarking system* to measure progress

- Conducted benchmark assessment at the seven Siderar plants
  - Developed a "Scoreboard for Maintenance Excellence" rating for each
  - Developed specific improvement opportunities for each
  - Defined improvement opportunities, gained value for an excellent ROI
- Developed improvement to existing planning and scheduling process
  - Introduced work measurement techniques
  - Measurement of planning effectiveness
  - Measurement of increased uptime of critical production constraint assets
  - Increased coordination/collaboration/cooperation with operations
- Defined improved *preventive/predictive maintenance* (PM/PdM) opportunities
  - Revision and upgrade of PM/PdM procedures
  - Improved support from manufacturing operations
  - Increased compliance to PM/PdM schedules
- Defined need for improved storeroom and shop operations
  - Developed need for strategic storeroom master planning/modernization
  - Implemented new storeroom procedures
  - Improved inventory control and accuracy
  - Increased opportunity for inventory reduction
- Developed a corporatewide performance measurement process
  - Developed corporatewide metrics
  - Performance goal for each metric established
  - Documented the process with written standard procedures
  - Established the Siderar *maintenance excellence index* (MEI)

The results

Siderar achieved its corporatewide maintenance goals and has in place a method to measure results. The key outcomes to provide these results were:

1. A corporatewide CMMS strategy was developed to integrate a "best of breed" CMMS with SAP R/3.
2. A Siderar *CMMS benchmarking system* in place to measure CMMS implementation progress.
3. An overall maintenance excellence strategy established as a result of the total maintenance operations assessment.
4. Best practices with potential savings of over $4 million were identified for implementation.
5. Maintenance performance measurement process established.

6. Validation of results via the MEI for Siderar
7. A world-class CMMS and standard best practices in place for implementation at all seven plants.
8. A method for greater accountability and productivity of all craft labor and material resources.
9. A renewed focus on effective planning, estimating, and scheduling coordination within the total steel making operations.
10. A renewed focus on storeroom modernization and improvement to include:
    - Removal of obsolete items
    - Establish bin locations
    - Use of bar code equipment
    - Improved material handling
    - Establish better physical layouts
    - Upgrade of the parts database
    - Measure improved inventory accuracy
    - Improved storage methods
11. A Siderar "Scoreboard for Maintenance Excellence" established to periodically evaluate overall maintenance excellence progress at each of the seven plants.

Part II: Details of the Justification Process for a New CMMS and Best Practice Implementation

Introduction

The following sections provide the methodology and rationale for determining the cost justification of a new CMMS for Siderar. The TMEI/Biasca team tried to make perfectly clear, the facts and understanding that implementation of just a new CMMS *will not* provide benefits. There must also be new business practices implemented concurrent with a new CMMS or expansion of SAP R/3. Their corporate engineering understood that from the start and there was an obvious commitment to proceed accordingly.

It is also important to note that the projected savings/benefits we projected for Siderar at this point were very realistic and conservative, but are of the nature that they could be measured. Therefore, Siderar had to commit to implementing valid measurement process such as the recommended Siderar specific MEI that we outlined in the final report. There was excellent justification available to provide a significant ROI for a new CMMS that is common to all of the current and future Siderar plants.

At this point it was critical that Siderar be confident with the range of magnitude of these projected benefits and the methodology being used by the TMEI/Biasca Allaince Team.

Primary areas of cost justification/savings

The following four areas were used to provide justification and will be discussed individually and savings projected for each.

1. Maintenance parts inventory reduction
2. Craft productivity improvement (craft utilization and performance)
3. Savings of staff time for user-friendly system
4. Increased uptime/utilization of critical steel making process

Maintenance parts inventory reduction. Inventory reductions related to CMMS implementation can be in the 25 to 35 percent range (per industry standards) in cases where the CMMS inventory module is a company's first step for parts inventory management. Siderar's inventory management process with SAP R/3 is in place and working as a valid system. Actual storeroom layout and operations of each plant storeroom was commendable and better than most seen by TMEI in U.S. plants in the late 1990s. However, improvements to provide a measurable and expected inventory reduction with better use of SAP R/3 or another CMMS such as MAXIMO did exist.

It is very conservatively estimated that a 10 percent inventory reduction can be accomplished across all seven Siderar plant sites with improved materials management practices and a fully integrated CMMS. Calculations are as follows:

Direct reduction in total inventory dollar value

$130,986,000	inventory value (a total of 135,805 stock keeping units/SKU)
× 0.10	a 10% inventory reduction
$13,098,600	direct reduction in inventory dollar value (one-time saving)

Note: A 10 percent inventory reduction is an accepted industry standard in our experience where an existing legacy inventory system is already in place.

Reduction in carrying costs for the 10 percent inventory reduction. So we used a carrying cost of 30 percent to determine the ongoing cost savings of not having $13,098,600 worth of parts inventory to maintain in inventory.

Typical Inventory Carrying Costs

	(%)
Cost of capital	8–13
Cost of obsolescence	6–8
Salaries/labor	3.5
Depreciation	1.5
Insurance	1.2
Record keeping	1.5
Material handling	1.8
Loss/shrinkage	3.5
Building/services	1.5
Total	25–40

$13,098,600 savings (one time)
× 0.30 estimated carrying costs

$392,580 per year savings on carrying costs after realization of the 10% inventory reduction

How new CMMS and best practices provide inventory reductions

- Spare parts listing for asset systems/subsystems.
- Spare part where used capability.
- Parts interchangeability capability.
- Identification of obsolete parts by type.
- Better capability for parts standardization.
- All sites have better visibility of total parts on hand at other sites.
- Standard Siderar process for adding and deletion parts.
- Improved engineering support for identification of "quality spares," standardization, and parts specification data.
- Corporate focus on measurable inventory reduction.

Craft productivity improvement

Craft labor utilization. Traditionally, from many surveys and personal TMEI experience, craft utilization is in the 30 to 40 percent range or below. Surveys at all seven plants indicated the Siderar wrench time to be above this typical baseline. Again craft utilization is a measure of pure hands-on wrench time compared to total craft time available and paid. Increases in craft utilization is measurable and improvements, therefore, provide real gained value; additional craft wrench time capacity output from the existing Siderar crafts workforce and contractor personnel.

Based upon site visits to all plant sites Siderar's level of craft utilization is higher than a pure reactive maintenance operation without planned positions which were being put into place. For the purpose of this benefit calculation Siderar's baseline craft utilization is established at 50 percent because of many best practices such as planning and scheduling and its PM/PdM processes.

Note: A detailed appendix was provided in the Siderar project report to explain in more detail, the methodology for calculating and understanding craft labor productivity improvements similar to Chaps. 13 and 14 in this book.

Improvement value of increased craft utilization. Siderar can expect a measurable 10 percent increase in craft utilization from 50 to 60 percent. In terms of potential benefits, the following calculations illustrate the potential gained value, assuming a base of 1000 craft positions only:

Note: In this case we used a very conservative wrench time base of 50 percent, where actually the typical baseline of 30 to 40 percent (shown below) is present for many, many operations. And for example, if Siderar wrench time was actually 40 percent, the benefits and gained value would be even greater than shown below.

- $1000 \text{ crafts} \times 40 \frac{\text{h}}{\text{wk}} \times 52 \frac{\text{wks}}{\text{yr}} = 2{,}080{,}000$ craft hours total available
- 2,080,000 craft hours available
  × 14.50 avg craft cost/h with fringes
  = \$30,160,000 craft labor cost/year with fringes for 1000 crafts base
- Total craft hours wrench time @ 50 percent utilization

$$1000 \text{ crafts} \times 40 \frac{\text{h}}{\text{wk}} \times 52 \frac{\text{wks}}{\text{yr}} \times 0.50 = 1{,}040{,}000 \text{ h wrench time @ } 50\%$$

$$= 1040 \text{ avg per craftsperson}$$

- Total craft hours wrench time @ 60 percent utilization

$$1000 \text{ crafts} \times 40 \frac{\text{h}}{\text{wk}} \times 52 \frac{\text{wks}}{\text{yr}} \times 0.60 = 1{,}248{,}000 \text{ h wrench time @ } 60\%$$

$$= 1248 \text{ avg per craftsperson}$$

- Total gain in wrench time @ 50 to 60 percent increase in craft utilization

  1,248,000 h @ 50%
  1,040,000 h @ 60%
  208,000 total hours of wrench time gained

- Total gain in equivalent craft positions:

$$\frac{208{,}000 \text{ h gained @ 60 percent utilization}}{1{,}040 \text{ average wrench time hours per crafts @ 50\%}}$$

$$= 200 \text{ equivalent craft positions}$$

- Value of 200 equivalent craft positions

$$200 \times \$14.50/\text{h} \times 40 \frac{\text{h}}{\text{wk}} \times 52 \frac{\text{wks}}{\text{yr}}$$

= $6,032,000 added gained value from better utilization of existing crafts for expanded work including contractors or the following possibility

= $6,032,000 potential reduction due to attrition, reduction in overtime, and the like for direct craft payroll reductions

Craft labor performance. Craft performance is related to actual time required (wrench time) to complete a repair job compared to a reliable planned time (or estimated standard) for the job. Typically, performance levels of well-trained and motivated craftspeople is in the 90 to 100 percent range for this metric. This projected increase of craft performance and craft utilization is included for illustrative purposes, but will not be used as a part of CMMS justification.

Note: Only the 10 percent increase in craft utilization was to be used for the CMMS justification process for Siderar. This projected increase of craft performance and craft utilization is included for illustrative purposes, but will not be used as a part of CMMS justification. Only the 10 percent increase in craft utilization will be used. The following set of calculations includes looking at gained value in term of "direct craft hours" which compares to "direct standard hours" when considering "direct production labor" cost within a standard cost system for manufacturing. Example follows with gains shown as direct craft hours.

Assume a conservative increase of 10 percent craft performance from an estimated Siderar baseline of 70 percent craft performance to a very realistic 80 percent craft performance level in the future. Typically a 95 percent craft performance is possible with well-planned work being performed by well-trained and motivated crafts people.

- Total craft hours available = 2,080,000
- 1000 crafts @ 50 percent utilization and 70 percent performance (assumed Siderar baseline)

$$1000 \text{ crafts} \times 40 \frac{\text{h}}{\text{wk}} \times 52 \frac{\text{wks}}{\text{yr}} \times 0.50 \text{ (util)} \times 0.70 \text{ (performance)}$$

$$= 728{,}000 \text{ direct craft hours per year}$$

(direct term used when both craft utilization and performance used in combination)

- 1000 crafts @ utilization and 80 percent performance (potential goal)

$$1000 \text{ crafts} \times 40 \frac{\text{h}}{\text{wk}} \times 52 \frac{\text{wks}}{\text{yr}} \times 0.60 \text{ (util)} \times 0.80 \text{ (perf)}$$

$$= 998{,}400 \text{ direct craft hours per year @ 60\% util @ 80\% perf}$$

- Total gain in direct craft hours

998,400 @ 60% util and 80% perf
728,000 @ 50% util and 70% perf
270,400 net gain in direct craft hours

1,248,000 h @ 50%
1,040,000 h @ 60%
208,000 total hours of wrench time gained

- Total gain in equivalent craft positions

$$\frac{208{,}000 \text{ h gained @ 60\% utilization}}{1{,}040 \text{ average wrench time hours per crafts @ 50\%}}$$

$$= 200 \text{ equivalent craft positions}$$

- Value of 200 equivalent craft position (shown as two options for using gained value)

  = $6,032,000 added value from better utilization of existing crafts for expanded work including bringing back contractor work in-house (A Very Important Note: If we measure in-house crafts we must for sure measure out house/contractor support too!)

  = $6,032,000 potential reduction in craft positions due to attrition, reduction in overtime, and so forth. for direct craft payroll reductions

- $1000 \text{ crafts} \times 40 \frac{\text{h}}{\text{wk}} \times 52 \frac{\text{wks}}{\text{yr}} \times 0.60 \text{ (util)} \times 0.80 \text{ (perf)}$

$$= 998{,}400 \text{ direct craft hours per year @ 60\% util @ 80\% perf}$$

- Total gain in direct craft hours

998,400 @ 60% util and 80% perf
728,000 @ 50% util and 70% perf
270,400 net gain in direct

- Baseline cost per direct craft hour @ 50 percent util and 70 percent perf

$$\frac{\$30{,}160{,}000 \text{ total craft labor cost (1000 crafts)}}{728{,}000 \text{ direct craft hours @ 50\% util and 70\% perf}}$$
$$= \$41.42 \text{ average cost per direct craft hour baseline}$$

- Cost per direct craft hour @ 60 percent util and 80 percent performance

$$\frac{\$30{,}160{,}000 \text{ total craft labor hour cost (1000 crafts)}}{998{,}400 \text{ direct craft hours @ 60\% util and 80\% perf}}$$
$$= \$30.21 \text{ average cost per direct craft hour}$$
$$\text{@ new levels of 60\% util and 80\% perf}$$

- Reduction in average cost per direct hour

$$\$41.42 \text{ (baseline)} - \$30.21 \text{ (new level)} = \frac{\$11.21}{\text{Direct craft hour reduction}}$$

- Craft labor productivity gain (hours) = 270,400 (i.e. 998,400 – 728,000) total direct craft hours

$$\frac{270{,}400 \text{ direct craft hours gained}}{728{,}000 \text{ baseline}} = 37\% \text{ gain in direct hours}$$

- Craft labor productivity gain ($value) = 998,400 direct craft hours × $11.21 gain/hour ($41.42 base/hour – 30.21) = $11,192,064 added value @ 60% craft utilization and 80% craft performance
- Gain in terms of equivalent craft positions

$$\frac{728{,}000 \text{ total direct craft hours (base)}}{1000} = \frac{728 \text{ h}}{\text{craft}}$$

$$\frac{\text{270,400 direct craft hours gained @ 60 and 80\%}}{\text{728 direct craft hours/craft @ base}}$$

$$= \text{330 equivalent craft positions}$$

How a new CMMS and best practices provide craft labor productivity improvements

Craft utilization. The estimated increase in Siderar's craft utilization is used at only 10 percent . . . And the value-added cost benefit of $6,032,000 will be the *only* value used within the cost/benefit analysis. As illustrated previously, the added value of both a 10 percent increase in craft utilization and craft performance provides value-added estimated benefits of $11,192,064 for a craft base of 1000 personnel.

Note: As you can see the cumulative impact of including both craft utilization and craft performance provides a much greater gained value.

The following capabilities of a new CMMS combined with best practices, provide the craft utilization improvements:

- Increased capability for planning/scheduling
- Improved kitting/reservation of parts linked to work order
- Better status information (for planning) on parts on order delivery dates
- Ability to measure craft utilization (wrench time) effectively
- Improved online parts information, as built drawings, and repair specification information
- Improved coordination with production and interface with process controls (QNX) to more effectively predict and schedule repairs for shutdown

Note: QNX was an acronym for their shop floor control and production planning and control system for steel making.

Craft performance. The combination of increasing both craft utilization and performance is not being used as part of the ROI for a new Siderar CMMS. However, Siderar can expect measurable benefits in this area with their new CMMS and by measuring the performance from both Siderar crafts and contractor employees. With extensive use of contractors, Siderar must commit to measuring both the utilization and performance of this expanding cost to the current maintenance operations.

The following capabilities will provide productivity gain for craft performance:

- Planners applying realistic planning time to all major planned PMs, inspections, and planned repairs with a methodology such as the ACE team process (details on the ACE System found in Chap. 15) or applicable maintenance standard data.
- Conduct craft job task analysis and continue or start training to increase overall skills proficiencies and existing technical skill gaps.
- Continued increases in motivation and work ethic of crafts (see PRIDE-in-Work.com) due to maintenance leadership, team-based processes and having an effective planning/scheduling process for proactive maintenance. Siderar was moving far away from a reactive, fire-fighting strategy for maintenance.
- Siderar crafts and contractors have a clear expectation of what is required for performance via reliable planning times.
- CMMS system will allow standard job plans and required actual times to complete them as part of the CMMS database.
- Planned time versus actual time provides one critical way to evaluate project planning effectiveness and progress status of major projects or shutdowns.

Savings of staff time for user-friendly system. The current legacy systems for CMMS at all seven plants suffer from not being user-friendly and not being directly interfaced with existing SAP R/3 inventory and purchasing processes. A client-server CMMS with windows-based *graphical user interface* (GUI) applications will improve Siderar staff productivity and more effective use of CMMS. Regardless of the final decision, Siderar must commit to effective training of the user—both the maintenance and production operations users. Surveys indicate along with interviews, that an effective focus on user training has not occurred. This must occur with a new CMMS implementation, whether SAP R/3, MAXIMO or if nothing is done and they keep the legacy CMMS. The power of existing functional capabilities of either one of these two new world-class systems will be wasted without effective user training.

Projected savings in this area of CMMS cost justification are as follows:

- 300 frequent users × 0.5 h/day saving in-time = 150 h staff time savings
- Value of frequent user staff time—150 h × 28.00/h (staff level) = $4200
- 2000 infrequent users × 0.10 h/day (i.e. 5–10 minutes) = 200 h staff time savings × $28.00/h (staff level) = $5600

  Total staff time savings are $9800.

Increased uptime/utilization of critical steel-making assets. This is a very specific and measurable benefit that can contribute to the bottom line. Selected Siderar assets are throughput constrains required at almost 100 percent capacity, such as, continuous casting operations, galvanizing, and so on. Increased uptime equates to increased capacity output and increased revenue and profits. Assets other than critical 24 hours per day operating assets also provide added value with increased uptime and increased availability.

The TMEI/Biasca team planned to use increases only for the value of increased uptime of the critical assets as part of the justification. During the review meeting with staff staff, the need for baseline information was requested. A specific request was made to the Director of Industrial Engineering for this information.

Baseline information to provide this calculation was not available as of our project report date and therefore is not included. With seven plants this direct cost savings would be very significant.

And as we stated, "the team of TMEI/Biasca can include this added information with revisions to the final report that Siderar request during final review scheduled on 12/15/97 and 12/16/97." With baseline on current uptime, another gained value from increased uptime can be easily added to an amended final justification.

Summary of projected benefits of new CMMS integrated with recommended best practices

- Inventory reduction = $13,098,600 one time
- Carrying cost avoidance = $392,580 @ carrying cost
- Craft labor utilizing = $6,032,000 improvement (10%)
- Improved use of staff time = $9,800
- Value of increased uptime = To be determined

Total range (without uptime $) is $19,000,000 to $21,000,000.

Part III: The Cost Evaluation Process for SAP R/3 and MAXIMO

Introduction

The following cost estimate assumes a base of 300 primary users of the new CMMS system and a total of 2000 infrequent users such as craftspeople, storeroom staff, production operators, and production leadership staff. This cost estimate provides only a preliminary estimate to support justification of changing to a new CMMS.

Hardware costs

The following hardware costs would be the same for all alternatives:

■ Expanded server capacity	$350,000
■ PCs for prime users ($1800/PC × 300 users)	$540,000
■ Printers (20 @ $1000 each)	20,000
■ Network installations ($200/each for 300 users)	60,000
	$970,000
Plus 10% contingency	+ 97,000
	$1,067,000

Training costs

The following estimated training costs include both the value of internal staff time away from current job for training on a new CMMS, plus the estimated costs of on-site training by the selected vendor. Training costs would be the same approximate total for all vendors.

Internal cost of training time

- 300 prime users @ 40 h each × $28.00/h = $336,000
- 200 infrequent users @ 8 h each × 14.50/h = $232,000

Total cost internal time = $568,000

External training costs

- 8 wks on-site time for training on new CMMS, train the trainer, and so forth.

8 wks × $6,000/wk/instructor = $48,000
Expenses @ $1000/wk = 8,000
Total = $54,000

- Company X staff travel to existing vendor sites + 20,000 = $74,000
- Total training costs = $640,000

Software costs

SAP R/3: Software costs are based on a total of 300 primary users (seats) for a new system

- SAP R/3 @ $3093/seat × 300 = $927,900
  5% maintenance fee first year + 46,395
  $974,295

■ MAXIMO @ $5,590/seat × 300	$1,677,000
■ SAP Maxlink interface	75,000
■ *Report writer (100 seats @ $400 ea)	40,000
■ *Analyzes (100 seats @ 1250 ea)	125,000
■ *Work manager (100 seats @ $1000 ea)	100,000
■ *Schedules (100 seats @ $2000 ea)	200,000
	$2,217,000

Cost summary comparison

Cost item	SAP R/3	MAXIMO
Hardware cost	$1,067,000	$1,067,000
Total training costs	640,000	640,000
Software costs	974,295	2,217,000
	$2,681,295	$3,924,000

Notes:

1. SAP R/3 software costs of $3093 per seat for plant maintenance module utilization. Software maintenance support cost: 5 percent for the first year, 10 percent for the second year, and 15 percent for the third and future years.

   MAXIMO software maintenance fee per year is 17 percent of software cost. 0.17 ($2,217,000) = $376,890 per year starting during the second year.
2. All additional fees for other interfaces of MAXIMO to SAP R/3 would require more specific definition and cost estimates, that is,
   - Interface to CAD engineering drawings
   - Interfaces to process control (QNX)
   - Interfaces to condition monitoring systems
3. SAP R/3 implementation costs would also have to be included.

So you can now take a look at the numbers presented in the case studies and develop your own judgment as to the final decision that was made.

*Assume that only 100 users will be primary users of the MAXIMO add-on modules listed above. MAXIMO costs are based upon a preliminary cost proposal on 11/26/97 Director of Latin American operations for MAXIMO.

Part

4

The Profit- and Customer-Centered Maintenance Operation

Chapter

13

Improving Craft Productivity: An Essential Strategy for Profit and Customer Service

The profit- and customer-centered maintenance operation must clearly understand productivity and be committed to measuring craft productivity. All the best practices we discuss in this book contribute either directly or indirectly to improving craft productivity. I know for sure that the profit- and customer-centered maintenance contractor such as Viox Services of Cincinnati and Akron, Ohio as part of EMCOR understand productivity and the need to take every means possible to improve craft productivity.

Defining overall productivity from the total maintenance process of the total operations can be a slippery task. Attempting to increase productivity can be even more difficult if managers mistake efficiency for productivity. A focus on increasing work-order completion rates may sacrifice quality of repair work. Cutting costs by specifying cheap, prone-to-fail equipment or parts yields the high cost of low-bid buying and maybe achieving short-term budget requirements.

Today's economic climate requires that we define and then improve maintenance and engineering productivity. Every employee must take ownership in the organizational mission and the maintenance mission.

Some Definitions of Productivity

The following illustrations reveal that productivity is still a mystery to many.

1. A director of maintenance operations/consulting services says the essence of productivity varies depending upon who is defining it. His own definition was "a formal and idealistic view defining productivity as time spent accomplishing the mission of the department," he says. "It's not just wrench-turning time."

   His view is that crafts that spend time reading manuals or enrolled in certification classes are being productive. In addition, one quick way to determine productivity and identify unproductive workers is with log sheets. If for an 8-h day a craft employee who lists his two 15-min breaks, plus a half-hour lunch, and lists the other 7.5 h as working, it is a red flag that he is not being very productive. Crafts should spend some time every day increasing their knowledge to help further boost productivity.

2. A school district has adopted a similar, if unconventional, definition. The director of facility services for the district rejects standard performance and productivity metrics such as maintenance costs per square foot or per employee, preferring instead to establish the maintenance mission of the district as a self-directed group. After giving a lot of thought to productivity for over 20 years his view borrows from the just-in-time phrase of materials delivery to help define his own productivity concept as "just enough."

3. Another vice president for university facility management chooses to define productivity as efficacy. He uses callbacks to determine productivity and whether craft workers have been effective. Managers cannot fully define productivity without considering the effectiveness of their workers. Work orders contain much information that is useful to gauging productivity. A scientific way through which we assess productivity is to compare "actual versus estimated work times," he says. We also stress the importance of training and certification because it is essential for productive workers. Investment in skills training can often be ignored, especially in a tight budget environment.

4. Another maintenance manager states that "If you've got a computerized maintenance management system, set it up to include all productive aspects of work," and, productive work should include education and reading about standards or design manuals, in addition to actual maintenance work.

5. Others state that your work-tracking module must help assess the productivity of workers. This does not require supervision of them in the traditional sense. Professional craft know they are not going "to be mothered." A critical aspect of employee productivity is having

each employee understand why he or she is here. Productive workers need to have a high level of mission understanding. I feel that here is where *pride*-in-maintenance really plays a key role: *pride* in the specific maintenance mission and the overall organizational mission.

6. A manager for the decision support for operations and maintenance program at the Pacific National Laboratory in Richland, Washington, categorizes productivity into four levels:

Survival. In a low-productivity environment, the goal is survival. Chronic operations and maintenance problems will plague such an organization, and low-reliability or outright equipment failures characterize this category.

Adequacy. Uncertainty might characterize this category, but the department is keeping equipment running. Low efficiency and low productivity are classic traits of this category.

Accuracy. Departments in this segment are secure in their knowledge of operating and maintaining their facilities, but would like better performance measurements and would want to know how their operations and maintenance affect facility processes.

Optimized. These departments know the slope of their performance curve. They search for ways to optimize the state of an already effective maintenance process.

What is productivity?

One definition of *productivity* is that it is a measure relating to quantity or quality of output compared to the inputs required to produce it. When we look at two elements of total plant productivity; production labor and production equipment productivity have the same three key factors in common. The same applies to craft employees performing repair and PM to support the production operation customer. My definition includes three key factors for productivity of craft labor:

1. The effectiveness factor—*Doing the right things.*
2. The efficiency factor—*Doing the right things and giving the best personal performance as possible.*
3. The quality factor—*Doing the right things giving the best personal performance as possible with high-quality results.*

These three factors apply to production labor productivity, to asset productivity (OEE), and to craft labor productivity as shown in Table 13.1.

TABLE 13.1 The Three Key Productivity Factors Compared for Three Elements of Total Plant Productivity

	Productivity: Three key factors		
Productivity factors	Production employees	Production equipment	Craft employees
Effectiveness	% time on direct labor adding value	% availability to add value	% true wrench time or craft utilization
Efficiency	% performance against a standard time	% performance against design speed	% craft performance against a standard planning time
Quality	% good compared to total produced	% good compared to total produced	% good repairs compared to total repairs

What is your craft productivity?

Surveys consistently show that wrench time (craft utilization) one element of craft productivity within a reactive, fire-fighting maintenance environment is within the range of 30 to 40 percent. A study using work sampling of a major corporation's maintenance operation revealed 26 percent. My surveys from speech and seminar attendees over the past 15 years shows some that think their true wrench time is 15, 20, or 25 percent and much lower than 30 percent. Therefore, if your baseline pure wrench time baseline is conservatively at 40 percent, this means that for a 10-hr day there is only four hours of actual hands-on, wrench time. You can do the math on other levels, 10, 20, or 30 percent.

Craft workforce not to blame. Typically, low-craft utilization is due to no fault of the craft workforce. Most of the lost wrench time is not the fault of the craft workforce. Lost wrench time can be attributed to the following reasons:

1. Running from emergency to emergency; a reactive, fire-fighting operation
2. Waiting on parts and finding parts or part information
3. Waiting on other information, drawings, instructions, and the like
4. Waiting for the equipment to be shut down
5. Waiting on rental equipment to arrive
6. Waiting on other crafts to finish their part of the job
7. Travel to/from job site

8. Lack of effective planning and scheduling
9. Make ready, put away, clean up, meetings, troubleshooting, and so forth

Should we measure craft productivity?

My answer to this is a resounding yes! A method that is easy to use and acceptable to the craft workforce such as the ACE team benchmarking process detailed in Chap. 15 should be used. Craft resources are becoming harder to find and in many areas there is a true crisis and shortage of craft labor. One question we want to answer for you is: "How can we get maximum value from craft labor resources and achieve higher craft productivity?" *First, we must be able to measure it.* Maintenance operations that continue to operate in a reactive, run-to-failure, fire-fighting mode, and disregard implementation of today's best practices will continue to waste their most valuable asset and very costly resource—craft time. Often true wrench time is within the 20 to 30 percent range for operations covering a wide geographic service area. Best practices such as effective maintenance planning/scheduling, preventive/predictive maintenance, more effective storerooms and parts support, delivery to job site, crafts skill training, and the best tools we can provide all contribute to proactive, planned maintenance, and more productive hands-on, "wrench time."

Craft productivity measurement and improvements in union plants. Craft productivity measurement and improvements in union manufacturing in the U.S. plants must occur. I said before that we could not perform maintenance offshore, except for maybe troubleshooting and online monitoring/operation of process control systems. Maybe we can send items offshore for rebuilding, and so on. But we can move plants offshore when current productivity of both production operators and craftspeople do not increase as union labor rates continuously go higher and higher. *At some point, higher wages without productivity gains will lead to plant closings due to profits or a move offshore for cheaper labor*, precisely the GM model of 2006. Or we might replace union maintenance with contractor maintenance. After seeing the Cooper Tool Group move seven unionized plants from the north (Plumb, Cresent, Wiss, Lufkin, Weller, and the Nicholson file and saw plants) to the south, there must be significant improvement in yes, performance measurement of unionized and nonunion maintenance operations. Craft productivity can be measured and improved. In addition, if we can do it for in-house maintenance we can surely do it for out-house maintenance, contractors as well. Therefore, what is your wrench time or craft utilization and wrench time as shown in Figs. 13.1 and 13.2.

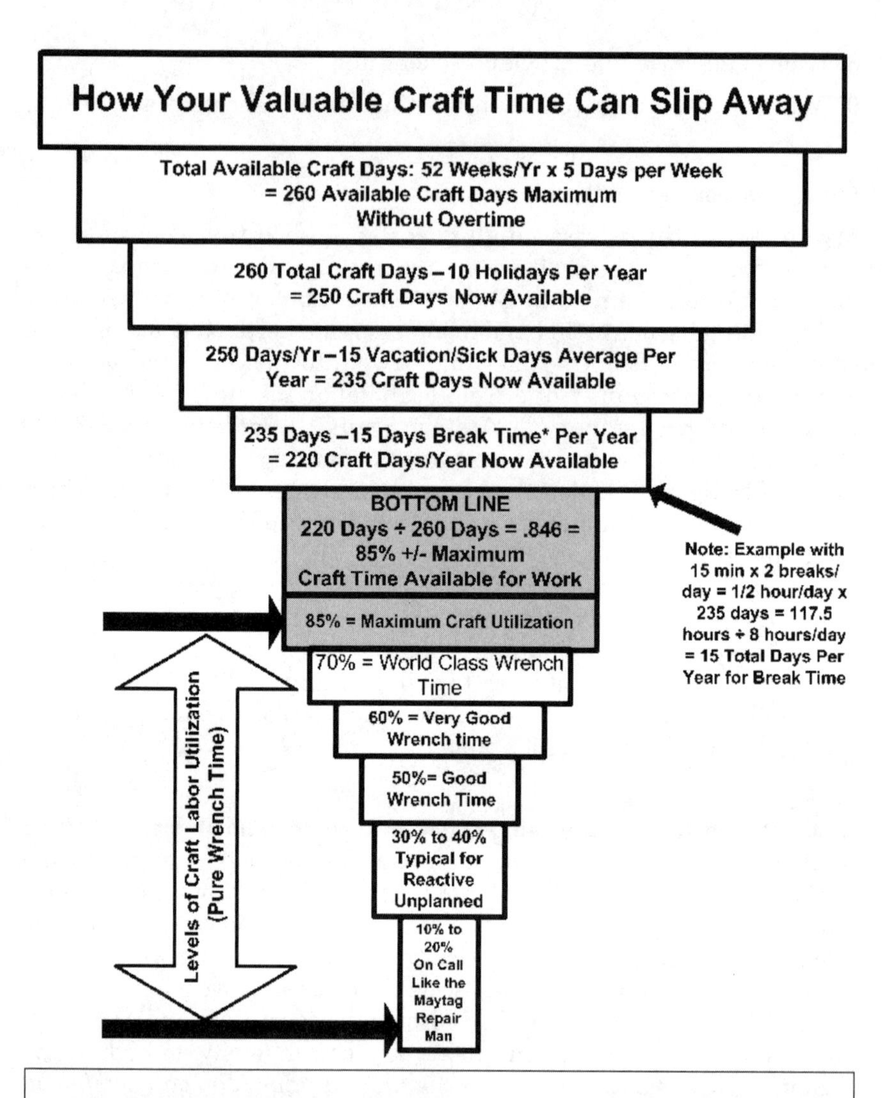

Figure 13.1 How valuable craft time can slip away.

Significant Gained Value Available from Increased Wrench Time

An improvement in actual wrench time from 40 to 50 percent represents a 25 percent net gain in craft time available and a significant gained value. When we are able to combine gains in wrench time with increased

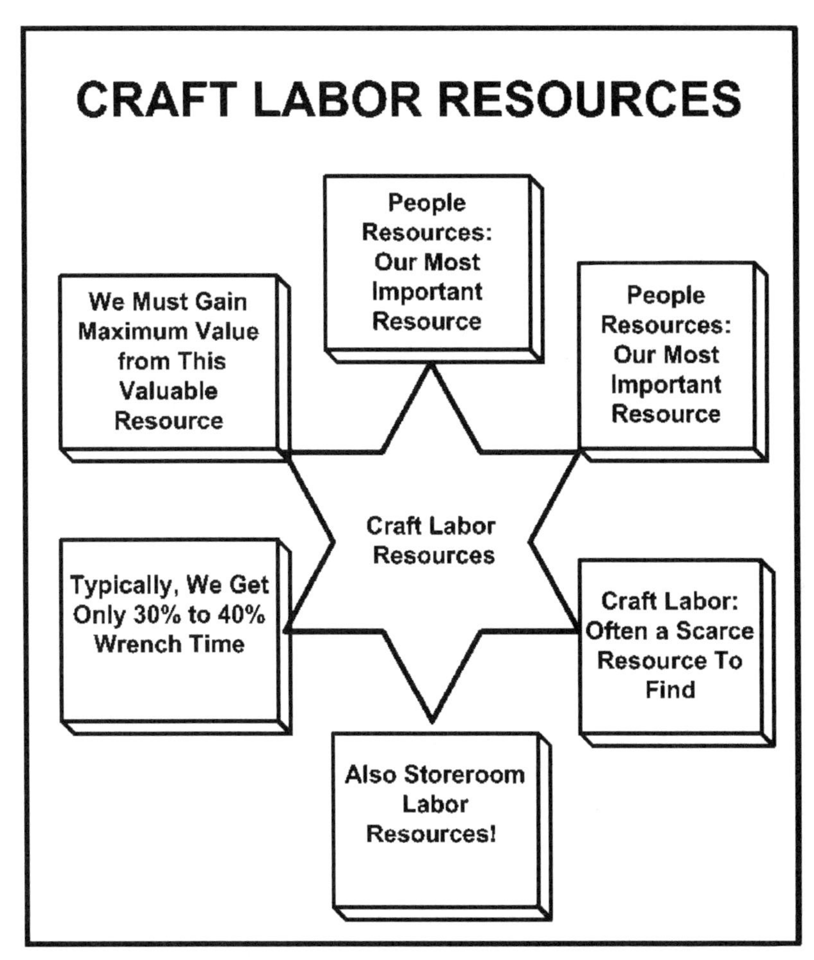

Figure 13.2 Craft labor resources—our most important maintenance resource.

craft performance when doing the job we increase our total gain in craft productivity. Measuring and improving overall craft productivity can be a key component to justify an effective *computerized maintenance management system* (CMMS) and other investments for maintenance improvement. We will now look at a new term *overall craft effectiveness* (OCE) in Chap. 14. As you will see, OCE has the same basic factors as *overall equipment effectiveness* (OEE) (availability, performance, and quality), but applied to craft productivity and people resources as compared to OEE that focuses upon critical physical assets.

Chapter

14

Introducing OCE as a New Buzzword: The Overall Craft Effectiveness (OCE) Factor

The profit- and customer-centered maintenance leader (in-house or contractor) must consider total asset management in terms of improvement opportunities across all maintenance resources. There are many questions to be asked about how we can improve the contribution that each of these six resources make toward your goal for maintenance excellence:

- Physical resources: equipment and facilities
- People resources: craft labor and equipment operators
- Technical skill resources: craft labor that is enhanced by effective training
- Material resources: *maintenance repair operations* (MRO) parts and supplies
- Information resources: useful reliability information, not a sea of useless data
- Hidden resources: synergy of team work as a true people asset multiplier

Measuring and improving *overall craft effectiveness* (OCE) must be one of the many components to *continuous reliability improvement* (CRI) process and total asset management. OCE includes three key elements very closely related to the three elements of the OEE Factor.

The OEE Factor = % Availability(A) × % Performance(P) × % Quality(Q)
An OEE factor of 85% is recognized as world-class.
Therefore OEE of 85% requires at least the 95% level for each of the 3 elements:
So if OEE = A × P × Q then if each factor is 95% or .95
OEE = 0.95 × 0.95 × 95 ≅ 85%

Figure 14.1 Overall craft effectiveness (OCE).

Overall Equipment Effectiveness (OEE)

We must clearly understand the elements of OCE and how the OCE Factor relates to better use of our craft workforce. Almost everyone recognizes and understands the world-class metric of *overall equipment effectiveness* (OEE) that measures the combination of three elements for the physical asset: equipment asset availability, performance, and quality output. OEE is about measuring asset productivity. The calculation of OEE is shown in Fig. 14.1.

The OCE factor focuses upon craft labor productivity and measuring/improving the value-added contribution that people assets make. Just like OEE, there are three elements to the OCE Factor:

- The effectiveness factor: craft utilization for OCE and *asset availability* for OEE
- The efficiency factor: craft performance for OCE and *asset performance* for OEE
- The quality factor: craft service quality for OCE and *quality of asset output* for OEE

All three elements of OCE can be as well defined as all three of the OEE Factors. We will now review the three key elements for measuring OCE and see how they very closely align with the three elements for determining the OEE factor for equipment assets. Table 14.1 provides a comparative summary and Fig. 14.2 defines how OCE is calculated.

TABLE 14.1 Summary Comparisons of OCE and OEE

Overall craft effectiveness (OCE)	Overall equipment effectiveness (OEE)	Elements of OEE and OCE
1. Craft utilization or pure wrench time (CU)	Asset availability/ utilization (A)	Effectiveness
2. Craft performance (CP)	Asset performance (P)	Efficiency
3. Craft service quality (CSQ)	Quality of asset's output (Q)	Quality

The OCE Factor = % Craft Utilization (CU) × % Craft Performance (CP) × % Craft Service Quality (CSQ)
Therefore OCE = % CU × %CP × %CSQ
Typically CU and CP can be easily measured.
Craft Service Quality (CSQ) is somewhat harder to measure and can be more subjective.

Later in Part II we will see how all three elements of OCE can be measured and how all three contribute to increased craft productivity.

Figure 14.2 Calculating overall craft effectiveness.

OCE focuses upon your craft labor resources

I strongly believe in basic maintenance best practices as the foundation for maintenance excellence. This is what I call CRI. CRI is about maintenance business process improvement that includes opportunities across all maintenance resources: equipment and facility assets as well as people resources—our crafts workforce and equipment operators. CRI must also include MRO materials management assets, maintenance informational assets, and the added-value resource of synergistic team-based processes. CRI improves the total maintenance operation and can start with measuring and improving OCE.

The *Maintenance Excellence Institute* (MEI) advocates, supports, and clearly understands the need for *reliability-centered maintenance* (RCM) and *total productive maintenance* (TPM) types of improvement processes. But out on the shop floor we see today's trend toward forgetting about the basics of "blocking and tackling" while going for the long touchdown pass with some new analysis paralysis scheme. RCM is not analysis paralysis when done correctly with true information and when it is not based upon "precisely inaccurate" data.

Build upon the basics

Your approach must be built upon the basics and then include, but go well beyond, the traditional RCM/TPM approaches to CRI. See Fig. 14.3.

Maintenance excellence can start with PRIDE in maintenance

Do not take a piecemeal approach that focuses only RCM-type processes on physical assets and equipment resources. Often the maintenance information resource piece, among others, is a missing link for the successful RCM-type process. RCM alone can often become analysis paralysis with no data or bad data. Your approach should be about improvement opportunities across all maintenance resources. There of course must be

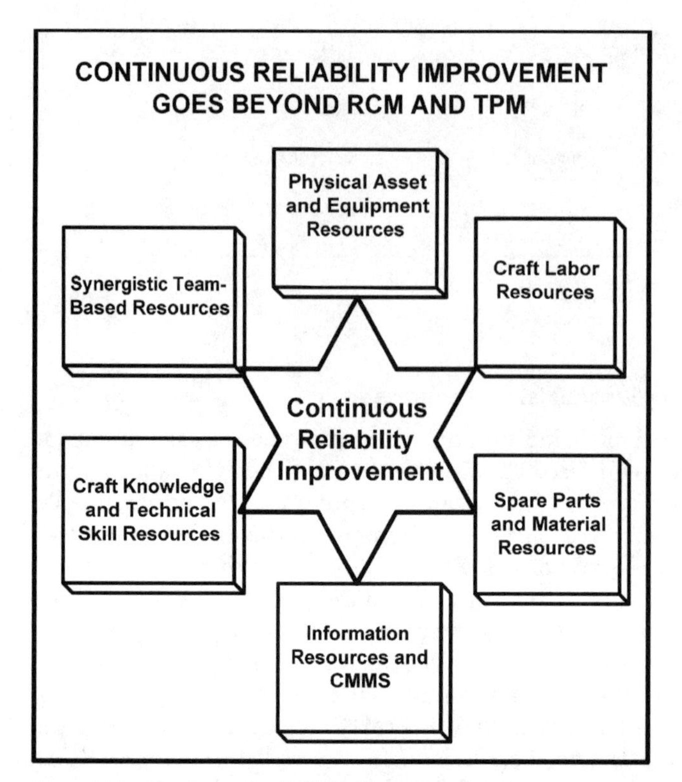

Figure 14.3 Continuous reliability improvements.

priorities as to where we start and where we make investments. For example with the craft labor resource, we can easily measure the three elements of OCE, as we will see later. However, we can start the journey toward maintenance excellence by just helping to achieve *pride* in maintenance from within the crafts workforce and among maintenance leaders at all levels and PRIDE in maintenance around the world as illustrated in Fig. 14.4.

A very important question for the craft workforce

I have seen maintenance operations and talked to craftspeople all around the world, starting in South Vietnam in 1970. Almost all want to do a good job and be appreciated for what they do to help achieve the mission of the organization. In addition, when asked the following question, I have received positive responses from 100 per cent of the thousands of dedicated professional craft people I have interviewed. Dedicated union leaders and union crafts are included in my personal sample.

Figure 14.4 PRIDE in maintenance around the world.

The question I always ask is: "How would you do this job or lead this crew if it were in fact your own maintenance business?" What if we could get attitudes that are more positive plus action from *all our crafts* focused on this important question? I think you can do it with a combination of many things from this book. So again I ask you, to apply what you think will work and then test other ideas that you may not fully agree with now.

Your own internal people can add greater value to your maintenance operation with a profit and customer-centered attitude about their job and the profession of maintenance. We feel strongly that maintenance excellence begins with *pride* in maintenance. In Part II we saw "How OCE Impacts Your Bottom Line" and how we can measure and improve the productivity of an important maintenance resource—craft labor.

Craft labor—a terrible thing to waste. Improving OCE is a key question we need to answer. Getting maximum value from craft labor resources and higher craft productivity requires measurement and knowing where you are now. Maintenance operations that continue to operate in a reactive, run-to-failure, fire-fighting mode, and disregard implementation of today's best practices will continue to waste their most valuable asset and very costly resource—craft time. Typically, due to no fault of the craft workforce, surveys and baseline measurements consistently show that only about 30 to 40 percent of an 8-h day is devoted to actual, hands-on wrench time. It is very important to understand, "How your valuable craft time can slip away" as illustrated previously in Fig. 13.1. Best practices such as effective maintenance planning/scheduling, preventive/predictive maintenance, more effective storerooms, and parts support all contribution to proactive, planned maintenance, and more productive hands-on, "wrench time." Measuring and improving OCE must be one of many components to CRI process and total asset management. OCE includes three key elements very closely related to the three elements of the OEE factor (Table 14.1) which we will discuss next in more detail.

Craft Utilization

The first element of the OCE factor is *craft utilization* (CU) or pure wrench time. This element of OCE relates to measuring how *effective* we are in planning and scheduling craft resources so that these assets are doing value-added, productive work (wrench time). Effective planning/scheduling within a proactive maintenance process is one important key to increased wrench time and craft utilization. It is also about having an effective storeroom with the right part, at the right place in time to do scheduled work with minimal nonproductive time on the part of the craftsperson or crew assigned to the job. Your valuable craft time can just slip away to a level of 30 to 40 percent, or below, when crafts service a large geographic area: a large plant, refinery, or facilities complex.

Pure wrench time is just that and does not include time caused by the following:

1. Running/traveling from emergency to emergency in a reactive, fire-fighting mode
2. Waiting on parts issue, actually finding parts, or getting parts information
3. Waiting on other asset information, asset drawings, repair instructions, documentation, and the like
4. Waiting for the equipment or asset to be shutdown to begin work
5. Waiting on rental equipment or contractor support to arrive at job site
6. Waiting on other crafts to finish their part of the job
7. Traveling to/from the job site
8. All other make-ready, put away, or shop clean-up time

9. Meetings, normal breaks, training time, and excessive troubleshooting due to lack of technical skills
10. Lack of effective planning and scheduling

Craft utilization (or wrench time) can be measured and expressed simply as the ratio of:

$$CU\% = \frac{\text{total productive (wrench time)}}{\text{total craft hours available and paid} \times 100}$$

Improve wrench time first

Go on the attack to increase wrench time in your operation even if you do nothing to improve the other two OCE factors: *craft performance* (CP) and the *craft service quality* (CSQ) level. As we will see in the following examples, very dramatic and significant tangible benefits can be realized with just focusing on increasing wrench time. Improvement from 10 to 30 percentage points from your current baseline of wrench time can typically be expected. In addition, often this can be achieved, just from more effective maintenance planning and scheduling. Let us now look at several examples showing the value of CU improvement within a 20-person workforce with an average hourly rate of $18.00 and see the significant benefits that a 10 percent increase in CU can provide.

Gained value of 10 percent in wrench time

What if through better planning and scheduling, good parts availability and having equipment available to fix it on a scheduled basis, we are able to increase actual wrench time by just 10 percent from a baseline of 40 percent? What is the gained value to us if we get wrench time increase across the board for a 20-person crew being paid an average hourly rate of $18 per hour? First let us look at what it is really costing us at various levels of wrench time in Table 14.2.

Total craft hours available and annual craft labor costs for crew of 20 crafts

$$20 \text{ crafts} \times 40 \text{ h/wks} \times 52 \text{ wks/yr} = 41{,}600 \text{ craft hours available}$$

$$41{,}600 \text{ craft hours @ \$18/h} = \$748{,}800 \text{ craft labor cost/yr}$$

Wrench time and actual costs per hour at various levels of craft utilization

Example: What If Baseline for Wrench Time Is 40 Percent With effective planning and scheduling we can achieve at a minimum a 10-point improvement in CU from our current baseline. Starting from a baseline of 40 percent and increasing wrench time up to a level of 50 percent, we in effect get a 25 percent increase in craft capacity for doing actual work.

TABLE 14.2 Actual Cost Per Hour at Various Levels of Wrench Time

Level of craft utilization (%)	Total wrench time (Hours)	Actual hands-on cost per hour ($)	Average wrench time hours per craft position
30	12,480	60.00	624
40	16,640	45.00	832
50	20,800	36.00	1040
60	24,960	30.00	1248
70	29,120	25.71	1456
80	49,920	22.50	1664
*85	35,360	21.18	1768
90	37,440	20.00	1872
100	41,600	18.00	2080

*Maximum possible craft utilization is 85 percent (as shown in Table 14.2 and in Fig. 13.1) considering paid holidays, vacation time, breaks, clean-up, employees meetings, craft training, and so forth.

- Total hours gained in wrench time: 4160-h gained

$$20{,}800 \text{ h @ } 50\% - 16{,}640 \text{ h @ } 40\% = 4160\text{-h gained}$$

- Total gain in equivalent number of crafts positions: 5 craft positions

$$\frac{4160 \text{ h gained}}{832 \text{ average wrench time hours @ } 40\%} = 5 \text{ equivalent craft positions}$$

- Total gained value of 5 equivalent positions: $187,200

$$5 \text{ equivalents} \times 40 \text{ h/wk} \times 52 \text{ wks/yr} \times \$18.00/\text{h} = \$187{,}200 \text{ gained value}$$

Valuable Craft Time Can Be Regained For the 20-person craft workforce, just a 10 percent improvement up to 50 percent wrench time is 4160 h of added wrench time. This gain represents a 25 percent increase in overall craft labor capacity. The maintenance best practice for planning and scheduling requires a dedicated planner(s). An effective maintenance planner can support and plan for 20 to 30 crafts positions. With only a 10 percent increase in CU for a 20-person craft workforce, we can get more than a 5 to 1 return to offset a maintenance planner position.

Example B: What if Wrench Time Is 30 Percent? For many operations, wrench time is only about 30 percent and sometimes below 30 percent. Again with effective planning and scheduling, good PM/PdM, and parts availability we can eliminate excessive nonwrench time. An improvement of at least 20 points in CU is very realistic. If we begin from a baseline of 30 percent up to a level of 50 percent, we are in effect getting a 67 percent increase in craft capacity for actual hands-on work.

The gained value goes from 30 up to 50 percent wrench time.

- Total hours gained in wrench time: 8320 h gained

$$20{,}800 \text{ h @ } 50\% - 12{,}480 \text{ h @ } 30\% = 8320\text{-h gained}$$

- Total gain in equivalent number of crafts positions: 13

$$\frac{8320 \text{ h gained}}{624 \text{ average wrench time hours @ } 30\%} = 13.3 \text{ equivalent craft positions}$$

- Total gained value of 13.3 equivalent positions: $497,952

$$13.3 \text{ equivalents} \times 40 \text{ h/wk} \times 52 \text{ wks/yr} \times \frac{\$18.00}{\text{h}} = \$497{,}952 \text{ gained value}$$

Valuable Craft Time Can Be Regained Tremendous opportunities are available for the 20-person craft workforce with wrench time currently in the 30 to 40 percent range. Just a 10 to 20 percent improvement up to 50 percent wrench time can be from 4,000 to 8,000 h of added wrench time. This gain represents a 25 to 67 percent increase in overall craft labor capacity. There are a number of best practices to help you regain valuable craft resources. The maintenance best practice for planning and scheduling requires a dedicated planner(s). An effective maintenance planner can support and plan for 20 to 30 crafts positions.

Use your CMMS/EAM as a mission-essential information technology tool that supports planning and scheduling, better MRO materials management, and effective PM/PdM. They are three best practices for improving craft wrench time. Bottom line results that give us 5 to 13 more equivalent craft positions and up to $500,000 in gained value of more wrench time with existing staff, can be dramatic proof that internal maintenance operations can be profit centered.

Craft Performance

The second key element affecting OCE is craft performance. This element relates to how *efficient* we are in actually doing hands-on craft work when compared to an established planned time or performance standard. CP is expressed as the ratio of:

$$\text{CP\%} = \frac{\text{total planned time (hours)}}{\text{total actual craft hours required}} \times 100$$

CP is directly related to the level of individual craft skills and overall trades experience as well as the personal motivation and effort of each craftsperson or crew. Effective craft skills training and technical development contribute to a high level of CP.

Craft performance calculation

The planned time for a minor overhaul or PM procedure is 10 h based on a standard procedure with parts list, special tools, permits required, and so forth.

- If the job is completed in 12 h, then $\text{CP} = \frac{10}{12} \times 100 = 83\%$
- If the job is completed in 9 h, then $\text{CP} = \frac{10}{12} \times 100 = 111\%$

An effective planning and scheduling function requires reasonable repair estimates. It is what I call reliable planning times and they should be established for as much maintenance work as possible. Because maintenance work is not highly repetitive, the task of developing reliable planning times is more difficult. However, there are a number of methods for establishing planning times for maintenance work including:

- *Reasonable estimates.* A knowledgeable person, either a supervisor or planner, uses their experience to provide their best estimate of the time required. This approach does not scope out the job in much detail to determine method or special equipment needed.
- *Historical data.* The results of past experience are captured via the CMMS or other means to get average times to do a specific task. Over time, a database of estimated time is developed which can be updated with a running average time computed for the tasks.
- *Predetermined standard data.* Standard data tables for a wide range of small maintenance tasks have been developed. Standard data represents the building blocks that can then be used to estimate larger, more complex jobs. Each standard data table provides what the operation is, what is included in the time value, and the table of standard data time for the variables that are included. The *Universal Maintenance Standards* (UMS) method used back in the 1970s and developed by H. B. Maynard Inc. represents a predetermined standard data method.

The ACE Team Benchmarking Process

As a means to overcome many of the inherent difficulties associated with developing maintenance performance standards, the ACE (*a consensus of experts*) team benchmarking process was developed. This process is detailed in Chap. 15 with complete forms in App. H and an example ACE team Charter in App. G. This process was developed back in 1978 by Ralph W. "Pete" Peters, founder of MEI as part of a Masters program; management information systems at North Carolina State University. This method is based upon principles of the Delphi technique. It relies primarily on the combined experience and estimating and methods improvement ability of a group of skilled craftspersonnel, planners, and foremen. The objective is to determine reliable planning times for a number of selected "benchmark" jobs. This team-based process which uses skilled craft people places a high emphasis on continuous maintenance improvement to reflect improvements in performance and repair methods and repair quality as they occur.

Generally, the ACE team benchmarking process parallels the UMS approach in that the "range of time concept" and the "slotting" technique is used once the work content times for a representative number of "benchmark jobs" have been established. The ACE team benchmarking process focuses primarily on two key areas: (a) repair methods improvement and (b) on the development of work content times for representative "benchmark jobs" that are typical of the craft work performed by the various craft work areas.

Once a number of benchmark job times have been established, these jobs are then categorized onto spreadsheets by craft and task area and according to workgroups which represent various ranges of times. Spreadsheets are then set up with four work groups/sheet with each work group having a time slot or "range of time." For example, workgroup E would be for benchmark jobs ranging from 0.9 up to 1.5 h and assigned a standard time (slot time) of 1.2 h. Likewise, workgroup F would be for benchmark jobs ranging from 1.5 up to 2.5 h and assigned a standard time of 2 h. Spreadsheets include brief descriptions of the benchmark jobs and represent pure wrench time. Work content comparison is then done by an experienced person, typically a trained planner to establish planning times within the 95 percent confidence range. A complete users guide, complete with step-by-step procedure, forms, and the recommended ACE team Charter for establishing the ACE team benchmarking process is available in App. G.

Planning times are essential

Planning times provide a number of key benefits for the planning/scheduling process. First, they provide a means to determine existing workloads for scheduling by craft areas and backlog of work in each area. Planning times allow the maintenance planner to balance repair priorities against available craft hours and to realistically establish repair schedules that can be accomplished as promised. Secondly, planning times provide a target or goal for each job that allows for measurement of CP. Due to the variability of maintenance-type work and the inherent sensitivity toward measurement, the objective is not so much the measurement of individual CP. The real objective is measurement of the overall performance of the craft workforce as a whole. While measurement of the individual craftsperson is possible, CP measurement is intended to be for the maintenance of craft labor resources (Fig. 13.2) doing skilled type work.

Craft Service Quality

The third element affecting OCE relates to the relative quality of the repair work. This element includes quality of the actual work, where certain jobs possibly require a call-back to the initial repair, thus requiring another trip to fix it right the second time. Recall the last time you had to call back a repair person to your home for a job not performed to your satisfaction. This is not a good thing for a real profit-centered maintenance company! However, *craft service quality* (CSQ) can be negatively impacted within a plant or facility complex due to no fault of the craftsperson when hasty repairs, patch jobs, or inferior repair parts/materials create the need for a call back.

We can measure call backs via the CMMS with special coding of call-back workorders. Typically, the CSQ element of OCE is of a more subjective value and therefore it must be viewed accordingly in each operation. However, the CSQ level does affect overall craft labor productivity and the bottom line results of the entire maintenance process. When reliable data is present for all three elements of OCE, then the OCE factor can be determined by multiplying each of these three elements:

$$\text{OCE} = \underset{\text{Craft utilization}}{\text{CU\%}} \times \underset{\text{Craft performance}}{\text{CP\%}} \times \underset{\text{Craft service quality}}{\text{CSQ\%}}$$

What OCE Can You Expect?

Since OCE is a rather new concept there are actually a limited number of case studies outside the real worldwide experiences of MEI staff and alliance members. Some organizations try to measure just wrench time and it is accepted that 30 to 40 percent is typical and 70 percent is great. Other organizations may measure and track CP if a sound planning process and reliable planning times are in place. Also other good consulting firms shy away from the often sensitive issue of measuring craft labor at all, especially within a union environment. The MEI does not even for a union shop, because at some point high wages, growing continuously without productivity will make the operation a candidate for movement outside the current U.S. borders.

MEI feels strongly that measuring and improving productivity of craft labor resources is essential to profit-centered maintenance and CRI. Measuring and improving OCE must be addressed by today's in-house maintenance operation. Likewise, we feel that the range of OCE element values shown in Table 14.3 represents the high, medium, and low

TABLE 14.3 Range of OCE Element Values

	Range of OCE element values		
OCE elements	Low (%)	Medium (%)	High (%)
Craft utilization	30	50	70
Craft performance	>80	90	95
Craft service level	>90	95	98
The OCE factor	22	43	65

combinations for OCE. Successful operations can expect an OCE factor in the high range of 65 percent or more.

All three elements of OCE are important

Maintenance craft labor may be very efficient with 100-percent CP and still not be effective if CU is low and craft service quality are poor.

- OCE = craft utilization × craft performance × craft service quality = 30% × 100% × 70%
- OCE = 0.3 × 1.00 × 0.7 = 0.21 × 100
- OCE = 21% OCE

Since the nature of determining the value of CSQ can be subjective, this element is typically not used for calculating OCE. When this element is not used in determining OCE, it is still an important part of effective planning and scheduling. One key part of planning is determining the scope of the repair job and the special tools or equipment that is required for a quality repair. A continuing concern of the maintenance planning function should be on improving existing repair methods whether by using better tools, repair procedures, or diagnostic equipment, and using the right skills for the job. Providing the best possible tools, special equipment, shop areas, repair procedures, and craft skills can be a key contributor to improving CSQ. And CSQ can often still be a *key performance indicator* (KPI) that is determined from periodic review of call backs, customer complaints, and customer surveys. Therefore MEI feels that the OCE factor is best determined by using just two elements for the OCE factor calculations:

$$\text{OCE} = \text{craft utilization} \times \text{craft performance}$$

The impact of improving both craft utilization and performance

Improved CU through more effective planning of all resources will increase available wrench time. Improved performance results from the fact that work is planned and the right tools, equipment, and parts are available made by planning the right craftsperson or crew for the job with the type of skills needed. Improving CP is a continuous process with a program for craft skills training and methods improvement to do the job right the first time in a safe and efficient manner. The ACE team benchmarking process mentioned earlier provides reliable planning times based upon "a consensus of experts" and a tremendous repair methods improvement effort as benchmark jobs are analyzed.

Example C: What if We Increase Wrench Time from 30 to 50 Percent and CP from 80 to 90 Percent? When we look at the combination of improving both CU and CP, we see an even greater opportunity for a ROI. Let us now look at a very realistic 20-point improvement in CU and a 10-point increase in CP for the same 20-person craft workforce shown back in Table 14.3 and having an average hourly rate of $18.00.

Example C Details

- *Baseline cost per direct maintenance hour @ 30% utilization and 80 percent performance:*

$$20 \text{ crafts} \times 40 \text{ h/wk} \times 52 \text{ wks/yr} \times 0.30 \text{ (CU)} \times 0.80 \text{ (CP)} = 9984 \text{ direct craft h/yr}$$
$$= 499 \text{ direct hours/craft position}$$

$$\text{Baseline cost: } \frac{\$748{,}800}{9984 \text{ direct hours}} = \$75 \text{ cost per direct craft hour}$$

- *Improved cost per hour with 50% CU and 90% CP:*

$$20 \text{ crafts} \times 40 \text{ h/wk} \times 52 \text{ wks/yr} \times 0.50 \text{ (CU)} \times 0.90 \text{ (CP)}$$
$$= 18{,}720 \text{ direct craft hours/year} = 936 \text{ direct hours/craft position}$$

$$\text{Cost per direct craft hour @ 50\% CU and 90\% CP} = \$40$$

$$\frac{\$748{,}800}{18{,}720 \text{ direct hours}} = \$40 \text{ cost per direct hour}$$

- *Total direct craft hours gained = 8736 total direct hours gained (87% increase):*

$$18{,}720\text{h} - 9984\text{h @ baseline} = 8736 \text{ direct craft hours gained}$$

$$\frac{8736\text{h gained}}{9984\text{h @ baseline}} = 87.5 \times 100 = 87\% \text{ gain in direct craft hours}$$

- *Total gain in equivalent number of craft position—17 equivalent craft positions:*

$$20 \text{ crafts} \times .87 \text{ (\% hours gained)} = 17.4 \text{ equivalent craft positions}$$

$$\frac{8736 \text{ h gained}}{499 \text{ h/craft baseline average}} = 17.4 \text{ equivalent craft positions}$$

- *Total gained value = $655,200*

$$\text{Gain of 8736 direct craft hours} \times \$75 \text{ baseline cost/direct hour}$$

$$= \$655{,}200 \text{ gained value } \frac{\$655{,}200}{\$748{,}800} = 87\% \text{ gain from}$$

$$\text{a baseline of 30\% wrench time and 80\% CP}$$

Summary of Our Previous Examples

The previous examples have illustrated that increasing OCE provides greater craft capacity and gained value from increased wrench time. Improving CP in combination with improving CU simply compounds our ROI: an astronomical amount of 87% as shown in Table 14.4.

Where can we apply OCE gained value

Maintenance operations that continually fight fires and react to emergency repairs never have enough time to cover all the work (core requirements) that needs to be done. Overtime, more craftspeople, or more contracted services typically seem to be the only answers. Improving CU provides additional craft capacity in terms of total productive craft hours available. In relation to OEE, OCE is increased people asset availability and capacity. It is gained value that can be calculated and estimated and then measured. The additional equivalent craft hours can then be used to reduce overtime, devote to PM/PdM, reduce the current backlog, and attack deferred maintenance which does not go away. Figure 13.2 shows "How your valuable craft time can just slip away."

Indiscriminate cutting of maintenance is blood letting

Typically, operations that gain productive craft hours desperately need to invest the time elsewhere. Likewise, we cannot automatically and indiscriminately reduce head count when we improve OCE. Indiscriminate cutting of maintenance is killing the goose that lays the golden egg. If an organization is not achieving core requirements for maintenance, cutting of craft positions to meet budget is like using blood letting as a new cure for a heart attack. It just will not work.

Just like the high cost of low-bid buying, gambling with maintenance costs can be fatal. Long-term stabilization and reduction of head count can occur. Attrition can absorb valid staff reductions that may result over the long term. We also may regain our competitive edge and get back some of the contract work we lost previously to low performance and productivity. We cannot indiscriminately cut craft labor resources when we increase OCE.

TABLE 14.4 Summary Comparisons of Previous Examples

Examples	Baseline	Improve to	Craft labor gain	Gained value
A	CU@40%	CU@50%	5	$187,200
B	CU@30%	CU@50%	13	$497,952
C	CU@30% and CP@80%	CU@50% and CP@90%	17	$655,200

Think profit centered

Today's maintenance leaders and craftspeople must develop the "maintenance-for-profit" mindset that the competition uses to stay in business. Measuring and improving OCE and the value received from improving our craft assets is an important part of total asset management. Profit-centered in-house maintenance in combination with the wise use of high-quality contract maintenance services will be the key to the final evolution that occurs. There will be revolution within organizations that do not fully recognize maintenance as a core business requirement and establish the necessary core competencies for the maintenance. The bill will come due for those operations that have subscribed to the "pay me later syndrome" for deferred maintenance. It will be revolution within those operations that have gambled with maintenance and have lost, with no time left before profit-centered contract maintenance provides the best financial option for a real solution.

Maintenance is forever

Contract maintenance will be an even greater option and business opportunity in the future. Again, we must remember—maintenance is forever! Some organizations today have neglected maintaining core competencies in maintenance to the point that they have lost complete control. The core requirement for maintenance still remains, but the core competency is missing. In some cases, the best and often only solution may be value-added outsourcing. Maintenance is a core requirement for profitable survival and total operations success. If the internal core competency for maintenance is not present it must be regained. Neglect of the past must be overcome. It will be overcome with a growing number of profit-centered maintenance providers that clearly understand OCE and provide value-added maintenance service at a profit.

Chapter

15

The ACE Team Benchmarking Process: A New Benchmarking Tool

This chapter outlines a new and highly recommended methodology for establishing team-based maintenance performance standards. The ACE (*a consensus of experts*) team benchmarking process (ACE system) is a team-based process that utilizes skilled craftspeople, technicians, supervisors, planners, and other knowledgeable people to do two things:

1. Improve current repair methods, safety, and quality
2. Establish work content time for selected "benchmark jobs" for planners and others to use in developing reliable planning times

Nearly every *computerized maintenance management system* (CMMS) allows a user to enter "planned" or "standard" hours on a work order, and then report on actual versus planned hours (the CP element of OCE) when the job is complete. This holds true for both preventive and corrective maintenance work orders as well a project-type work for renovation, major overhauls, and capitalized repairs.

Determining the standard hours an average maintenance technician will require to complete a task under standard operating conditions provides everyone involved a sense of what is expected. The standards provide maintenance leaders with valuable input for manpower planning, scheduling, budgeting, and costing. Labor standards also form the baseline for determining labor savings for improved methods.

Many maintenance managers have observed firsthand the negative effect that setting a reliable planning time can have on the workforce. As methods improve, reliable planning times tighten and technicians perceive a noose that's getting tighter with each improvement. This situation discourages people who must do the work from coming forward with innovative ideas that might further improve efficiency.

At the other end of the hierarchy, many managers fear that a reliable planning time might set an upper limit on worker productivity. The thought is that people will work only as hard as they must. Therefore, managers do not want to set reliable planning times that are too loose.

Some companies attack the problem by setting ideal, unachievable planning times. No allowance for the realities in the environment such as having to walk to a remote location, equipment that is difficult to access, and other less-than-ideal conditions. Unachievable planning times can be disastrous because technicians, and their supervisors for that matter, are continually frustrated by not being able to meet, no matter how they try.

Set reliable planning times. The goal is not meeting unrealistic management-imposed standards. Wherever possible, achieve a series of reasonable short to longer duration planning times, set by or with the help of the experienced technicians. This avoids both the frustration of never attaining ideal planning times and the problem with easily achievable, loosey goosey production standards as we used to call "measured daywork" plans at Crescent/Excelite and around the Cooper Tools plants.

Not for discipline. Use standards primarily for accounting for your maintenance requirements and scheduling purposes, not for disciplining an individual who cannot achieve the so-called reliable planning time. If a craftsperson consistently misses target times, determine the cause, use other means to prove a problem exists, and then deal with it. For example, the reason someone cannot meet a reliable planning time can be a lack of training. Some say it could be as nasty as sleeping on the job or conducting union business. In either case, the problem will be identified first via planned-versus-actual reports the CMMS generates. However, maintenance leaders should investigate and confront a person only after gathering evidence from other means such as "management by walking around." We can observe a lot by just watching as Yogi Berra has often said: pure laziness and lack of skill and effort are all three fairly obvious.

Focus on improvement. Again the goal is not simply meeting current standards in the form of a reliable planning time. As you will see the ACE Team Benchmarking Processs is about improving craft productivity for a repair process or task by using methods improvement, better tools, focused training, or simply expecting improved performance. Various built-in CMMS analysis tools can determine problem areas, their causes, and the cost/benefit of alternative solutions. The most effective method improvements not only result in reliable planning times, but also improve working conditions and increase the level of craft responsibility and bottomline personal craft productivity

Standards are much easier to accept when crafts are allowed to influence some of the decisions that affect their jobs. In turn, job satisfaction improves, as does productivity. Basic industrial engineering, value engineering, chartered leadership-driven, productivity improvement teams, cost improvement initiatives, and serving on cross-functional teams to improve processes, all help build pride in ownership of the maintenance process. Likewise avoid a Six Sigma effort I witnessed in 2003 that had already eliminated facilities maintenance and was being directed to eliminate plant maintenance and then . . . share the savings across Six Sigma team members. It was an event that I witnessed personally while conducting an assessment and asked "Why don't you sit in on our first Six Sigma team meeting to improve plant maintenance?

Competition Between craft areas. Having reliable planning times is not about measuring the individual craftsperson. When coupled with a productivity improvement program such as the ACE Team process it can involve competitiveness between craft areas and/or first-shift, second-shift, or third-shift teams. Not only are there individual targets for various processes or tasks, the various craft teams can be responsible for establishing craft performance, craft utilization, and craft service quality targets for a craft, shift, department, or even a plant or a division. Here teams can be competitive against each other and achieve the highest productivity level possible. Peer pressure and group dynamics are powerful elements as well as both help to identify and minimize bottleneck processes or people.

Technology is available for support. CMMS provides an excellent vehicle for entering, analyzing, and reporting standard and actual performance data. CMMS-based analysis tools, such as Pareto analysis, root-cause analysis and trend analysis, can switch the emphasis from a "sea of analysis paralysis data" into data you can see that can identify improvement opportunities.

As an extension to the CMMS, personal digital assistants (PDA) are fast becoming very valuable and cost-effective technology tools to support your most valuable people (MVP) doing the real work. Mobile technology forces maintenance planners to assign tasks and download them to individual maintainers' PDAs based on planned hours. Each PDA then automatically tracks the time spent on each assigned task for comparison to plan and standard.

More sophisticated implementations of mobile technology can result in productivity gains of as much as 20 percent or more. Some of the advanced PDA features contributing to this gain include:

- Validating inspection readings
- Capturing data as work is completed, instead of at the end of the day
- Issuing workorders on the fly
- Alarming capability
- On-the-job access to workorder history and analysis tools to assist in PM and PdM task completion

With properly implemented mobile technology, crafts feel they have a portal into the main CMMS system. They feel empowered to get the work done effectively, without the burden of additional paperwork.

Monetary incentives. Financial incentives for individuals or groups can provide significant motivation for improving productivity. One important incentive is to get the maintenance job tasks analyzed correctly and get your maintenance wage plan at least recommended to your top leaders. Examples of incentives include profit sharing, gain sharing, suggestion plans with financial rewards, and performance-based bonuses. Again a Six Sigma suggestion plan that I witnessed is not the *answer*! Establish incentives that simultaneously encourage individual creativity, teamwork at the line or departmental level, and net positive gains across the division or company. Bottomline direct involvement allows the craft workforce to be a part of answering the big question asked preciously; "How would I do this job if I in fact owned the maintenance operation?"

Nonmonetary incentives. Money talks and good maintenance people want you to "show me the money". Nonmonetary or low-cost incentives also can work well but they do not buy milk and bread, but maybe pizzas. Examples include providing pizza as a reward when the department or team reaches a milestone performance target or resolves a major problem. The easiest and lowest-cost incentive is recognition, such as a write-up in the company newsletter or a wall chart showing performance targets that were met or exceeded.

Summary of The ACE Team Benchmarking Process: The ACE system uses skilled craft, planners, and even skilled supervisors as the ACE; for a consensus of experts who know the jobs and can also help improve them. In turn, using the ACE Team lets you use a relatively small number of representative "benchmark jobs" from the major work areas/types within the operation. Benchmark jobs are then arranged into time categories ("time slots") on spreadsheets for the various craft work areas. By using spread sheets to do what is termed "work content comparison" or "slotting." A trained planner then is able to establish reliable planning times for a large number of jobs using a relative small sample of "benchmark jobs." This chapter provides the step-by-step process on using the ACE system. Most importantly it will illustrate how this method supports *continuous reliability improvement* (CRI) and quality repair procedures for all types of maintenance repair operations.

The ACE System Supports Continuous Realiability Improvement

Maintenance work by its very nature seldom follows an exact pattern for each occurrence of the same job. Therefore, exact methods and exact times for doing most maintenance jobs cannot be established as they can for production-type work. However, the need for having reliable performance measures for maintenance planning becomes increasingly important as the cost of maintenance labor rises and the complexity of production equipment increases. To work smarter—not harder—maintenance work must be planned, have a reasonable time for completion, use effective and safe methods, performed with the best personal tools and special equipment possible, and have right craft skill using the right parts and materials for the job at hand. With an investment in maintenance planners there must be a method to establish reliable planning times for as many repair jobs as possible. The ACE system provides that method as well as team-based process to improve the quality of repair procedures.

Various methods for establishing maintenance performance standards have been used, including reasonable estimates, SWAGs, historical data, and engineered standards such as *Universal Maintenance Standards* (UMS) using predetermined standard data. These techniques generally require that an outside party establish the standards, which are then imposed upon the maintenance force. This approach often brings about undue concern and conflict between management and the maintenance workforce over the reliability of the standards.

The ACE System: A Team-Based Approach

Rather than progressing forward together in a spirit of continuous improvement, the maintenance workforce in this type environment often works against management's program for maintenance improvement. The ACE system overcomes this problem with a team-based approach involving craft people who will actually do the work that will be planned later as the planning and estimating process matures. As shown later, the ACE system is truly a team-based process that looks first at improving maintenance repair methods, the reliability of those repairs to improve asset uptime, and then to establish a benchmark time for the job.

Gaining Acceptance for Maintenance Performance Standards

To overcome many of the inherent difficulties associated with developing maintenance performance standards, the ACE system is recommended and should be established as the standard process for modern maintenance management. Other methods such as the use of standard data (RS Means) can supplement the ACE system. The ACE system methodology relies primarily on the combined experience and estimating ability of a group of skilled craftspeople, planners, and others with technical knowledge of the repairs being made within the operation.

The objective of the ACE team benchmarking process is to determine reliable planning times for a number of selected "benchmark" jobs and to gain a consensus and overall agreement on the established work content time. This system places a very high emphasis on improving current repair methods, continuous maintenance improvement, and the changing of planning times to reflect improvements in performance and methods as they occur. The ACE system is a very progressive method to developing maintenance performance standards; a very hard area in itself to develop reliable and well-accepted planning times for maintenance.

The Basic Approach: A Simple 10-Step Procedure That Works

Generally, the ACE system parallels the concepts of the UMS approach. For both UMS and the ACE system, the "range of time concept" and "slotting" are used once the work content times for a representative number of "benchmark jobs" have been established. The ACE system focuses primarily on the development of work content times for representative "benchmark jobs" that are typical of the craft work performed by the group. An example of an actual UMS benchmark job that has been analyzed with standard data to establish work content time is included as Figs. 15.1 and 15.2.

From the example illustrated in Fig. 15.1, we see that through the use of UMS standard data, the eight elements of the job including oiling of the parts, has been analyzed and assigned time values that total 1.07 h. Since the time value for this benchmark job falls within the time range of 0.9 to 1.5 h (see Fig 15.3) it is assigned a standard work content time of 1.2 h.

What this implies is that the actual work content for this benchmark job will generally be performed within the time range for work group G (0.9 to 1.5 h) with a confidence level of 95 percent. When we refer to "work content" the following applies: The work content of the benchmark job *excludes* things such as travel time, securing tools and parts, prints, delays, and personal allowances. The benchmark time that is estimated does not include the typical "make ready" and "put away" activities that are associated with the job. Therefore, a number of allowances must be added to work content time by a planner to get the actual planning time for the job being done.

Benchmark Analysis Sheet

Decription: Remove and reinstall 3 oil wiper rings	**B.M. No:**
split type of air compressor 1950 cfm at 100 psi	**Craft:** Mech
	Dwns: N/A
	No. of Men: 1 **Sh.** 1 **of** 1
	Analyst: JEB **Date:**

Line	Men	Operation Description	Reference Symbol	Unit Time	Freq.	Total Time
1	1	Remove and reinstall 1 crankcase cover	PWN-10-10	.030	2	.060
2		Remove and reinstall 2 bolts	PWN-10-1	.011	4	.044
3		Slide gland off and on	PWN-10-7	.012	2	.024
4		Unfasten 3 garter springs and refasten	PWN-10-8	.023	6	.138
5		Remove and reinstall 9 wiper ring	PWN-10-9	.012	18	.216
		segments				
6		Fit 9 ring segments to piston rod	PWN-5-2	.040	9	.360
7		Clean 12 springs and rings	PWN-8-1	.016	12	.192
8		Oil 12 parts	PWN-3-9			
		2 squirts per pint				
		.0023 + .0012(N2)		.0023	1	.002
		N2 = number of application		.0012	24	.029

Notes:	**Benchmark Time**	1.07
	Standard Work Group	E

C.

Figure 15.1 An example of a benchmark job analysis using UMS standard data.

ACE Team Benchmark Job Analysis Sheet						
Benchmark Job Description			Benchmark Job No.: MECH-AC-5			
Remove and reinstall 3 oil wiper rings, split type air compressor 1950 cfm @100 psi			**Craft:** Mechanical			
			Ref. Drawing: AC-9999			
			No. of Crafts: 1			
			Analyst: JEB/ACE Team			
Line No.	**No. of Crafts**	**Operation Description**	**Ref. Code**	**Unit Time**	**Freq**	**Total Time**
1	1	*Remove 2 bolts to remove crankcase cover*				.10
2		Slide gland off and unfasten 3 garter springs				.10
3		Remove 9 wiper ring segments				.10
4		Clean 12 springs and rings and properly oil all 12 parts per Lube Spec #AC-2000	CLN-1			.25
5		Reinstall 9 wiper ring segments and fit to piston rod				.50
6		Slide glide on and fasten 3 garter springs				.10
7		Replace. crankcase cover and fasten with 2 bolts				.10
Notes: CLN-1 = Average benchmark time established to clean and oil this # of small parts per lube specification noted above.			Benchmark Time for **Work Content**			**1.25**
			Standard Work Group			**E**

Figure 15.2 Example of an ACE team benchmark analysis.

Important note

The estimated time for a benchmark job is for pure work content time and is made under these conditions:

a. The right craft skills and level of competency is available to do the job.

b. An average skilled craftsperson, two-person team, or crew is doing the job giving 100 percent effort, that is, a “fair day’s work for a fair day’s pay.”

Work Group	ACE System Time Ranges		
	From	Standard Time (Slot time)	To
A	.0	.1	.15
B	.15	.2	.25
C	.25	.4	.5
D	.5	.7	.9
E	.9	1.2	1.5
F	1.5	2.0	2.5
G	2.5	3.0	3.5
H	3.5	4.0	4.5
I	4.5	5.0	5.5
J	5.5	6.0	6.5
K	6.5	7.3	8.0
L	8.0	9.0	10.0
M	10.0	11.0	12.0
N	12.0	13.0	14.0
O	14.0	15.0	16.0
P	16.0	17.0	18.0
Q	18.0	19.0	20.0
R	20.0	22.0	24.0
S	24.0	26.0	28.0
T	28.0	30.0	32.0

Figure 15.3 ACE system standard work groupings and time ranges.

c. The correct tools are available at the job site or with the craftsperson or crew.

d. The correct parts are available at the job site or with the craftsperson or crew.

e. The machine/process/asset is available and ready to be repaired.

f. The craftsperson or crew is at the job site with all of the foregoing points and proceeds to complete the job from start to finish without major interruption.

Once a sufficient number of benchmark job times have been established for craft areas and work types, these jobs are categorized onto spreadsheets. They are estalished on spreadsheets by craft and task area and according to the standard work groups (Fig. 15.3), which represent various ranges of time. This is exactly the concept behind the UMS

Craft: ______________ Code: ______________

Task Area: Task areas would be mechanical, electrical, hydraulic, etc or by major areas such as fork lift, conveyor systems or building systems types of repairs.			
Group E **1.2 hours**	**Group F** **2.0 hours**	**Group G** **3.0 hours**	**Group H** **4.0 hours**
Job Description A	Job Description E	Job Description I	Job Description M
Job Description B	Job Description F	Job Description J	Job Description N
Job Description C	Job Description G	Job Description K	Job Description O
Job Description D	Job Description H	Job Description L	Job Description P
Note:			

Figure 15.4 Example of spreadsheets for work groups E, F, G, and H.

approach. Figure 15.4 provides an example of a spreadsheet for workgroups E, F, G, and H with jobs that have benchmark times of 1.2, 2.0, 3.0, and 4.0 h respectively.

A complete set of ACE team spreadsheets for work group A (0.1 h) up to workgroup T (30.0 h) is available from *The Maintenance Excellence Institute* (TMEI).

ACE system work groupings and time ranges

Figure 15.3 includes a listing of the ACE system work groupings, the respective time ranges for each workgroup from A to T. Likewise, spreadsheets for work groups A, B, C, and D would be developed with benchmark jobs having work content time below 1.2 h. Spreadsheets for workgroups I, J, K, and L for benchmark jobs having work content time from 5.0 to 9.0 h respectively would also be developed, as they were needed.

Spreadsheets provide means for work content comparison

After sufficient spreadsheets have been prepared based on the representative benchmark jobs from various craft/task areas, a planner/analyst now has the means to establish planning times for many different maintenance jobs using a relatively small number of benchmark jobs as a guide for work content comparison. By using work content comparison (or *slotting* as it is called) combined with a good background in craft work and knowledge of the benchmark jobs, a planner now has the tools to establish reliable performance standards consistently, quickly, and with confidence for a large variety of different jobs.

Since the actual times assigned to the benchmark jobs are so critical, it is very important to use a technique that is readily acceptable. The ACE system provides such a technique since it is based on the combined experience of a team of skilled craftspeople and others. It is their consensus agreement on the range of time for the benchmark jobs—a consensus of experts who know the mission-essential maintenance work that is to be done.

The 10-Step Procedure for Using the ACE System

1. *Select "benchmark jobs."* Review past historical data from work orders and select representative jobs that are normally performed by the craft groups. Special attention should be paid to determine the 20 percent of total jobs (or types of work) which represent 80 percent of the available craft manpower. Focus on determining repetitive jobs where possible in all craft areas.

2. *Select, train, and establish team of experts (ACEs).* It is important to select craftspeople, supervisors, and planners who as a group have had experience in the wide range of jobs selected as benchmark jobs. All craft areas should be represented in the group. In order to ensure that this group understands the overall objectives of the maintenance planning effort, special training sessions should be conducted to cover the procedures to be used, reasons for establishing performance measures, and the like. A total of 6 to 10 knowledgeable team members is the recommended size for the team.

   Develop ACE team charter—At this point it is highly recommended that a formal ACE team charter be established. Appendix G provides a sample charter format that can easily be tailored for each site. The ACE team has an important task that will take time to accomplish. However, the task of the ACE team will be important and in turn their success as a team can contribute significantly to CRI and increased asset uptime.

3. *Develop major elemental breakdown for benchmark jobs.*

   a. For each benchmark job that is selected, a brief element analysis should be made to determine the major elements or steps for completing the total job. Another example is shown in Fig. 15.2. Here the elements of the same job that we illustrated in Fig. 15.1 are used but arranged in a more logical sequence of the actual repair method. In this example a standard time allowance for cleaning and oiling a small group of parts had already been established, so reference to CLN-1 task was made. This task referenced back to a standardized lube specification (No. AC-2000).

b. This listing of the major steps of the job should provide a clear, concise description of the work content for the job *under normal conditions*. It is important that the work content for a benchmark job be described and viewed in terms of what is a *normal* repair and not what may occur as a rare exception. All exceptions along with make ready and put away time are accounted for by the planner when the actual planned time is completed.

c. An excellent resource to consider for doing the basic element analysis for each benchmark job are the craftspersons (ACEs) that are selected for doing the estimating or even other craftspeople within the operation. Brief training on methods/operations analysis can be included in the initial training for the ACEs. Very significant methods improvements and methods to improve reliability can be discovered and implemented as a result of this important step.

d. The ACE team process must include and also lead to getting answers to the following questions:
   - Are we using the best method, equipment, or tools for the job?"
   - "Are we using the safest method for doing this job?"
   - "Are we using the best quality repair parts and materials, or is this a part of our problem?"
   - "What type of preventive task and/or predictive task would help identify or eliminate the root cause of the problem?"
   - "Where can we work even smarter, not necessarily harder?"

e. Major exceptions to a routine job should be noted if they are significant; generally an exception will be analyzed as a separate benchmark job along with an estimate of time required for such repair.

f. This portion of the ACE team process ensures that the work content of each benchmark job is clearly defined so that each person/planner doing the estimating has the same understanding about the nature and scope of the job. When the benchmark jobs are finally categorized into "spreadsheets," the "benchmark job description" information developed in this step is then used as key information about the benchmark job on the ACE team spreadsheets.

4. *Conduct first independent evaluation of benchmark jobs.*

   a. Each member of the group is now asked to review the work content of the benchmark jobs and to assign each job to one of the

UMS time ranges or slots. Each member of the group provides an independent estimate, which represents an unbiased personal estimate of the "pure work content" time for the benchmark job. It is essential that each team member does an independent evaluation of each benchmark job and not be influenced by others on the team with their first evaluation.

b. Focus on work content time—It is important here for each member of the team to remember that only the work content of the benchmark job description is to be estimated and not the "make ready" and "put away" activities associated with the job. This part of the procedure is concerned only with estimating the pure work content *excluding* things such as travel time, securing tools and parts, prints, delays, and personal allowances.

c. The estimate should be made for each job under these conditions:
   - An average skilled craftsman is doing the job giving 100 percent effort, that is, a "fair day's work for a fair day's pay."
   - The correct tools are available at the job site or with the craftsperson.
   - The correct parts are available at the job site or with the craftsperson.
   - The machine is available and ready to be repaired.
   - The craftsperson is at the job site with all of the foregoing points and proceeds to complete the job from start to finish without major interruption.

d. The work accomplished under these conditions therefore represents the "pure work content" of the job to be performed. Establishing the "range of time" estimate for this "pure work content" is the prime objective of the first evaluation.

e. It is important for each ACE team member to remember that developing a planning time requires "pure work content" time plus additional time allowances to cover "make ready" and "put away" type activities associated with each job as illustrated in Fig. 15.5. The "make ready time" and "put away time" will be accounted for as the planner adds time and allowances for these elements as the actual planning time is completed. Make ready and put away times are established specifically for each operation and added to the work content time to get the total planned time for the job being estimated.

Make ready +	Work time +	Put away =	Planning time

Figure 15.5 Planning time elements.

5. *Summarize first independent evaluation.*
   a. Results of the first evaluation are then summarized to check the agreement among the group as to the time range for each benchmark job. A coefficient of concordance can be computed from the results if required, but normally this level of detail is not needed. A coefficient of concordance value of 0.0 denotes no agreement while a value of 1.0 denotes complete agreement, or consensus among the ACEs. Generally, a consensus can be reached by the ACE team in one, two, and at the most three rounds of evaluations.
   b. Define high and/or low estimates—Team members who are significantly higher or lower than the rest of the group for a particular benchmark job are then asked to explain their reasons for their respective high or low estimates. They explain their reason for their estimate to the group, discussing the method, condition, or situation for their initial time estimate. This information will then be used during the second evaluation to refine the next round of time estimates from the entire group.

6. *Conduct second independent evaluation of benchmark jobs.*
   a. A second evaluation is conducted using the overall results from the first evaluation as a guide for the entire team. Various reasons for high or low estimates from the first evaluation are provided to the group prior to the second evaluation. Normally this can be done in an open team discussion with team members making personal notes to use in their second independent evaluation.
   b. The second round allows for adjustment to the first estimates if the other ACE team member's reason for a higher or lower estimated time is considered to be valid. In other words, results from the first evaluation plus reasons for highs and lows will allow each team member to reconsider their first estimate. In many cases, a review of the repair method or scope of work will be more clearly defined, causing a change to the time estimate for the second evaluation.

7. *Summarize second evaluation.*
   a. Results of the second evaluation are then summarized to evaluate changes or improvements in the level of agreement. The goal is a consensus among the ACEs as to the time range (work group) for each benchmark job. The second round should bring an agreement as to the time range.
   b. The second independent evaluation should produce improved agreement among the group. If an extreme variance in time range estimates still exists, further information regarding the work

content, scope, and repair method for the job may be needed. Here those with high/low estimates should again review their reasons for their estimates with the team describing the scope of work that they see is causing differences from the rest of the team.

8. *Conduct third independent evaluation if required.* This evaluation is required only if there remains a wide variance in the estimates among the group.
9. *Conduct review session to establish final results.* This session serves to finalize the results achieved and to discuss any of the high or low estimates, which have not been resolved completely. A final team consensus on all time ranges is the objective of this session.
10. *Develop spreadsheets.*
    a. The benchmark jobs with good work content descriptions and agreed upon time ranges can now be categorized onto spreadsheets. From these "spreadsheets," which give work content examples for a wide range of typical maintenance jobs, a multitude of individual maintenance performance standards can be established by the planner through the use of work content comparison.
    b. The basic foundation for the maintenance planning system is now available for generating consistent planning times that will be readily acceptable by the maintenance workforce that developed them. Attachment B to these appendices provides a graphical illustration of the ACE system.

The ACE team approach combines the Delphi technique for estimating along with a proven team process plus the inherent and inevitable ability of most people to establish a high level of performance measures for themselves. As used in this application, the objective for the ACE team process is to obtain the most reliable, reasonable estimate of maintenance-related "work content" time from a group of experienced craftspeople, supervisors, and planners. This process provides an excellent means to evaluate repair method, safety practices, and even to do risk analysis on jobs that leads to improved safety practices. The ACE team process can contribute significantly to continuous reliability.

The ACE team approach allows for independent estimates by each member of the group, which in turn builds into a consensus of expert opinion for a final estimate. The final results are therefore more readily acceptable since they were developed by skilled and well-respected craftspeople from within the work unit. Application of the ACE system promotes a commitment to quality repair procedures and provides the foundation for developing reliable planning times for a wide range of maintenance activities.

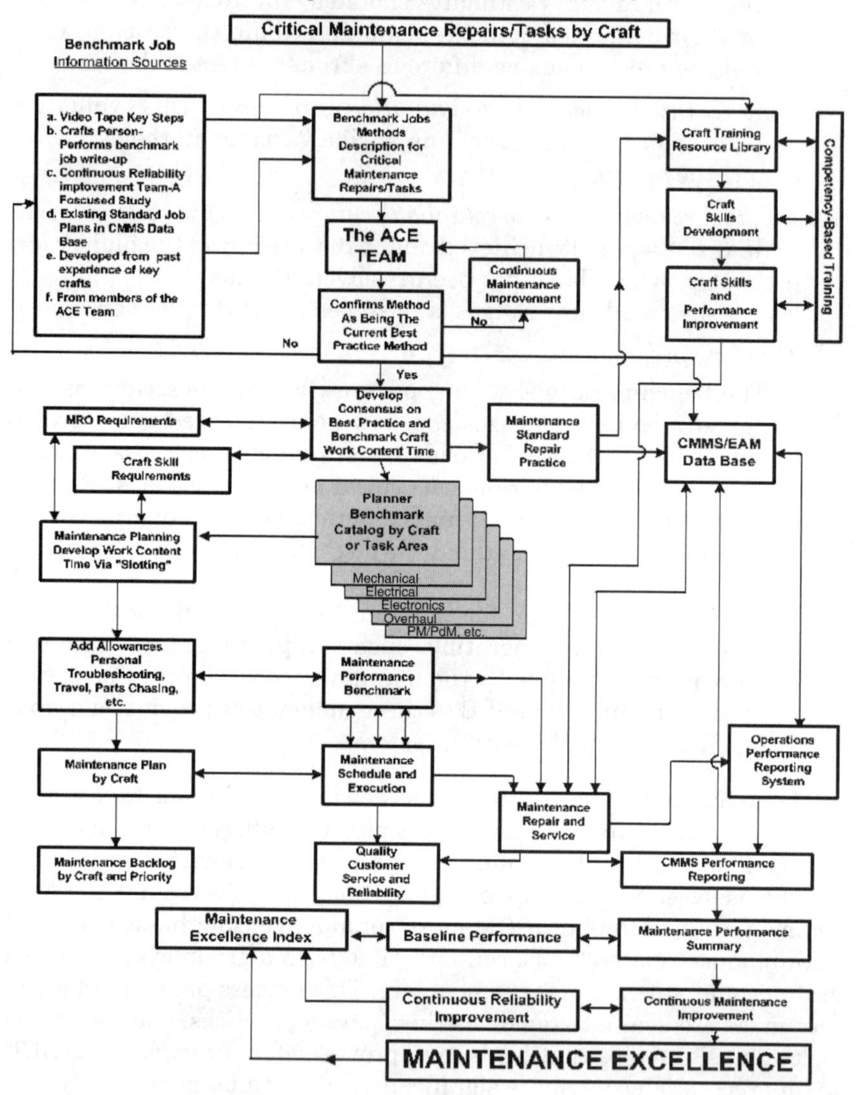

The ACE team benchmarking process for reliability and quality maintenance customer service.

Chapter

16

Profit- and Customer-Centered Best Practices

Continuous Reliability Improvement (CRI)

Continuous reliability improvement (CRI) is an overall strategy committed toward improvement of the total maintenance operation as a business process. CRI is the application of the traditional *reliability centered maintenance* (RCM) processes, using the best from *total productive maintenance* (TPM) and going beyond these traditional approaches. CRI is not focused just upon the physical assets within a plant manufacturing environment. It encompasses an improvement process to include all maintenance resources: equipment/facility, craftspeople/ operators, storeroom operation, *maintenance repair operations* (MRO) materials management, and maintenance information resources. CRI includes effective team processes to create synergy that serves as people-asset multipliers. CRI is about uncovering and expanding the hidden assets of people within your maintenance operation. CRI also includes maximizing the capability of the physical asset through comprehensive asset facilitation that does combine the best preventive/predictive and reliability technologies available.

CRI is a total maintenance operations improvement process to support the total operational success within a manufacturing plant, facilities complexes, fleet operations, and specialized facilities such as hospitals and healthcare operations. CRI covers maintenance operations in the "green industry": golf course maintenance, landscape operations, and maintenance within national and municipal parks systems. CRI focuses not just on teams but true leadership-driven, self-managed teams that are formally chartered and focused on continuous improvement. CRI focuses team processes and maintenance improvement strategies on CRI opportunities of six maintenance resource areas as shown in Fig. 14.3.

1. *Physical assets.* Use of reliability improvement technologies: reliability centered maintenance, *preventive/predictive maintenance* (PM/PdM), and knowledge-based/expert systems for maintenance of the physical asset. Asset facilitation to gain maximum capacity at the lowest possible life-cycle cost. Figure 16.1 provides key points for this important maintenance resource area.
2. *MRO parts and material resources.* Effective MRO parts, supplies, and materials for quality repair with effective storeroom and procurement processes. Figure 16.2 provides key points for this important maintenance resource area.
3. *Information resources.* Quality information resources for maintenance management and control from *computerized maintenance management systems* (CMMS), EAM, ERP, vendor, and customer. Figure 2.4 provides key points for this important maintenance resource area.

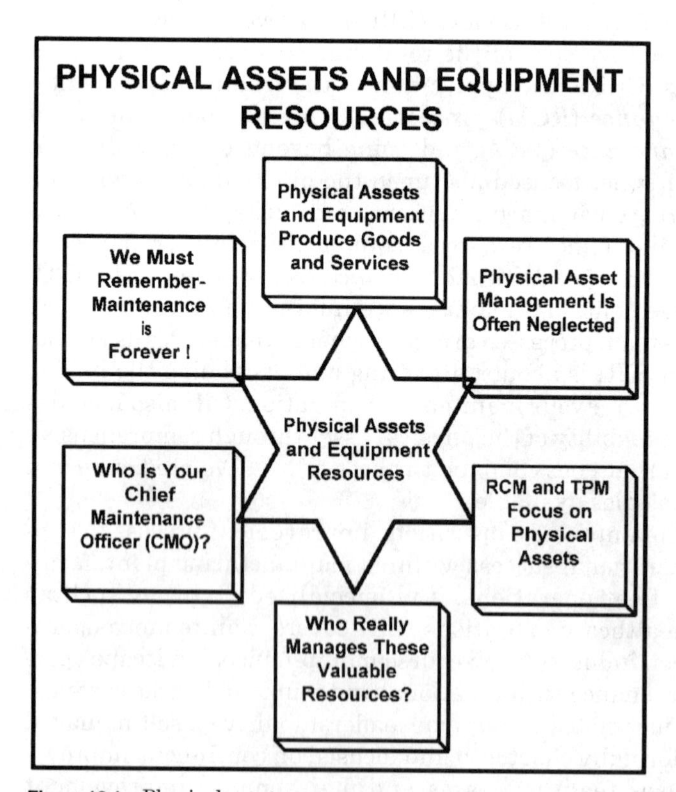

Figure 16.1 Physical asset resource area.

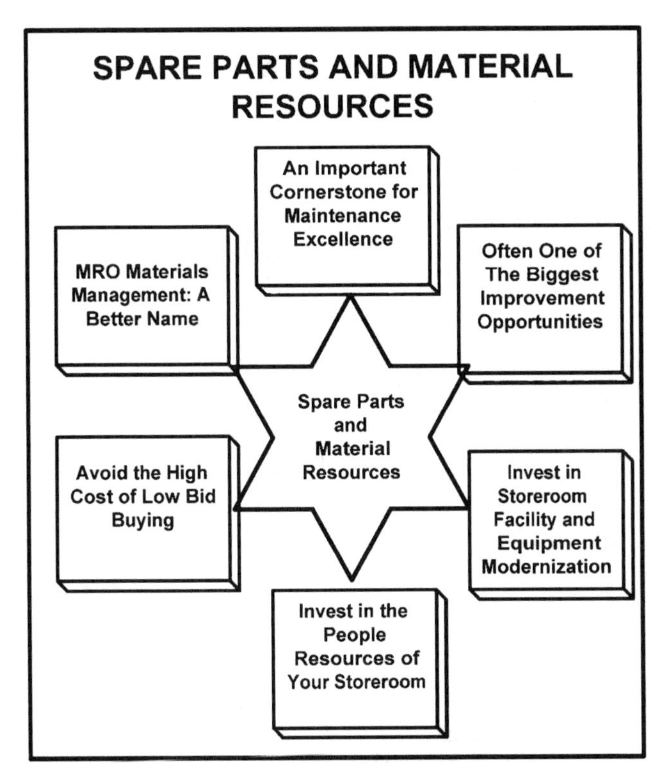

Figure 16.2 MRO parts and material resources.

4. *Craft, storeroom, and operator resources.* Quality skill improvement by craftspeople assets for the maintenance process to support customer service to the total operation. It also includes operator resources and the added value of equipment operators instilled with pride in ownership at the most important level—the shop floor to support the maintenance process with *operator-based maintenance* (OBM). Figure 2.5 provides key points for this important maintenance resource area.

5. *Technical skill resources.* The added value of crafts skills development and the continuous improvement of technical capabilities of craft resources. Figure 16.3 provides key points for this important maintenance resource area.

6. *Synergistic team processes that serve as multipliers of people assets.* Formal and informal teamwork that can be "mustered" to directly support the maintenance leaders with new ideas, attitudes, and positive actions, which bring about positive change within the total maintenance team to support the total operations team. Figure 16.4 provides key points for this important maintenance resource area.

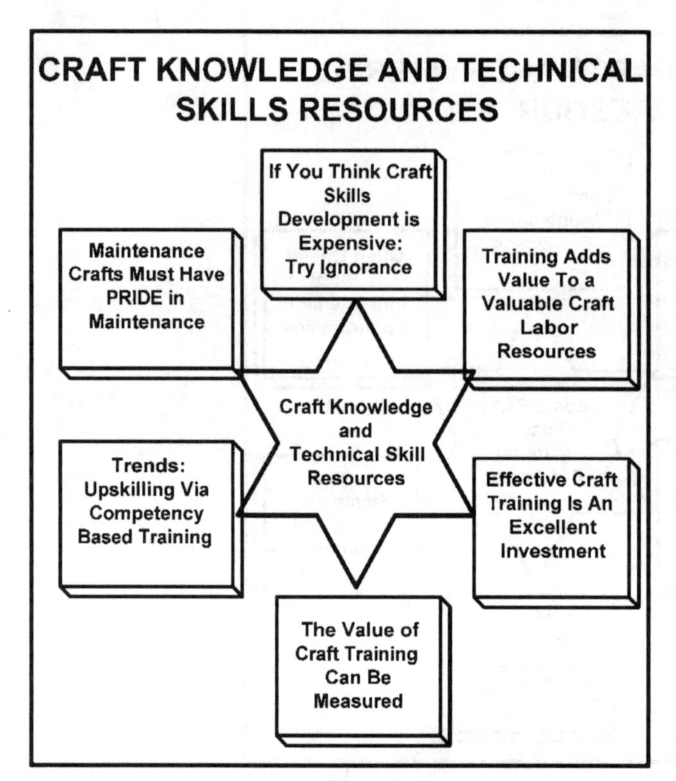

Figure 16.3 Craft knowledge and technical skill resources.

Successful CRI must include all six key areas and all areas must be supported by the many best practices in this book and the profit-centered best practices that follow.

Critical Asset Facilitation

This very comprehensive process is a component of CRI and is Category Z [critical asset facilitation and *overall equipment effectiveness* (OEE)] on the "Scoreboard for Maintenance Excellence." It is conducted for critical assets and typically discrete manufacturing equipment. The objective is to maximize the capability of the asset to perform its primary functions. It takes a complete look at the requirements for maintenance, setup, operations, changeover, and shutdown; and documents these into standard operating and maintenance procedures. The results of a typical asset facilitation process conducted by *The Maintenance Excellence Institute* (TMEI) would include the following.

Development of a "quality operations and maintenance guide" for the asset that provides complete operating and maintenance guidelines, procedures, and documentation for:

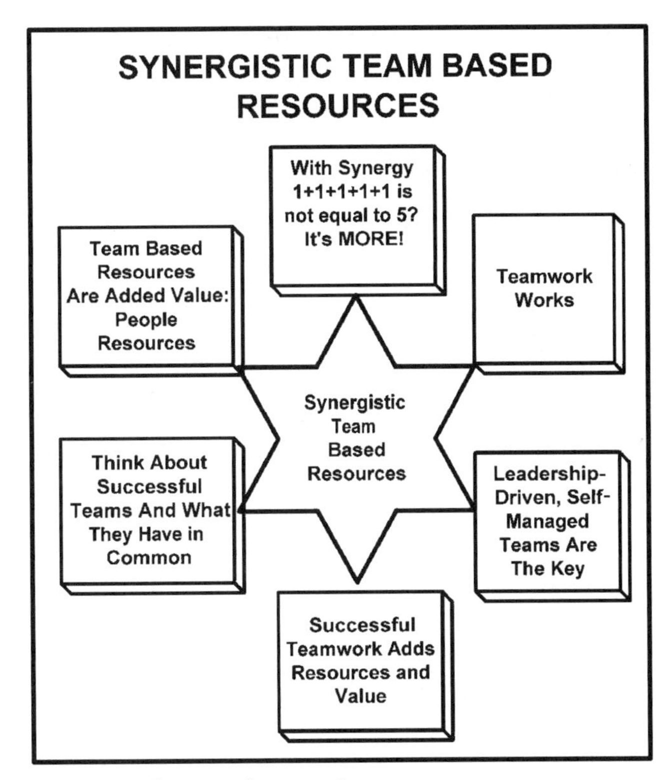

Figure 16.4 Synergy of teamwork.

- Equipment safety and housekeeping requirements
- Defines operator and operations departmental requirements for the 5S process
- Defines requirements for subsequent 5S evaluations and ratings
- Asset documentation references and drawings, either manual or electronic medium, using electronic document imaging capabilities
- PM and PdM requirements
- Setup/validation/changeover/operator inspection procedures
- Start-up, operating, and shutdown procedures for the asset
- Quality control requirements
- OBM tasks with details on:
  - Recommended operator-level PM tasks and lubrication services
  - Other well-defined maintenance tasks to be performed by the operators

- Operator training and certification requirements—production-related tasks
- Operator training and certification requirements for maintenance related (OBM) tasks
- Develop visual management to support OBM, production tasks and pure maintenance department tasks
- Quality requirements for ISO/QS compliance

Preventive Maintenance (PM)

PM is an interval-based surveillance method in which periodic inspections are performed on equipment to determine the progress of wear in its components and subsystems. When wear has advanced to a degree that warrants correction, maintenance is performed on the equipment to rectify the worn condition. The corrective maintenance work can be performed at the time of the inspection or following the inspection as part of planned maintenance. The decision of when to perform the corrective repair related to PM inspections depends on the length of shutdown required for the repair.

Consideration is given to availability of parts and the impact of the shutdown on the operation versus how immediate is the need for repair. If it is judged that the worn component will probably continue to operate until a future repair can be scheduled, major repairs are postponed until they can be planned and scheduled.

Periodically, a PM inspection is made. If the inspection reveals serious wear, some maintenance operation is performed to restore the component or subassembly to a good state of repair and reduces the probability of failure. A PM system increases the probability that the equipment will perform as expected without failure until the next inspection due date or the planned corrective repair.

Determining the interval between inspections requires considering the history of maintenance for the equipment in each unique operation. Time between PM inspections (intervals) will ultimately be guided by a number of resources. These include manufacturer's recommendations, feedback information from repair history of breakdowns, and the subjective knowledge of the maintenance craftsmen and supervisors who maintain the asset daily. Equipment operators may also be a good source of information in some operations.

A central characteristic of PM is that in most major PM applications, the asset must be shutdown for inspection. For example, a heat exchanger must be shutdown and isolated when a nondestructive eddy current inspection is made on its tubes. The inspection process would require a discrete amount of downtime for the unit, which is typical for PM.

The loss of operational time when significant PM inspections are made is one of the reasons that PM programs are often less than successful. This is especially true in applications where there are few redundant units and equipment must operate at 100 percent of capacity. In some situations the loss from shutdown is considered too high a penalty and PM inspections are resisted. Chapter 20 iprovides a more detailed look at this best practice area and a strategy to help you make PM a success in your organization.

Predictive Maintenance (PdM)

In contrast to PM, PdM is a condition-based system. PdM measures some output from the equipment that is related to the degeneration of the component or subsystem. An example might be metal fatigue on the race of a rolling element bearing. The vibration amplitude produced by the rolling element as it passes over the degenerating surface is an indicator of the degree of severity of wear. As deterioration progresses, the amplitude of vibration increases. At some critical value, the vibration analyst concludes that corrective action should be taken if catastrophic failure is to be avoided.

PdM usually permits discrete measurements which may be trended compared to some predefined limit (baseline), or tracked using statistical control charting. When an anomaly is observed, warning is provided in sufficient time to analyze the nature of the problem and take corrective action to avoid failure. Thus, PdM accomplishes the same central objective as PM.

Confidence in the continued "on specification" operation of the asset is increased. By early detection of wear, you can plan for and take corrective action to retard the rate of wear, or to prevent or minimize the impact of failure. The corrective maintenance work restores the component or subassembly to a good state of repair. Thus, the equipment operates with a greater probability of trouble-free performance.

The enhanced ability to trend and plot numbers collected from PdM measurement give this method greater sensitivity than traditional PM methods. The technique yields earlier warning of severe wear and thus provides greater lead-time for reaction. Corrective actions may be scheduled so that they have minimum impact on operations.

A principal advantage of PdM is the capacity it offers the user to perform inspections while the equipment is operating. In particular, in order to reflect routine operating conditions, the technique requires that measurements be taken when the equipment is normally loaded in its production environment. Because the machine does not need to be removed from the production cycle, there is no shutdown penalty. The ability to conduct machine inspections while equipment is running is especially important in continuous operations such as in utilities, chemical, and petrochemical manufacturing.

Another advantage of PdM is that the cost of surveillance labor is much less than the cost of PM activities. Although the technical knowledge required for PdM inspections is usually higher than those for PM, the inspection time required per machine is much less. With PdM, the machines do not have to be disassembled for inspection. For example, with vibration analysis, 50 to 60 machines may be inspected in a single day using modern computer data collectors. When comparing cost advantages of PdM over PM, consider production downtime costs, maintenance labor costs, maintenance materials costs, and the cost of holding spare parts in inventory.

If PdM methods are superior to PM, why use PM at all? The answer is simple. The nature of your operation will determine which methods are most effective. In actual practice, some combination of PM and PdM is required to assure maximum reliability. The degree of application of each will vary with the type of equipment and the percent of the time these machines are operating.

Pumps, fans, gear reducers, other rotating machines, and machines with large inventories of hydraulic and lubricating oils lend themselves to PdM surveillance methods. On the other hand, machines such as those which might be involved in high-speed packaging may be better inspected using traditional PM methods. Machines which have critical timing adjustments, which tend to loosen and require precision adjustments, or have many cams and linkages which must be reset over time, lend themselves to PM activities.

The strategy for selecting the appropriate preventive or predictive approach involves the following decision process:

- Consider the variety of problems (defects) that develop in your equipment.
- Use the predictive method if a predictive tool is adequate for detecting the variety of maintenance problems you normally experience. One or a combination of several PdM methods may be required.
- Use PM if it is apparent that PdM tools do not adequately apply. Inspection tasks must be developed that reveal the defects not adequately covered by PM.
- After you have decided the combination of inspection methods, then determine the frequency at which the particular inspection tasks must be applied.

Some equipment will be satisfactorily monitored using only PdM. Other equipment will require PM. Ultimately, some combination of methods will provide the required coverage for your operation to assure reliable performance. In most operations, it is wise to apply a combination of methods to ensure that equipment defects do not go undetected.

The following provides a brief summary of some of the leading PdM technologies.

Vibration analysis. Today, electronic instrumentation is available that goes far beyond the human limitations with which the old time craftsmen had to contend when trying to interpret vibration signals with a screwdriver-handle-to-the-ear method. Today's instruments can detect with accuracy and repeatability, extremely low-amplitude vibration signals. They can assign a numerical dimension to the amplitude of vibration and can isolate the frequency at which the vibration is occurring. When measurements of both amplitude and frequency are available, diagnostic methods can be used to determine the magnitude of a problem and its probable cause. When you use electronic instruments in organized and methodical program of vibration analysis you are able to:

- Detect machine problems long before the onslaught of failure
- Isolate conditions causing accelerated wear
- Make conclusions concerning the nature of defects causing machine problems
- Execute advance planning and scheduling of corrective repair so that catastrophic failure may be avoided
- Execute repair at a time which has minimum impact on operations

Shock pulse. Shock pulse is a method of surveillance that is specific to rolling element bearings. Since rolling element bearings (sometimes referred to as antifriction bearings) are so common in machines, the method has many applications. A secondary, but important, feature of the shock pulse method is that it permits maintenance workers to judge the adequacy of the lubrication program applied to this type of bearing.

Spectrometric oil analysis. Oil in machines carries the products of deterioration resulting from wear and mechanical failure. By analyzing the oil resident in a machine, or the debris the oil carries, predictions can be made about the state of health of the machine. The critical measurement reflecting the condition of machine wear is the amount of microscopic metal wear particles that are suspended in the oil system of the machine. The spectrometric oil analysis process is a laboratory technique, which uses various instruments to analyze a used oil sample from a machine. The spectrometric result is compared to a baseline level of metal found to be typically suspended in the oil under normal operating conditions. When the wear is meaningful, the sample will show high levels (in parts per million) of wear metals compared to the baseline oil sample.

Standard oil analysis. In addition to the spectrometric analysis, the oil laboratories also check the oil using common oil analysis techniques. For example, the oil is usually checked for viscosity. Other types of standard oil analysis might include total acid number, percent moisture, particle count (for hydraulic systems), total solids, percent silicon (representing dirt from the atmosphere in the form of silicon dioxide or perhaps just from an additive).

Ferrographic oil analysis. Ferrography provides the maintenance manager with two critical sets of decision-support information—condition monitoring that prevents unnecessary maintenance and precise trend information that allows maintenance to initiate repairs before equipment failure. This is accomplished by the analysis of wear particles in lubricants to determine their size, distribution, quantities, composition, and morphology—form and structure. Ferrography is a technique that provides microscopic examination and analysis of particles separated from fluids. Developed in the early 1970s as a PdM technique, it was initially used to precipitate ferrous wear particles from lubricating oils magnetically. This technique was used successfully to monitor the condition of military aircraft engines. That success has led to the development of other applications, including testing of fluids used in vacuum pumps within the semiconductor industry.

Infrared thermography. The use of infrared thermography has grown significantly in the past 10 years. Equipment is easier than ever to use and more effective. The real power of thermography is that it allows quick location and monitoring of problems. It then presents critical decision-making information in *visual* form, making it easy for management to understand. Infrared imaging systems, as they are generally called, produce a picture, either black and white or color, of the invisible thermal patterns of a component or process. These thermal patterns, when understood, can be used to monitor actual operating conditions of equipment or processes.

For instance, viewing a thermogram (heat picture) can clearly show the heat of a failing bearing or a pitted contact on a disconnect switch. Today's sophisticated imaging equipment is capable of acquiring a thermal video. This allows us to see *dynamic* thermal patterns—of a casting process or a belt-wear pattern in real-time for instance. When teamed with the power of computer-based analysis systems, it can go one step further to compare and trend the critical thermal changes that often precede equipment failure or loss of production quality.

Thermography can be used to quickly locate and prevent recurrence in many equipment and process problems such as the following:

- Catastrophic electrical failures
- Unscheduled electrical outages or shutdowns
- Chronic electrical problems in a piece of equipment or process
- Excessive steam usage
- Frozen or plugged product transport lines
- An inability to predict failures accurately
- Inefficient use of downtime maintenance opportunities
- Friction failures in rotating equipment
- Poor product quality due to uneven hearing or cooling or moisture content
- A fire in a wall or enclosed space
- Inability to locate or verify a level in a tank
- Replacement of refractory in a boiler, furnace, or kiln
- A leaking flat roof
- Uneven room temperature affecting product quality or employee productivity
- Trouble locating underground water, steam, or sewer lines

Ultrasonic detection. A variety of tools using airborne ultrasound technology (commonly called ultrasound) have revolutionized many maintenance programs, allowing inspectors to detect deteriorating components more accurately before they fail. Ultrasound, by definition, is beyond the limits of normal human hearing, so an inspector uses a sophisticated detector to transpose ultrasonic signals to the range of human hearing. Fluid and gas systems and other working machinery have constant ultrasound patterns. Changes in the "sonic signatures" can be recognized as wear in components. An ultrasonic detector senses such subtle shifts in the "signature" of a component and pinpoints potential sources of failure before they can cause costly damage.

Maintenance Storeroom Operations

MRO parts, materials, and supplies are the key material resources necessary for the execution of the maintenance process. Often the physical storage, control, and procurement of MRO items never get recognized as to its value to the total operation. The effective operation of a maintenance storeroom is a cornerstone for maintenance excellence, but it is often a neglected area for management attention.

Best practices for maintenance storeroom operations parallel those for a finished good warehouse in some cases, but nonetheless are distinctively different. For more information on storage and warehousing, material and information handling systems, and warehouse management, readers should also refer to the respective chapters for these best practices that have general application to the maintenance storeroom operation. The following key strategies should be applied to the maintenance storeroom operation:

- *Professionalism.* The company needs to view the maintenance storeroom as an important activity and not as a necessary evil. Both the dollars invested in maintenance storeroom materials and the impact of downtime have highlighted the need for a more professional approach to maintenance storerooms.
- *Customer awareness.* Successful maintenance storerooms will have a high regard for the customer, know the customer requirements, and consistently meet these requirements. The right materials will be available at the right time, in the right quantity, at the right location.
- *Measurement.* The maintenance leader will establish storeroom standards, performance will be measured against these standards, and timely actions will be taken to overcome any problems. Performance reports will be distributed to management on a monthly basis.
- *Operations planning.* Systems and procedures will be put into effect that allow the storeroom manager to proactively plan the storeroom operations as opposed to reactively respond to external circumstances.
- *Materials planning.* Systems and procedures will be put into effect that will ensure having the right materials on-hand in the right quantity at the right time. The materials planning systems will provide for good inventory rotation, a minimum of stockouts, the elimination of obsolete materials, and the addition of new items when new equipment and systems are installed.
- *Adaptability.* Maintenance storeroom facilities, operations, and personnel must become more adaptable. The pace of the storeroom will continue to increase; reduction of lead times, shorter equipment lives, increased inventory turns, more *stock-keeping units* SKUs, and more customer demands require that storeroom adaptability be present to satisfy customers.
- *Uncertainty.* All certainty must be minimized; all interactions with the maintenance storeroom must be based on meeting expectations. No surprises.
- *Integration.* The activities within the maintenance storeroom will be integrated (from storeroom item identification to item issue), and the

maintenance storeroom will be more integrated with the overall organization.

- *Material control.* Maintenance storeroom facilities, procedures, and systems will be designed to provide for the control of all materials. The importance and enforcement of maintenance storeroom security will be widely understood and accepted.
- *Maintenance information system.* The maintenance information system will support exceeding customer expectations. The information system will include the maintenance catalog system and the management of inventory.
- *Inventory accuracy.* Cycle counting will be used to manage inventory accuracy, and accuracy above 98 percent will be the norm.
- *Space utilization.* Space will be more efficiently and effectively utilized.
- *Housekeeping.* Quality housekeeping will be a priority. These will be an acceptance of the fact that there is efficiency in order.
- *Human resources.* A priority in the maintenance storeroom will be establishing a positive culture and the training and education required to achieve quality maintenance.
- *Team players.* Everyone associated with the maintenance storeroom and MRO procurement functions within the organization must be integrated into a single service-providing activity.

Storeroom inventory management

It is also important to establish an inventory planning methodology. As part of this, a determination must be made of what to stock, what the inventory policy will be and how the inventory will be managed. A barcoding system could be just the solution for improving accountability and accuracy. This would be just one part of a strategic plan to introduce many new technologies for a major upgrade of the maintenance storeroom.

Inventory accuracy is a must for a maintenance storeroom. It must be 98 percent or more. If it is not, the craftspeople will bypass the storeroom and order a new part. It is critical that they have confidence in the accuracy. In order to attain this level of accuracy, it is necessary to cycle count. Timely detection of errors and correction of causes for the errors is essential to good control. All storerooms must have a proper cataloguing system as a permanent record of all storeroom items and as a tool for identifying and locating items.

A maintenance storeroom strategic master plan is a prerequisite for success. There is efficiency in order. We must know what is to be done and what is the proper order to do it. It all starts with a plan, and successful planning requires teamwork. Phillip Slater provides his thoughts on the five levels to world's best practice inventory management.*

The five levels to world's best practice inventory management

For many companies the key MRO inventory issue is availability, and so long as they have an item when it is required they care little about the cash investment. However, this approach will not maximise your ROI and, in almost all cases, cannot be financially justified on any level. This is because the excess inventory investment that this approach generates provides little or no value to your business. The excess is invested in inventory that does not move or becomes obsolete.

World's best practice inventory management demands that the "management system" is optimized, not just the inventory. It is in this field that best practice can be both easily identified and readily achieved.

Each level on the ladder to world's best practice provides a greater degree of control and management but is only at level 5—system optimization that the management system is optimized. By reaching this level companies can reduce their inventory investment, freeing up cash, and achieve their desired availability levels.

The five levels to world's best practice inventory management are:

Level 1—ad hoc. Purchases are made on an "as needed" basis. At this level there is little control necessary as inventory is expensed when purchased and used immediately. While this may seem to reduce the cash investment it may not reduce the total cash expenditure. This approach can only be viable if the items are available "instantly" and the cost of a "stock out" is negligible.

Level 2—storage. Inventory is expensed when purchased and stored for use but not strictly controlled. Similar to above except that items are stored because of the cost of a stock-out. This approach appears to solve one problem but it raises two others. Firstly, total expenditure is likely to increase as items are purchased in "economic quantities." Secondly, without controls there is little opportunity for review and development.

Level 3—capitalization. Inventory is capitalized and is subject to some level of control, either manual or software based. This approach is by

---

*Phillip Slater is the author of the book *Sustainable Inventory Reduction and the developer of the Inventory Cash Release*™ System a best practice approach to inventory management and reduction. For more information visit his website at http://www.InitiateAction.com, Initiate Action Pty Ltd, 49 Vida St, Aberfeldie Vic 3040 Australia, +61 3 9331 3181, Email: info@InitiateAction.com far the most popular as it appears to provide the required mix of availability and control. Unfortunately, most organizations use their software solely for counting and accounting. There is a strong reliance on human calculation of inventory requirements, but often little review of outcomes. The result is likely to be good availability but a significant overinvestment in inventory and high levels of obsolescence.

Level 4—software optimization. Inventory is capitalized and stock levels are optimized based on a risk/return algorithm. This is the basis of most software solutions. Most software packages will incorporate the ability to automatically adjust the required stock levels based on the history of demand and supply. Very few companies actually use this feature because they know that they cannot trust the results. This is not due to a software flaw, but because the supply and demand may not represent typical usage.

Level 5—system optimization. Inventory management minimises the overall cash investment without an increase in risk. This is the world's best practice. At this level, all of the factors that influence the actual inventory investment are reviewed on a regular basis. This review is manageable because it is limited to the "vital few" items that have a real impact on the level of investment. Inventory levels are adjusted to take account of changing needs and this minimizes the likelihood of obsolete inventory.

Any company that already has the software required for level 3 can achieve level 5—world's best practice. What is needed is the knowhow, policy development, measures, and reporting required to take a company to level 5, not more software. Once these key issues are addressed you are implementing a true management system. Software only goes to level 4, it is the management system that provides the bridge to level 5.

MRO Materials Management and Purchasing

Excellence in corporate-level and plant-level purchasing is an important contributor to success within your total maintenance and asset management operations. Likewise, total operations success must have high-quality, mission-essential maintenance to survive profitability. The purchasing and procurement function is essential at all levels. Corporate, division, or plant site purchasing professionals must understand their key role and strive for purchasing excellence for raw materials, finished goods for distribution and MRO-related (maintenance repair operations) parts, supplies, equipment, and third-party services.

At one point during my career, I had direct responsibility for overall materials management and plant purchasing within a world-class, hand tool-manufacturing plant—Crescent-Xcelite. Never once during this time did I have to refocus attention toward expediting MRO items or checking delivery status of spare parts. We instead focused on achieving just-in-time delivery from sister plants, specialty steel mills and other raw material vendors. Expediting large forgings, which we got from our sister plants, however, was a severe challenge that took all the focus we could find.

At this plant, we were progressing rapidly toward world-class status in both maintenance and manufacturing. Our progress at this point in time did not require the typical purchasing support needed for a purely reactive, fire-fighting maintenance operation.

- Our engineering manager and his maintenance team in turn allowed valuable purchasing resources to focus on our closed-loop MRP II implementation.
- Our maintenance leaders had their acts together in regard to MRO vendors and what they needed through MRO vendor contracts, blanket orders, and consignment stocking of MRO items.
- The storeroom was reliable and we trusted the inventory stocking decisions and accuracy.
- We knew and trusted our vendors for spare parts for our precision machine tools and automated forging equipment.
- We knew that Chambersburg would service their "impactors" and provide JIT parts service for their number one, worldwide site for automated forging.
- We also trusted our reliable contract PdM service provider to collect accurate vibration data.
- We believed the PdM findings when they were bad and took action to address recommendations to avoid catastrophic failures.
- We were not perfect but we did not waste valuable purchasing power and time because of the MRO materials management and storeroom process.

The Storeroom: The Foundation for Effective Maintenance

Our plant's maintenance storeroom was not a planning afterthought, stuck in a dark corner of the plant, tucked strategically in available attic space, or put on a one-half story mezzanine built high, dry, and tight against the plant ductwork system. Our proactive maintenance strategy eliminated expediting "hot shot" MRO requisitions so we did not help UPS and FedEx build their business in rural South Carolina.

This picture is not normal. In many operations, purchasing can be perceived as the villain, someone else to blame where reactive run-to-failure strategies for mission-critical maintenance are present. Purchasing must be a solutions provider working with maintenance for effective asset management. Purchasing must also understand that MRO inventory items and MRO procurement are different in many ways than procurement of raw materials to make a finished product. For instance, see the following.

The high cost of low-bid buying

There is a high cost associated with a low-bid buying strategy for purchasing MRO items as well as contract maintenance services. Some of the tragic airline accidents of recent years can be attributed directly or indirectly to this strategy for using low-cost contracted maintenance service. Very simple but important, low-grade nuts and bolts have contributed to serious catastrophic failures in refineries and chemical processing plants. The mechanical integrity of normally "Class C" inventory items-fasteners must be "Class A" quality in corrosive, high-pressure systems. Purchasing must listen, take the proper actions and do the right thing when the maintenance leader pleads for quality MRO parts and materials. We gamble with all three elements of overall craft effectiveness (1) craft utilization, (2) craft performance, and (3) craft service quality when we compromise the quality of maintenance parts, materials, and tools.

MRO items and raw material procurement are different. Purchasing must understand the basic differences related to MRO items and raw materials. The following are key areas related to MRO items:

- Generally more MRO SKUs with higher relative value
- Lower inventory turn rates in range of 2 to 3 times per year
- Clearly defined technical specifications not understood by the typical nonmaintenance-oriented buyer
- Quality inspection of inbound parts and materials can be required
- Shelf life can attribute to obsolescence for milk, fruit, belts, and other MRO items
- *Just in time* (JIT) vendor service essential if a run-to-failure culture exists
- Total operations quality and through-put suffers without quality parts to support quality maintenance service and physical asset care
- Low-bid buying of subquality parts and material can be fatal
- Maintenance can and must be a key player on the MRO purchasing team

- Purchasing must get technical help from maintenance to support MRO procurement
- Maintenance must pull its share of the MRO database challenge

The life-cycle information loop

Information links between purchasing, the MRO storeroom, the maintenance planning process, MRO vendors, equipment designers/manufacturers, and the craftspeople should be developed. Purchasing must be a part of what I call the "life-cycle information loop" for equipment assets and component parts of the respective assets we must maintain. For example:

- Equipment designers/manufacturers need feedback information on reliability and maintainability issues. Basic warranty information often fails to be captured at the shop-floor level. Root-caused failures related to design problems do not flow easily back to the *original equipment manufacturers* (OEMs).
- Purchasing and maintenance play key roles in the "life-cycle information loop."
  - Supporting CMMS databases with reliable vendor performance data, quality information, lead-times reliability, costs, alternate sources, and the like
  - Developing partnerships with MRO vendors and OEMs with key maintenance leaders on the team
  - Requiring via the initial vendor or OEM contract that:
    - Parts manuals, equipment drawings, and specification packages come with each new asset or major subsystem modification
    - Recommended spare parts lists are also included
    - Recommended PM/PdM procedures are included

Teamwork works for achieving purchasing excellence

Two key internal team relationships are very important contributors to purchasing excellence: the maintenance storeroom and maintenance planning and scheduling functions.

- *The MRO storeroom.* Effective purchasing is one of the keys to an effective storeroom. Purchasing must get the right MRO items into the inventory management system. Purchasing and the MRO storeroom must share and maintain a common inventory management database. The inventory management module of a CMMS or of an enterprise system must be responsive to both purchasing and to maintenance users. Vendor data, lead times, and status of purchase orders are typically a purchasing challenge to maintain. Most of the rest falls in the maintenance arena. The MRO inventory database item master must be accurate and up-to-date regarding MRO parts used on physical assets, the accuracy of on-hand balances, assess control, and the day-to-day storeroom operations. For run-to-failure operations, purchasing can often stay on the "hot seat" expediting emergency MRO purchases at high UPS and FedEx shipping rates. Within a proactive planned operation, purchasing can achieve its function hand-in-hand with the planning and scheduling process with parts and materials staged and kitted for large and small planned jobs.

- *The maintenance planning and scheduling process.* All maintenance repairs are planned one way or another. Response to a reactive, run-to-failure strategy can cost three to five times more for procurement, craft labor, and unplanned capacity loses. Systematic planning and scheduling by qualified planners is a world-class maintenance best practice in small, medium and large maintenance operations. Proactive maintenance requires effective planning and reliable support from MRO purchasing and vendors. Purchasing and maintenance planners must work as a close-knit team to anticipate MRO parts and materials for projects and shutdowns. Planners must work with the internal storeroom and purchasing to get the right parts at the right time for day-to-day planned repairs, as well as major projects or shutdown-related work.

There are two key external team relationships that are essential to purchasing excellence for MRO items.

- *MRO vendor partnerships.* Partnering is still a key buzzword today and is a strategic initiative for smart organizations within a peaceful global society. Today's *information technology* (IT) gives us the Internet (and Intranets) to make sharing information convenience from anywhere. Vendors give electronic parts manual and catalogs. *Electronic commerce* (EC) and *electronic data interchange* (EDI) help share information of all types: purchase requests, receipt payments upon delivery, inventory level data sharing with automatic replenishment by vendors, vendor stocking plans, and vendor consignment stocking. Many of these progressive partnering practices are now becoming standard practices, enhanced by today's IT advances. Purchasing, maintenance, and the respective vendors must work as a total team to make these practices happen at the shop-floor level. As we enter the new millennium, there are few excuses for not having effective partnerships and strong strategic alliances with MRO vendors.
- *Partnering with OEMs.* Many organizations view the original equipment manufacturer as a key part of the total team already. OEMs must provide a reliable source of repair parts, service, technical

advice, recommended PM procedures, recommended spare part stock levels, and warranty support. OEMs must be a key part of the team providing reliable equipment and reliable engineering drawings, electrical diagrams and schematics, parts manuals, and repair and troubleshooting manuals. Purchasing must fully understand the support maintenance required from its base of OEMs. If your organization has extensive custom equipment, designed internally by corporate engineering in technical cooperation with subsystem OEMs, you must have a sound internal partnership with corporate engineering too.

Develop win-win alliances with vendors

The marketplace and maintenance operations of the future will be dominated by organizations that successfully build partnerships and strategic alliances today. Partnerships with MRO vendors, OEMs and yes, corporate- or plant-level engineering are all essential ingredients for operational success and CRI. True partnering is people working together, internally and externally, for a mutual benefit and a two-way profit picture. Companies create partnership agreements, but people create the partnership. True partnerships can shift the focus of the customer/supplier relationship from adversarial to a win-win situation.

Maintenance and purchasing need stronger win-win relationships. There must be a desire for a long-term relationship based on trust. Purchasing must understand the challenges and technical requirements that the maintenance leader faces each day. Both must understand today's IT technology, and the purchasing and maintenance best practices that are available to support total operations success.

Never before than in today's trend toward downsizing and "dumbsizing" has maintenance needed to assume the attitude that it is truly a "profit center." Also:

- Purchasing must accept its role for achieving MRO materials management success.
- Purchasing must understand the technical challenge of asset management and maintenance.
- Maintenance must accept and go by the purchasing rules and systems.
- Purchasing and maintenance will both be profit centers if continuous improvement achieves reliable internal/external partnerships.
- Total operations excellence requires a true partnership and win-win alliance between maintenance and purchasing.

The purchasing and procurement function are essential at all levels. Quality vendors and OEMs must be trusted and reliable team members. Corporate, division, or plant site purchasing professionals must understand their key role in total asset management. Each level must strive for purchasing excellence of raw materials, finished goods for distribution, MRO-related items, and third-party services to support maintenance. When purchasing wins, maintenance and the total operation win. The total operation wins with long-term profitable survival and better total asset management.

Planning and Scheduling

Planning for maintenance excellence requires planning at the strategic level and at the shop-floor level. This section introduces the need for implementing the maintenance best practice for planning and scheduling at the operational, shop-floor level. Surveys consistently show that only about 30 to 40 percent of an 8-h craft day is devoted to actual hands-on wrench time. Without effective planning and scheduling, maintenance operations continue to operate in a reactive, fire-fighting mode wasting their most valuable resource—craft time. Gambling with maintenance costs is not an option for today's organization that wants long-term survival and profitability. Achieving world-class status requires a world-class maintenance operation. World-class maintenance requires strategic planning, especially at the shop-floor level. The effective planning and scheduling of the most valuable maintenance resource, craft skills, and labor can provide an important step forward in a strategic maintenance plan.

Effective planning and scheduling improves the *overall craft effectiveness* (OCE)—the factor which focuses upon measuring and improving the value-added contribution that people assets make to total asset management. There is also a very real concern within many areas of the United States and the world about the availability of craft skills. Scarce technical resources and craft skills are terrible things to waste because they are so hard to find and keep. Review the three key elements for measuring OCE, and see how they align very closely with the three elements for determining the OEE factor for equipment assets.

The OCE factor. Measuring and improving OCE is one of the key benefits from maintenance planning and scheduling. The OCE factor includes three key elements very closely related to the three elements of the OEE factor.

Overall craft effectiveness (OCE)	Overall equipment effectiveness (OEE)
Craft utilization (CU)	Availability/utilization
Craft performance (CP)	Performance
Craft methods and quality (CM&Q)	Quality

An effective planning and scheduling function requires that reasonable estimates and planning times be established for as much maintenance work as possible. Because maintenance work is not highly repetitive, the task of developing planning times is more difficult. However there are a number of methods for establishing planning times for maintenance work.

Planning times are essential. Planning times provide a number of key benefits for the planning function. First, they provide a means to determine existing workloads for scheduling by craft areas and backlog of work in each area. Planning times allow the maintenance planner to balance repair priorities against available craft hours and to realistically establish repair schedules that can be accomplished as promised. Second, planning times provide a target or goal for each planned job that allows for measurement of craft performance. Here we are not as concerned with measuring individual craft performance, but rather with the overall performance of the craft workforce as a whole. Individual craft performance can be determined by comparing the group performance to individual performance over a period of time. Training needs are normally identified when individual craft performance is consistently below the group norm.

Getting started with planning and scheduling. Select from within the maintenance workforce the best-qualified candidate possible. Normally 1 planner per 25 to 30 craft personnel is sufficient. The planner position requires knowledge of existing equipment repair needs and a strong craft skill background. It also requires new skills for planning/scheduling, estimating, parts coordination, computer use, personal relations, customer service, and so forth. Sufficient time must be invested in formal training of the planner(s) and for on-the-job training to get the planning function off to a smooth start. Since the planning function is often the focal point for successful CMMS utilization, planners must understand all functions of the CMMS and be capable of helping to train others in the organization.

Focus on customer service. The ultimate success of maintenance planning and scheduling will be determined by whether or not the customer is satisfied. All preliminary work to develop a plan and to coordinate the scheduled repairs is wasted if execution of the schedule does not occur as promised. The customer (operations) will determine the true success of the planning process. The entire maintenance workforce must understand their service role to operations. As a formal planning process is implemented, an increased focus on customer service must be established. Operations will expect better service, and maintenance must commit to providing it.

Measure effectiveness of the planning function. The measurement process should start first within the planning function. Develop and use performance measures to evaluate the return on investment and the effectiveness of the planning function.

Reliability-Centered Maintenance (RCM)

The evolution of RCM

In the early 1960s, the developmental work for RCM was done by the North American civil aviation industry; the airlines began to see that many of their maintenance philosophies were not cost-effective. But most importantly they did not achieve the best possible conditions for safety. The airline industry then put together a series of "*maintenance steering groups*" (MSG) to reexamine all aspects of aircraft maintenance operation. Representatives from aircraft manufacturers, airlines, and the Federal Aviation Authority were members of the MSG team. The Air Transport Association formulated maintenance strategies in 1968 known as MSG 1, later refined to MSG 2 1970.

In the mid-1970s, the U.S. Department of Defense commissioned a report written by Stanley Nowlan and Howard Heap of United Airlines with the title "Reliability Centered Maintenance" to define state-of-the-art strategies for maintenance within the airlines industry. Published in 1978, it is still one of the most important documents in the history of physical asset management.

A considerable advance in thinking, well beyond MSG 2 was achieved in Nowlan and Heap's report. It was a traditional value engineering approach. It was used as a basis for MSG 3 in 1980 that was revised in 1998 (Rev 1) and in 1993 (Rev 2). It is still used to develop prior-to-service maintenance programs for new aircraft types such as the Boeing 777 and the Airbus 330/340. RCM was progressing fairly well for Boeing when our scoreboard assessment was performed

Nowlan and Heap's report and MSG 3 have since then been used as a basis for various military RCM standards, and for nonaviation derivative programs with acronyms such as FMECA, MSG3, TPM, RCA, RBI, and RCM2. CRI uses the best from RCM, TPM, and the total operations approach to leadership-driven, self –managed teams.

Overview of the RCM process

The key elements of the RCM process include the following:

- Analysis and decision on what must be done to ensure that any physical asset, system, or process continues to do whatever its users want it to do. Includes essential information gathering.

- Define what users expect from their assets in terms of primary performance parameters such as output, throughput, speed, range, and carrying capacity.
- As applicable, the RCM 2 process defines what users want in terms of risks, process and operational safety, environmental integrity, quality of the output, control, comfort, economy of operation, customer service, and the like.
- Identify ways in which the system can fail to live up to these expressions (failed states) and failure consequences.
- Conduct *failure modes and effects analysis* (FMEA) to identify all the events, which are reasonably likely to cause, each failed state.
- Identify a suitable failure management policy for dealing with each failure mode in the light of its consequences and technical characteristics. Failure management policy options include:
  - Predictive maintenance
  - Preventive maintenance
  - Failure-finding
  - Change the design or configuration of the system
  - Change the way the system is operated
  - Run-to-failure (if preventive tasks are found)

TPM is a maintenance improvement program concept and philosophy that resembles *total quality management* (TQM). The TPM movement that has excellent objectives:

- Zero unplanned downtime
- Zero defects
- Zero machine capacity losses
- Zero accidents
- Minimum life-cycle asset care cost

It requires a total commitment to the program by upper-level management and employees to be empowered to initiate corrective action. It is a long-term strategy, so a long-range outlook must be accepted. TPM may take more than a year to implement and is an ongoing process. Changes in employee mind-set toward their job responsibilities must take place as well. TPM brings maintenance into focus as a necessary and vitally important part of the business. It is no longer regarded as a nonprofit activity.

To successfully apply TPM concepts to plant maintenance activities, the entire workforce must first be convinced that upper-level management is committed to the program. Typically a TPM coordinator is hired or recruited to sell the TPM concepts to the workforce through an extensive training program. To do a thorough job of educating and convincing the workforce that TPM is just not another "program of the month," will take time, perhaps a year or more. Once the coordinator is convinced that the workforce is sold on the TPM program and that they understand it and its implications, the first study and action teams are formed. These teams are usually made up of people who directly have an impact on the problem being addressed. Operators, maintenance personnel, shift supervisors, schedulers, and upper management might all be included on a team. Each person becomes a "stakeholder" in the process and is encouraged to do his or her best to contribute to the success of the team effort.

Usually, the TPM coordinator heads the teams until others become familiar with the process and natural team leaders emerge. The action teams are charged with the responsibility of pinpointing problems areas, detailing a course of corrective action, and initiating the corrective process. Recognizing problems and initiating solutions may not come easily for some team members. They will not have had experiences in other plants where they had opportunities to see how things could be done differently. In well-run TPM programs, team members often visit cooperating plants to observe and compare TPM methods, techniques, and to observe work in progress. Publicity of the program and its results are one of the secrets of making the program a success.

The initial stages of TPM will include taking the machine out of service for cleaning, painting, adjustment, and replacement of worn parts, belts, hoses, and so forth. As a part of this process, training in operation and maintenance of the machine will be reviewed. A daily checklist of maintenance duties to be performed by the operator will be developed. Factory representatives may be called in to assist in some phases of the process. After success has been demonstrated on one machine and records began to show how much the process had improved production, another machine is selected, then another, until the entire production area has been brought into a "world-class" condition and is producing at a significantly higher rate. This is one of the basic innovations of TPM. The attitude of "I just operate it!" is no longer acceptable. Routine daily maintenance checks, minor adjustments, lubrication, and minor part change out become the responsibility of the operator. Extensive overhauls and major breakdowns are handled by plant maintenance personnel with the operator assisting in some cases.

Operator-based Maintenance (OBM)

Pride in ownership and the process of OBM started early in American history. It is also a key element of TPM where it is called *autonomous maintenance.* OBM is also a key element in the very successful and effective maintenance program of all U.S. armed services.

OBM began with the entrepreneurial spirit of the cottage industry in the United States' free enterprise system. In most cases, the owners fixed the saw mills, the cotton gins, the printing presses, and the wagon wheels. Owners were the skilled craft people working within a free enterprise culture and system. For many Americans, OBM started with Henry Ford's Model T that came with tools: a Crescent wrench, slip joint pliers, screwdriver, and hammer, for the do-it-yourself repairs.

Early American culture did not normally employ the philosophy of, "I Own—You Fix—You Operate." Back then it was an attitude of, "We are all responsible for the equipment." However simplistic it may have been prior to the industrial revolution, it was a matter of necessity and survival, just as it is in a combat zone.

Ironically, we have returned to the world-class attitude toward maintenance in many organizations being, "We are all responsible for our equipment." The key word in the TPM world-class attitude is "our"; *our* equipment not *the* equipment. These objectives can and will be achieved in varying degrees for organizations that include OBM as part of the total operations/maintenance strategy. The following is a basic strategy for developing successful OBM:

- Start with an overall strategic maintenance plan, and include defined goals/objectives for OBM within this plan down to the operational or shop-floor level.
- Understand that OBM is a deliberate process for gaining commitment by operators toward:
  - Keeping equipment clean and properly lubricated
  - Keeping fasteners tightened
  - Detecting symptoms of deterioration
  - Providing early warning of catastrophic failures
  - Making minor repairs and being trained to do them
  - Assisting maintenance in making selected repairs
- Start with the first-things-first. Provide the necessary communication between maintenance operations and the rest of the total operation to gain the commitment and internal cooperation needed to start OBM.
- Clean the equipment to like-new condition, make minor repairs, and develop a list of major repairs for the future.
- Utilize leadership-driven self-managed teams with whatever team names that evolve. Do not use self-directed teams with true leadership as a missing link. Names, for example, can be "equipment improvement team," "SWAT team," or "CRI team."

- Develop a written and specific team charter (examples included in App. F and G).
- Avoid using self-directed teams with no technical leadership "driver" for the process.
- Have teams evaluate/determine the best methods for operator cleaning, lubrication, inspection, minor reapers, and level of support during major maintenance repairs.
- Develop standard written OBM procedures for operators and include in the "quality operations and maintenance guide." We will discuss in the "Boeing Commercial Airplane Group" case study in Chap. 18.
- Evaluate your current PdM and PM procedures and include those that the operator can do as part of OBM.
- Document start-up/operating/shutdown procedures along with set-up/changeover practices.
- Consider quality control and safety requirements.
- Document operator training requirements and what maintenance must do for OBM support.
- Develop operator certification to validate operator-performed tasks.

There are many organizational roadblocks to effective OBM. However, the roadblocks to the world-class attitude, "We are responsible for our equipment" exist only by self-imposed limitations we create by our less than optimal attitude toward maintenance. It has been said many times, throughout history, starting with the red lettered words in the Bible; "whatever the mind of man can conceive, can be achieved." So as per your choice, it becomes a successful OBM implementation.

Chapter

17

Maintenance Quality and Customer Service

Maintenance must be just as concerned about customer service and quality as companies such as Ford and Toyota. First, to look at the definition of quality, let us look at a couple of examples from the countries of these two car manufacturers. For that small island nation southeast of Russia, the definition of quality in the 1990s was very simply "satisfying the customer." The United States' most consistent definition is "the conformance to requirements." The combination of these two definitions provides yet another definition of quality: Quality is the conformance to customer requirements.

The customers for maintenance are everywhere in the typical plant operation and facility complexes. And customers are at all levels. They include the equipment/machine operators and supervisors of the operation who depend on equipment for production or service. They include all employees in offices and operational areas that depend on maintenance, students in dormitories, and fans in the stands at this year's NCAA Final Four basketball tournament and the masters.

Let us now use a brief analogy and take a look at some of the basic elements of quality for a car and see how these same items, presented in David Gavin's book *Managing Quality: The Strategic and Competitive Edge,* apply to maintenance.

For example, if we look at the basic elements of quality for a car, the customer's concern might sound something like this:

1. *Performance.* Does it have good acceleration, handling, and good gas mileage? Do I really need a sports car or a sport utility vehicle?

2. *Features.* Does it have air conditioning, four doors, four-wheel drive, or power windows? Is it big enough for the family? Do I really need a sunroof?
3. *Reliability.* Does it have an extended warranty, will it start in cold weather, and will the engine hold up? Should I really buy an American car this time?
4. *Conformance.* Has this car been assembled well and thoroughly inspected, or was it put together on the last shift prior to the Christmas holidays?
5. *Durability.* Will it last longer than my last car payment? Will the paint hold up? Will it outlast my teenager's learning permit/license?
6. *Serviceability.* How far will I have to take it for repairs? Will they have really qualified mechanics and the right parts to keep it serviced? Could my wife change the oil on this one, too?
7. *Aesthetics.* Does the styling of the automobile fit me and do I like it? Will the body style change soon?
8. *Perceived quality.* What do I think about the overall quality of this car? Over the years I have come to think of the names Cadillac and Honda as meaning quality in cars. When I think of quality in trucks, I think of Ford. But, was I brainwashed on quality about my last car through advertising?

Now let us focus on maintenance quality from the customer's point of view using these same eight elements of quality.

1. *Performance.* The ability of maintenance to make planned repairs with a firm schedule that is responsive, reasonable, and reliable. Repairs are performed within a reasonable period of actual time that is safe, productive, and effective.
2. *Features.* The maintenance craftsperson is multiskilled, versatile, and capable of performing repairs in more than one craft area. Maintenance is adaptable to support new technologies, new equipment, variable workloads, and emergency situations.
3. *Reliability.* Maintenance is able to do the job right the first time. It is able to anticipate and prevent unexpected breakdowns through effective *preventive/predictive maintenance* (PM/PdM). Well-trained and reliable craftspeople are able to troubleshoot nonroutine equipment problems for primary causes. Maintenance information is reliable, accurate, and timely; most of all, it is able to improve the level of service to maintenance customers.
4. *Conformance.* Maintenance is able to conform to its established repair schedules. It can achieve the total scope of service required of the

PM/PdM program. It conforms to the recommended best practices for lubrication, PM services and equipment condition monitoring. In general, maintenance continually seeks new ideas and conforms to the best practices within the maintenance profession. Maintenance conforms to all local, state and federal safety, environmental, and energy regulations.

5. *Durability.* Maintenance realizes the high cost of low-bid buying of parts and materials. Quality parts and supplies that are cost effective are obtained from the best sources. Parts usage and reliability are evaluated continuously. Maintenance insures that the right parts are available at the right time and place to meet the repair schedules and the critical needs of specialized equipment.
6. *Serviceability.* Maintenance and operations/operators work as a team to improve equipment effectiveness. The factors of maintainability, redundancy, modularity, and reliability are considered for new and rebuilt equipment to increase uptime and serviceability.
7. *Aesthetics.* The craftsmen are proud of themselves as well as their profession, tools, shop, and storeroom operations. They strive to keep all clean and in good order. They clean up the customer area after all repairs and set the standard for a clean, safe work environment.
8. *Perceived quality.* The attitude, appearance, and actions of maintenance leaders and craftspeople set the tone for the perceived quality of maintenance, either good or bad. Results from good (or bad) maintenance in the past leads to good expectations (or bad) from the customer about future maintenance service. If the quality of maintenance is perceived to be good, then chances are that there is, in fact, good maintenance and quality service.

The customer forms the perception of maintenance quality as either good or bad. Maintenance quality can go from a questionable perception to a positive expectation with a strategy of continuous reliability improvement. With dedicated actions, positive expectations can become "maintenance excellence."

All successful programs to achieve maintenance excellence must involve the maintenance customer. Positive interactions with the customer to define improvement opportunities result in joint cooperation to implement solutions. Ideas to improve maintenance customer service must start with the customer.

Maintenance leaders must understand and continuously communicate with the customer. They must listen and they must act to satisfy customer needs. They must become quality conscious and customer-service oriented just as if they were a business. In fact, maintenance is a business within a business. Let us look closer at one of my favorite areas—the maintenance storeroom and MRO materials management. Here Phillip Slater provides five reasons why it takes more than software to make best practice in this important area.*

It Takes More Than Software to Reach Best Practice

Businesses around the world spend millions of dollars on software and inventory management systems in an effort to maximize their *return on investment* (ROI) from inventory. Until now even the most sophisticated of these systems left businesses way short of best practice. In fact most of these systems institutionalize excess inventory.

The problem is that most software relies on optimization and this limits the opportunity to reduce inventory because it ignores external influences. Software can only optimize the values it has, not what could be.

World's best practice inventory management demands that the "management system" is optimized not just the inventory. Most inventory software takes today's data and runs an algorithm to optimize holdings. What they miss are the changes in the management system that could further reduce the total level of investment. This flaw makes software systems self-limiting in their results.

Inventory management is much more than just the software system. Inventory management is the combination of know-how, process, measures, and reporting that together provide the opportunity for maximizing availability while minimizing cash investment.

The five reasons why your inventory management is not best practice and is costing you money are:

1. *The responsibilities are misaligned.* The people that make the day-to-day decisions will typically not be responsible for the working capital outcomes; they will be responsible for availability. The problem is that if you run out of stock all hell breaks loose, but if you overstock there is no repercussion. This is especially the case with indirect inventory that is not subject to the usual planning scrutiny. Given this, what do you think most people do? That's right, they overstock!
2. *The optimization is incomplete.* Sophisticated software can track all sorts of data, and in many cases the software can make optimization decisions based on that data. This can reduce your inventory but it

*Phillip Slater is the author of the book *Sustainable Inventory Reduction and the developer of the Inventory Cash Release* ™ *System*, a best practice approach to inventory management and reduction. For more information visit his website at http://www.InitiateAction.com. Initiate Action Pty Ltd, 49 Vida St, Aberfeldie Vic 3040 Australia, +61 3 9331 3181, Email: info @InitiateAction.com.

is self-limiting. The problem is that software optimizes only on known data and ignores process and behavioral changes that can impact that data. This is software optimization not system optimization. The software should only be a tool within a bigger process of optimization.

3. *It is managed reactively.* Inventory is often seen as "set and forget," that is, once the item is optimized for the current situation the requirements are not systematically revisited. It is often only when there is a "cash crunch" or some other emergency that action is taken. Yet, even indirect inventory can represent millions of dollars of investment and deserves frequent attention. When action is taken it usually addresses the highly visible items rather than the real "cash burners."
4. *There is a significant time lapse before problems emerge.* The number one question asked about inventory is "what do I do with slow moving or obsolete stock?" Depending upon the accounting policies in your company this stock has taken three to five years to reach the point where that question is asked. By this time it often seems irrelevant to revisit the original decision or processes that produced this result. No one would accept this approach to quality management! No one ever asks "how do I prevent the accumulation of slow moving or obsolete stock?"
5. *It is painful to fix and easy to ignore.* In most cases the removal of obsolete inventory will result in a "hit" to the profit and loss account. However, if a reason can be found to justify it for another year then few will argue. Eventually someone is going to have to make a decision and it will be painful. For this reason, obsolete inventory decisions are often driven by the opportunism of results reporting rather than good management principles.

To truly achieve best practice your organization must review these issues and develop systems that will minimize their impact or eliminate them altogether.

Summary: The elements of quality in this chapter's brief analogy represents only a part of the indicators for evaluating quality and customer service in today's maintenance operation. Take the time to develop your own definition of maintenance quality in your operation. Develop your own specific indicators for customer service. Set high standards for maintenance customer service and develop your own measures for quality.

At the same time, involve your customers in developing the indicators and measures. Keep track of results, publicize the successes, and work as a team with your customer to correct the weaknesses. The customer will be a valuable partner in your path forward to greater customer service and maintenance excellence. A team-based approach to continuous reliability improvement with a strong partnership with the customer will succeed.

Chapter

18

Case Study—Critical Asset Facilitation: A Lesson Learned at Boeing Commercial Airplane Group

This case study will provide some important lessons learned for the reader to consider when conducting the asset facilitation process for critical plant equipment. Asset facilitation is enhanced tremendously when a mature and effective *computerized maintenance management system* (CMMS) is in place. On the other hand, asset facilitation will help build strong support and expanded use of CMMS functional capability for a new system. We first saw asset facilitation at the *Boeing Commercial Airplane Group* (BCAG) in Wichita, KN. Dennis Westbrook their focal engineer for their *advanced maintenance process* (AMaP) was leading this effort along with many more Boeing staff in support. Our role included the total assessment of all BCAG maintenance operations in the five manufacturing regions at that time. So we evaluated the asset facilitation process when we conducted "The Boeing Scoreboard for Maintenance Excellence" assessment with the custom Boeing benchmark evaluation items included in this customized Scoreboard. As a result we also tried this Scoreboard customizing process for a large three-plant operation fabricating an important maintenance tool—the ladder. This chapter shares those results and recommendations as a case study at XYZ Company. Sergio Rossi, an experienced electrical engineer from Argentina, now living in Dallas, Texas was the project engineer who coordinated this effort. Sergio along with Roland Newhouse were on the team for the Boeing "Scoreboard for Maintenance Excellence" benchmark assessment discussed previously in our lesson learned.

Summary of Overall Continuous Reliability Improvement Savings and Gained Value				
Extrusion	Plant A	Plant B	Plant C	Savings and Gained Value
Extrusion utilization cost savings (5%)	$369,066	$205,616	$453,666	$1,028,347
Pultrusion	Plant A	Plant B	Plant C	
Pultrusion utilization cost savings (3%)	N/A	$191,767	$56,389	$248,156
Extension ladders	Plant A	Plant B	Plant C	
Ext ladder utilization cost savings (3%)	$122,187	$159,430	$147,628	$429,245
Summary of savings:				
Asset utilization improvements:				$1,705,748
Original asset utilization savings estimated by XYZ company				–$174,380
Net asset utilization savings due to CRI improvements				$1,531,367
Estimated savings obtained from facilitated asset process				$132,000
Total craft & asset utilization savings from asset facilitation process & CRI				$1,663,367

Figure 18.1 Summary of benefits for XYZ Company.

The *asset facilitation process* at the XYZ Company included three metal fabrication plants with aluminum extrusion machines as the very first manufacturing process. At this point in time, maximum production was needed and the extruders (four to six per plant) were the primary capacity constraints. Also ladder assembly was another critical production operation. Two pilot assets were selected for each of the three plants for a total of six mission-essential assets: one extruder and one ladder assembly operation per plant. The 20-week duration of project time that it took to implement this process provides a good indication of the effort that the XYZ Company team committed to complete the undertaking. The asset facilitation process for two critical asset per plant was just one of the many manufacturing improvements underway as part of an overall *continuous reliability improvement* (CRI) program. Figure 18.1 provides a summary of overall benefits for the XYZ Company.

A Strong Internal Technical Team Is Required

This case study is not just an illustration of support, where an external consultant did all the work. Most of the work was done by XYZ Company people at all levels from production operators and craftspeople on up the ladder to engineering and quality staff. Results achieved at the XYZ Company were a total team effort and is just another example where a strong team effort was needed. It was originally intended to implement this process in a 4-week span at each site. Based on heavy production schedules and engineering projects, it was determined critical to spread the 4 weeks evenly and rotate the visits among the different sites to allow XYZ Company more time to complete the different and numerous tasks assigned.

Understand benefits, purpose, and results

The initial challenge for the XYZ Company was to understand the purpose that the *quality operations and maintenance guide* (QO&MG) provides to each and all departments and to each operator, craft and operations supervisor, and to the craft supervisor. Once the asset facilitation process was understood and how the QO&MG book related to greater output with better quality, the cooperation and ideas for improvement began to flow. This process also creates a change in culture. Some of the old "keep-the-secret-for-myself" attitude slowly disappeared and the process took off almost immediately.

A 43-item project task list was provided to track process' progress and to obtain a better understanding of those areas in need of improvement. The project list was used as a status reporting tool to manage the asset facilitation process details while it was being simultaneously started at all three sites. The progress for each site was documented, trended, and communicated to those responsible for the process development and completion. Attention was focused on the growth rate of each plant's *facilitated asset benchmarking baseline curve*. Because an evaluation like this was a subjective process, a single consultant, Sergio Rossi evaluated the progress to compensate for normal process variations.

Asset facilitation process results

Once completed, the QO&MG were taken to the shop floor in Plant B where it was tested for content, helpfulness in running the line, and troubleshooting. The feedback we received was very positive. The guidebook was very much appreciated by extruder operators, maintenance and quality technicians, supervisors, and managers who valued the worth and many applications of this process. A testimonial was even presented in our report to the XYZ Company that showed how the benefits of this process are perceived and highly appreciated by the XYZ Company personnel. The following results at the XYZ Company were directly related to the asset facilitation process:

1. *Visual management (VM) implementation.* This included a broad spectrum of visual management tools to improve the communications between different groups at the plant such as electronic boards, marquis, and others. VM included labeling of main piping systems, labeling of all controllers and all disconnects for safety reasons. VM techniques can be applied to sensors, gauges with upper and lower

ranges, *programmable logic controller's* (PLC's) *inputs and outputs* (I/O), cabinets, motor phases inside control cabinets labeled according to phases A, B, C, and ground or color coded (National Electric Code) to produce savings.

2. *Preventive maintenance* (PM) analysis completed and PMs updated.
3. Asset facilitation process for pilot extruder completed and can be easily applied to others at each plant.
4. Asset facilitation for pilot ladder machines completed and can be easily applied to other ladder machines at each plant.
5. Developed "QO&MG" for each site for both pilot assets.
6. Provided total operations and maintenance specific guidelines and procedures for:
   a. Equipment safety and housekeeping requirements
   b. 5S started for each facilitated asset (5S form in App. I)
   c. Asset documentation and drawings references were all located and or updated for use by all
   d. PM procedures renewed
   e. *Predictive maintenance* (PdM) opportunities identified and reliability equipment requirements identified
   f. Setup/changeover requirements established
   g. Setup/changeover validation process initiated
   h. Operator-based inspection renewed
   i. Start-up requirements by maintenance refined and reestablished
   j. Operator training included in QO&MG
   k. Shutdown procedures established as part of operator training
   l. All quality control requirements documented
7. *Operator-based maintenance* (OBM) tasks established to include:
   a. PM tasks and lubrication services
   b. Maintenance tasks to be performed by the operators
   c. Operator training for OBM tasks
   d. Certification process for production tasks
   e. Operator training/certification for maintenance of OBM tasks
8. *Reliability centered maintenance* (RCM) seminar tailored and presented at all three plants to the asset facilitation team.

Review of the PM analysis process

The PM analysis process served as the basis for developing a comprehensive PM program to include the power of machine reliability improvement predictive technologies and their applications to machines under the QO&MG process. The technologies for improving the reliability of machines were being contracted out by the engineering department and were not included in the PM list.

PM review teams. The PM review teams at each site were in the process of updating the PMs while Sergio Rossi was implementing the QO&MG process. This allowed us to evaluate both, the past tasks and the updated tasks, as they were being changed. We had the chance to analyze the PM tasks completed by Plant C and by Plant A for use at Plant B that was the first plant to begin. The PM analysis was tailored for each plant to comprehend the balance of task workload being performed by operators and maintenance personnel.

A complete review and close inspection of current PM tasks was done for each of the two-pilot assets to better understand the type of PMs performed and their effectiveness. Experienced XYZ Company maintenance personnel recommended and demonstrated how each new PM was to be performed. At this point we defined how the selected PM tasks could be transferred to production operators. This gave us a better understanding of current maintenance practices, allowed us to assess the needs for training, and permitted us to suggest some generic PdM tasks to improve the efficiency and effectiveness of work performed.

After evaluating the existing PM procedures, we found that revised PM tasks were much improved from the original ones. PM procedures and task descriptions that use to be broad based and generic, "go check tasks" were turned into a more detailed and specific "go do action," which could then be utilized to build asset history in the CMMS.

PdM tools. A large number of PdM tools could be utilized. Some, such as the *man-machine-interface* (MMI), were currently being used at XYZ Company to a certain extent to define punch press PM based on actual use of the asset. Parameters such as temperature and pressure were also being tracked via very extensive process controls for each extruder. Other key process factors were obtained only sparingly and not frequently enough to make them valuable. The PM-analysis phase of the total asset facilitation process considered the application of today's best reliability technology. Simple reliability technology instruments such as oil-condition analyzers, ultrasonic data, vibration overall meters were available at a very low cost and required little training. These tools have been implemented very successfully with training implemented properly and their applications understood.

OBM became an additional best-practice option

One very important best-practice application came as a result of the asset facilitation process. OBM was definitely a possible strategy for the XYZ Company where operators could be trained on well-defined PM tasks. Many other world-class operations have involved operators to cope with the increased number of daily tasks demanded by more complex and sophisticated machines. OBM is not the total maintenance of the asset but very well-defined tasks. When trained properly it was felt that extruder and ladder assembly operators could carry out a number of well-defined tasks with continued support from maintenance as required. With operators taking responsibilities for daily PM tasks, maintenance personnel then would be able to concentrate on more technically oriented tasks for this critical task.

Ten Important Lessons Learned

The role of the operator did change to more of a "pride in ownership" attitude as they progressed with a more involved, OBM-oriented role. This change must be properly addressed and implemented jointly between maintenance and production to obtain the highest benefits. Ten important lessons learned about OBM, during this asset facilitation project at XYZ Company, are included below:

1. Training must be launched simultaneously with the development of the program. We must keep in mind that, the more complex the tasks transferred, the more training will be required.
2. Operator's training must be performed at different levels and everyone should be involved.
3. Management must understand the benefits as well as the new responsibilities of the operators and craft people.
4. The corporate- and plant-level environment must be set for the OBM events to occur almost naturally.
5. Management must support the transferal of knowledge from one group to another and emphasize training for those people who do not welcome change.
6. Management must communicate the purpose of OBM to all employees and support must be shown at the floor level to demonstrate and prove that management is not just providing lip service.
7. Maintenance must lead and coordinate the training to ensure a smooth transition. Supportive external classes may be necessary to supplement in-house training. Continuous assessment of operator's understanding of new tasks is extremely important. This classroom training must be enforced with *hands-on-training* (HOT) and *competency-based development* (CBD). The QO&MG book should contain all the information needed for this training.
8. Operators must be trained to the point that they feel comfortable transiting to the new assignments. It is only after they reach this level of confidence and comfort that results are obtained as expected. A clean working environment is the first step for successful OBM.

9. Ideas and suggestions by operators must be accepted by all with open communications between operators and maintenance.
10. Operators must feel and appreciate the improvements and OBM results obtained from their efforts.

Asset Facilitation Supports QS9002 Compliance and Technical Imaging

The XYZ Company has achieved QS9002 compliance. In order to satisfy rigorous quality control requirements for documenting manufacturing procedures, the XYZ Company has established corporate-wide manufacturing, quality, engineering, and maintenance procedures. Central control of process modifications by corporate engineering is a keystone for QS9002 compliance in the product design or redesign phases. These procedures have taken different shape throughout each site. While the quality of change procedures were consistent at the three plants, there were some differences at all the three plants for maintenance and engineering written documents and procedures.

The QO&MG process contains information on training and troubleshooting on many manufacturing processes. In order to comply with ISO and QS9000 procedures, the QO&MG process must include a controlled change process for document and drawings. The asset facilitation project helped corporate engineering to increase justification and to fully fund their corporate wide program for the *technical document imaging* (TDI). The paper-based asset documentation reference within the QO&MG book later became available throughout the XYZ Company via the TDI system.

Summary

Asset facilitation helps in moving benefits of CMMS to the next level. It requires a total team effort. If a team-based continuous improvement effort for production is in progress with maintenance improvement, asset facilitation flourishes. In our case, we were blessed by having a plant engineer like Trevor Hartland and corporate engineering staff like Tim and Robin Varner located at this plant site. Asset facilitation will produce bottom savings via increases in uptime, throughput, capacity compliance to ISO/QS requirements, operator and craft productivity, quality, and safety. A production asset is a terrible thing to waste when the production orders are there, but capacity is constrained by equipment downtime. Operator and craft assets are also terrible things to waste, and asset facilitation improves their capability to operate and maintain mission-essential production assets.

Chapter

19

PRIDE in Ownership with Operator-Based Maintenance

In July of 1970, *operator-based maintenance* (OBM) first became personally important to me in a very harsh and challenging maintenance environment—South Vietnam. Dump trucks, graders, dozers, asphalt plants, paving machines, jeeps, and weapons did not perform without OBM. So I say that the OBM process is not new, and it did not begin in that "small island nation" as autonomous maintenance under the total preventive maintenance (TPM) banner. OBM was important in "The Nam," just as it was at Gettysburg, and before and after the Normandy Invasion. Today, airline pilots, tank drivers, truck drivers, and equipment operators in progressive organizations throughout the world practice OBM with a sense of personal pride in ownership.

Pride in ownership and the process of OBM started early in American history. It began with the entrepreneurial spirit of the cottage industry. In most cases, the owners fixed the sawmills, the cotton gins, the printing presses, and the wagon wheels. Owners were the skilled craftspeople working within a free enterprise culture and system. For many Americans, OBM started with Henry Ford's Model T that came with tools: a real *crescent* wrench, slip joint pliers, screwdriver, and hammer, for the do-it-yourself repairs.

Early American culture did not normally employ the philosophy of, "I Own—You Fix—You Operate." Back then it was an attitude of, "We are all responsible for the equipment." However simplistic it may have been prior to the Industrial Revolution, it was a matter of necessity and survival, just as it is in a combat zone.

Ironically, we have returned to the world-class attitude toward maintenance in many organizations being, "We are all responsible for our equipment." The key word in the TPM world-class attitude is "our": *our* equipment not *the* equipment.

The TPM movement, which emerged from America's direct support after World War II, has excellent objectives:

- Zero unplanned downtime
- Zero defects
- Zero machine capacity losses
- Zero accidents
- Minimum life-cycle asset care cost

These objectives can and will be achieved in varying degrees for organizations that include OBM as part of the total operations/maintenance strategy. The following is a basic strategy for caring for our equipment assets with OBM:

I. *Understanding, commitment, and communications*
- Start with an overall maintenance strategic master plan, and include defined goals/objectives for OBM within this plan.
- Understand that OBM is part of the continuous maintenance improvement process.
- Understand that OBM is a deliberate process for gaining commitment by operators toward:
  - Keeping equipment clean and properly lubricated
  - Keeping fasteners tightened
  - Detecting symptoms of deterioration
  - Providing early warning of catastrophic failures
  - Making minor repairs and being trained to do them
  - Assisting maintenance in making selected repairs

  Why let an operator sit drinking coffee watching a craftsperson work under extreme pressure? If there are no other machines to operate, let them gain valuable *on-the-job training* (OJT) knowledge helping repair their machine. If this operator owned the machine, you can bet that they would already have it fixed or be working frantically to help maintenance fix it.
- Start with the first-things-first. Provide the necessary communication between maintenance operations and the rest of the total operation to gain the commitment and internal cooperation needed to start OBM.
- Communicate clearly to the labor organization and to your own human resources staff.

II. *Getting started*

- Clean the equipment to like-new condition, make minor repairs, and develop a list of major repairs for the future. This is like getting a vehicle ready to sell yourself so you can then buy a newer truck.
- Utilize teams with whatever team name that evolves. For example, "equipment improvement team," "SWAT team," or "continuous reliability improvement (CRI) team."
- Develop a written and specific team charter. Apps. F and G provide a charter format that can be used.
- Avoid self-directed teams with no technical leadership "driver" for the process—Leadership-driven (and lead), self-managed teams only!
- Have your team evaluate/determine the best methods for operator cleaning, lubrication, inspection, minor reapers, and level of support during major maintenance repairs.
- Your team should also be included and involved with the initial equipment cleaning.

III. *Document the quality process*

- Develop standard written OBM procedures for operators and include in your "quality operations manual."
- Evaluate your current *predictive and preventive maintenance* (PdM/PM) procedures and include those that the operator can do as part of OBM.
- Document start-up/operating/shut-down procedures along with set-up/changeover practices.
- Consider quality control and safety requirements.
- Document operator training requirements and what maintenance must do for support.
- Develop operator certification to validate operator-performed tasks.

IV. *Establish internal benchmarks to measure results*

- Determine the best method to evaluate *overall equipment effectiveness* (OEE) within your organization.
- Measure the OEE factor on the most critical equipment or work center first.
- Establish an OEE-factor baseline. The average OEE-factor baseline is 40 to 55 percent. World class is about 85 percent.
- The OEE factor measures how well a machine is in regards to three critical categories and six major losses associated with the three categories (Fig. 19.1):

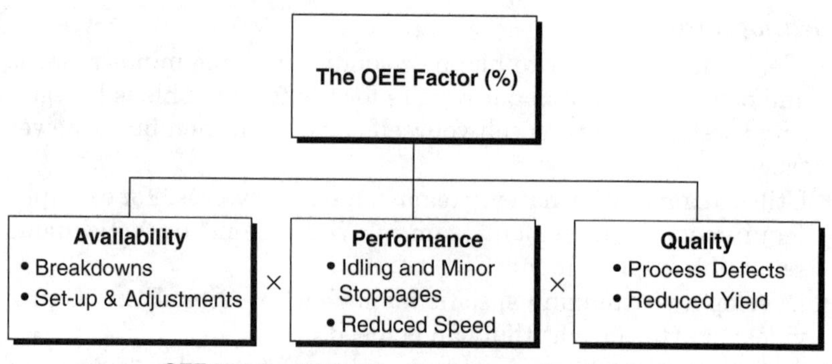

Figure 19.1 The OEE factor.

V. *Continuously measure OEE results*
- Measurement of OEE must be continuous to validate continuous improvement.
- Use the OEE factor as an internal benchmark along with other key performance indicators for measuring your journey to maintenance excellence.
- Develop valid and reliable methods to validate results of OBM and other maintenance improvement activities to leaders in your organization.

VI. *Focus team process on CRI considering*
- Use of reliability improvement technologies, reliability centered maintenance, PdM, knowledge-based/expert systems for maintenance
- Material resources—quality parts, supplies for quality repair
- Information resources—quality information from your CMMS, your vendor, your customer
- Craft resources—quality skills by maintenance to support total operations
- Operator resources—adding value to existing operators, paying additional if justified, and creating *pride* in ownership at the most important level—the shop floor

Forging, cold forming, stamping, die casting, surface mount technology for chip making, and injection molding are some of the most important manufacturing processes. How would the stock racing and the trucking industries survive without die-cast engine blocks, crescent wrenches, and other forged tools? Your organization—whether large, medium, or small—can survive and prosper by doing OBM and overall maintenance one way: the right way.

There are many organizational roadblocks to effective OBM. However, the roadblocks to the world-class attitude, "We are all responsible for *our* equipment" are for the most part self-imposed limitations. They are limitations that we create by our personal attitudes toward maintenance and our attitude toward achieving maintenance and total operations excellence and success. We do have a choice with our personal attitudes, which can be switched immediately and be changed to positive expectations and actions to achieve effective operator-based maintenance.

Ben Gibbs, Founder of GEMS (*Gibbs Equipment Management Services*), now living on the North Carolina side of Lake Gaston, provides some very personal lessons on pride in ownership and pride in maintenance. Ben has over 35 years of fleet management experience including direct responsibility for the largest centrally managed highway equipment fleet (NC DOT) management operation in the United States. He has been recognized internationally and all across the United State for his ability to implement practical solutions to complex and challenging fleet management improvement opportunities.

One area that is often missed is getting your mechanics to have pride in their work. Pride in their work can be assisted by providing the proper test equipment and necessary training to use it. The facility needs to be clean, and parts and supplies readily available. The mechanics need to have some say in improving their work environment and support from their supervision. Mechanics have pride in their ability to perform tasks that their fellow mechanic cannot seem to get, encourage the in-house training aspect of your operation. Establish a tool complement for each mechanic and buy tools for your valuable craft people. That will solve the problem of hiring a mechanic who does not have tools necessary to accomplish your required repairs. Establish a training program calling on the manufacturing department to furnish training in your purchase of equipment and make that part of the contract. Have a suggestion box and recognition program to reward the mechanic for cost improvements, and the like. If the company furnishes uniforms make sure that they are a first-class uniform and do not leave anyone out such as helpers or housekeeping staff, if you do have this in-house support to your shop and office areas. The workplace tools and how well the upkeep of the shop is maintained is very important to developing pride in work. Companies have to supply the mechanic with the parts, supplies, and facility necessary to perform first-class work. Also the supervisors and support people need to feel part of the finished product.

What about OBM! Operators need to have an incentive to look after their equipment, to be part of the maintenance, requirement of that unit of equipment. They have to be part of your total support operation. Example is the Department of Transportation (DOT); they are given a daily checklist to perform every morning before starting that day's work.

They are also required to fill out for performance problems and turn it in to their supervisor.

Another good illustration/story is UPS (*United Parcel Service*) with the "brown trucks." In the past some of their regional maintenance supervisors would arrive in the equipment storage yard early and place a $ 20 bill in the dip stick before the UPS trucks left the yard. Then they would go to the gate and wait until the trucks came to the gate to start their workday. They would stop the trucks they placed the $20 bill in and remove it if they had not checked their oil level. So when they were in the area, the drivers checked their oil, that is, after a few visits by the supervisor placing $20 bills around like the "tooth fairy."

Also while on a visit to a France Construction Company to support training equipment operators, an interesting true story came to light. The trainer told me about his experience with the operators performing maintenance. He said when it was time to go home, a crew of men working with Jackhammers sat down at the job site and started cleaning and servicing the hammers. He questioned the operator why they were doing this work on his time. He was told "If this hammer does not work tomorrow we go home until it is repaired." That goes for the roller and the crawler tractor, and all those heavy equipment operator. Each is responsible for their unit of equipment. If the equipment does not operate there is no pay for the operator. A pretty good incentive for OBM it seems to me!

For those focused on pure fleet maintenance, for those that have a large internal fleet such as the Idaho National Labs near Idaho Falls, ID, Ben also provides some key points on fuel management, an important area for larger fleets and another area related to our scoreboard benchmark category on energy management. Operators can definitely contribute to profit with profit-centered fuel management.

Profit-Centered Fuel Management*

Higher cost—not lower fuel cost

With the rising cost of fuel, management needs to consider an automated fuel control system. There are numerous systems on the market all these systems are not equal, but all offer a true record of fuel use. If you are using in-house fuel, your tanks need to be large enough to purchase tank wagon loads 8000 gallons are better. You need to separate off road fuel for your earth moving equipment and construction equipment. A system that records both mileage and hour-meter reading is preferred. Today you can order fuel electronically from your office and then can track performance of your vehicles with an automated reporting

*This section is contributed by Ben L. Gibbs, Lake Gaston, NC and founder of GEMS.

system. Management can develop fuel consumption averages for the vehicles and run comparisons to monitor performance.

The Smart Card was the way of the future and now is commonplace. These cards (or pens) hold asset information that can be read at the fuel island during dispensing of fuel. The vehicle life history can be imbedded on the chip to include the repair history. This smart card concept is well on its way where your operation will have hours of operation fuel used and mileage of the vehicle transferred to your computer as fuel or service is used—thereby allowing you to schedule services by fuel usage or actual operating time, rather than by calendar days.

Illustrative Example: With fuel cost over $2.50 per gallon if you could save 20% of the cost to purchase an automated system would pay for its self in months. Boeing Commercial Airplane Group purchased an automated fuel control system for its Western Regions operations around Ramond, California. The first week it was in operation the Snack Vender for the Boeing complex made a special visit to the West Region Director of the total complex. He had a complaint about not being able to get fuel like he had been doing for 10 or more years. The new fuel management system only let company vehicles have fuel, not vendors anymore. As they say on TV and radio here in the United States: "Crooks are really stupid!"

Another thing I learned long ago was to monitor fuel deliveries. A clever trick of the truck drivers was to park on an incline which would allow the tank to have 100 or more gallons left on the truck. With a tank monitor you can see what was delivered, not what the bill of laden had recorded from the load point. The carrier tanker truck for the fuel does not work for the fuel company in most cases—he is just a contractor. Crook truck drivers are smarter than most others!

Fuel management enhances preventive maintenance

The first thing to remember about Preventive Maintenance (PM) or what I call Prevent Major repairs is that keeping equipment from running for one hour of PM, may save you days of waiting for repair parts. Repair standards (reliable planning times) are needed to set up a good PM program as well as to keep track of your equipment actual cost.

The North Carolina DOT, Equipment has 500 mechanics working under job standards set up under a MTM-based work measurement system called Universal Maintenance Standard (UMS). This system was implemented in 1973–1976 by myself and Pete Peters.These standards are now maintained on a computer-based system and both PM and repair jobs are given a completion time which is continuously averaged on the computer from 50 jobs of that type. Example if you have 500 one-half ton pick-up trucks, being repaired and PMed, in a few weeks you will have a base period for your job standard and a "reliable planning time" as was discussed previously. This process can be used for all types of maintenance including road maintenance such as (1) Installation of drive way pipe, (2) Paving half mile of secondary road, (3) installing center line reflectors etc. With today's computer technology the life expectancy chart for your equipment assets can also be established. Profit-Centered Fuel Management can capture Fuel Cost, Maintenance Cost, Repair History, and PM History. Below are ten key benefits (things you may want) when considering the purchase of a fuel management system.

1. Wanting accountability for 100% of fuel cost for operating budget
2. Storing and dispensing company-owned fuel environmentally safely
3. Having a management information system (CMSS/EAM) to use data that gives you information by linking the fuel management system to your CMMS/EAM system
4. Wanting to track and then trigger equipment PMs through fuel system and use hours or miles rather than calendar time periods
5. To develop supply-chain partnerships with fuel suppliers that will support and integrate to your system
6. Wanting to use your management information system to track asset performance and life-cycle costs through fuel use
7. Desire for 100% fuel accountability and not having fuel loss due to poor reporting as a continuing write-off to your bottom line (pure profits)
8. Wanting to have rental equipment fuel charged to a specific project
9. Establish the ability to price jobs based on estimated fuel use
10. Having a relatively easy method of environmental reporting of your fuel tank security

Ben Gibb's contribution fits well here just as well as it fits into Chap. 4 in the energy management category. Our son Brian Peters, for example, is the golf course superintendent at a world-class course, The Preserve at Jordon Lake, just west of Raleigh, North Carolina. His recent solution fuel management was an enclosed box at his fuel pump station with a simple form that everyone getting fuel has to fill out. This includes both the course maintenance vehicles as well as all the gas-powered golf carts. Now with the fuel budget under only one budget code, all will be able to see which way the fuel goes: for the customer (the players) or to the golf course maintainers. Again think about OBM with the positive expectations that it will work for your unique maintenance operation. As we said in Chap. 7 about a self-assessment using the "Scoreboard for Maintenance Excellence" . . . *go ahead and just do it*!

Chapter

20

Case Study: Developing an Effective Preventive Maintenance Strategy

Introduction

Due to the unique nature of each organization, it is virtually impossible to apply an "off-the-shelf" program for *preventive maintenance* (PM). We have many CMMS systems to choose from, but when it comes to a PM program we must normally start from ground zero. However, a systematic approach to PM should be taken. We must ensure that there is top leaders' commitment at all levels and be sure we use the basic principles and practices that are tailored to our specific operation. These best practice guidelines are presented as an approach for effectively starting a new PM program. They can also be used for reviving or expanding an existing program as a new CMMS is being evaluated and selected for implementation.

What Are the Choices for a Maintenance Strategy?

Top leaders have two basic choices for an overall maintenance repair strategy. For each asset and its subsystems there are other strategies focused on specific tasks which might well be run to failure if failure consequences are minimal via a "hardcore RCM type analysis." But for simplicity of this chapter on PM, there are two main maintenance strategies, both with shades of grey and black in between. They are as follows:

1. *Run to failure*. Continue to operate with a high level of uncertainty by running equipment or operating facilities at required capacity and shutting down only for emergency repairs. Continue to gamble with maintenance cost and the cost of unexpected downtime.
2. *Planned proactive maintenance with PM and CRI of all maintenance resources.* Reduce the uncertainty of unplanned downtime, emergency repairs with a program of planned inspections, adjustments, lubrications, and testing/monitoring with *predictive maintenance* (PdM) tools and techniques such as CRI of the six major resources areas we have discussed previously.

Planned maintenance with PM and CRI are two obvious first step choices, but they are not always selected. Nor is managing maintenance as an internal business as we have discussed many times in previous chapters. The results of effective PM enhanced by CRI and with an improved *computerized maintenance top leaders system* (CMMS) can be a key ingredient for *return on maintenance investment* (ROMI) for the right maintenance strategy and for your CMMS decision. The approach provided in this chapter can help your operation realize measurable benefits. This chapter is based on an actual case study and it will also help you along the way to use your CMMS as an effective tool for PM and PdM success.

The Recommended Approach

The maintenance leader who is faced with the challenge of changing a top leader strategy of "run to failure" must develop, present, and implement a plan of action, based on well-documented savings and results. The approach must be systematic, comprehensive in scope and technically detailed, and include the following six phases. And very important to success is that all must be tailored to the specific needs of the maintenance operation and operating environment.

Phase I	Top leader awareness
Phase II	Top leader commitment
Phase III	Pilot program design
Phase IV	Evaluate pilot program
Phase V	Expand and operate the total program
Phase VI	Continuous improvement and evaluation of total program

Phase I—Top leader awareness of PM benefits

This is the launch phase of the overall plan and requires that all levels of top leaders and operating personnel be involved. It requires that the maintenance leader step forward and create awareness at all levels for the potential benefits of an effective PM program. It consists of four basic steps.

Step 1: define PM. Maintenance leader clearly defines PM for the specific operation to ensure a complete understanding of what it is and what it is not. PM must be viewed as a key element in both the top leaders and leadership of the maintenance function. Generally, the PM inspection activity is a major source of information for developing planned and scheduled work. In turn, unscheduled and emergency work is minimized, craft labor is better utilized, and downtime costs are reduced.

Step 2: determine downtime level. Current downtime costs and the costs by maintenance to correct breakdowns should be established in as much detail as practical from existing equipment repair records. If possible, a two-year period should be established and compared to the potential savings as if a PM program had been in place. An alliance team member of TMEI, Dr David Worledge, is doing just this with a software program that provides important "what ifs." (If I do the required PMs; what are the benefits? And what if I do not do PM as required, what will be my additional cost? So basically we must be able to develop a valid listing of potential benefits and establish savings projections where applicable. If a program like Dr. Worledge's applies to you, then consider buying and using it for step 2. Then develop the concept of the overall plan of action for new or improved PM and prepare for top leaders' presentation.

Step 3: top leader presentation. Create greater top leader awareness for the potential of PM, review possible savings in downtime and maintenance cost, overall benefits, and the overall plan of action. Focus on gaining a firm commitment from top leaders to proceed to Phase II where more detailed costs/benefits will be presented for a decision on the commitment of resources.

Step 4: establish PM leadership team. A cross-functional team made up of representatives from operating departments, engineering, key top leaders' staff, and representatives from maintenance, including selected craftsmen should be established. This team provides both direct and indirect support to PM development. It monitors and reviews progress and approves selected actions completed by the PM team.

The PM leadership team should also develop and establish a method for measuring PM compliance and a method for measuring the overall success and results of the program.

Phase II—Top leader commitment to PM

This phase involves basic research activities that provide the primary foundation of the PM program. It also includes a more in-depth review of the scope, benefits, and costs. More specific information on costs is provided in this phase so that a firm top leader commitment on resources can be made.

Step 1: determine equipment for pilot program. Approximately 10 percent of all equipment should be identified as top priority equipment for the pilot program. Factors such as criticality of the equipment, level and frequency of downtime, and types of repairs should be used to determine the priority of equipment. In many operations, 10 percent of the most critical equipment will cover 50 to 70 percent of the emergency breakdown problem areas.

Step 2: finalize equipment for pilot program. PM leadership team reviews list of critical equipment, finalizes list, and approves equipment for pilot program.

Step 3: determine level of PM for pilot equipment. If available, review CMMS database and compile available historical repair data on pilot equipment. Develop scope of PM inspection or condition monitoring for pilot equipment. Consideration should be given to questions such as: Should PM include only inspection and no disassembly? Should only completely worn or broken parts be replaced? Is the component to be disassembled for inspection? Is the component to be disassembled for planned replacement of parts? What level of PM can be performed by the operator? The concept of *operator-based maintenance* (OBM) including related PM tasks should be addressed early since this will impact the total craft time required for PM inspections and future planning.

Step 4:determine manpower requirements. Based on the initial estimate of the scope and level of PM for pilot equipment, maintenance establishes its initial manpower estimates. All options for additional manpower are reviewed to include new staffing, overtime, and better utilization of existing craftsmen. If manpower is not available, look to reduce pilot equipment initially to a level that can be handled with existing manpower. Start with something and do not use the excuse "we do not have enough staff."

Step 5: determine projected costs/benefits. With the initial scope and level of PM established along with manpower requirements, a projected cost should be established for the pilot program. Benefits should also be established and/or reconfirmed for the pilot equipment.

Step 6: gain top leader commitment. This step involves conducting a top leader presentation related to activities completed in Phase II and, in turn, gaining a commitment of required resources. Sufficient details about the program are presented and available for a firm commitment of "go" or "no go." Decisions on manpower must be made. A qualified person should be appointed to coordinate completion of the PM design and to administer the program.

Phase III—Pilot PM program design and implementation

This phase involved finalizing the details of items such as PM instructions, inspection intervals, scheduling, and record-keeping methods. The program was designed and tailored to fit the organization and the equipment selected for the pilot.

Step 1:establish PM team. A well-qualified member of the maintenance staff, in this case the maintenance supervisor selected as team leader. As the PM program evolves into full operation, this person would become the overall (PM) supervisor/coordinator. The PM team was a very active, working team that designed the program details and procedures. It included:

- Two well-qualified craftsmen to provide multicraft technical support in mechanical and electrical
- Qualified maintenance planner (if available)
- Engineering support from in-house or from an out-house consultant
- One representative from operations (It could be an operations' supervisor and an operator)
- Also, the team included craftsmen selected to do initial visual inspections (Step 2-B) and the actual inspections as PM instructions/checklists were completed in Step 6

Step 2-A: develop PM instructions/checklists. Information from all available sources was brought together to determine what needs to be included in the instruction/checklist for each piece of equipment. A number of sources were used to include:

a. *Review past experience.* Complete review and analysis of the equipment history file to see what has failed in past and why. Also, the team tried to capture any information from operators and maintenance craftsmen related to past equipment problems and to use any recommendations on what should be included. Information on any previous equipment modifications should also be considered.

b. *Manufactures' recommendations.* This is a valuable source of information but must be viewed in terms that vendors may tend to over-PM to maintain machine reputation or zero downtime. However, this is an excellent source but may not cover all the variable conditions that equipment is subjected to in its current operating environment.

c. *Local distributors.* Most distributors of bearings, pumps, mechanical seals, and lubricants are normally willing to help in providing recommendations in setting up PM programs. Some oil companies will provide a plantwide survey of recommended lubrication services provided their products will be used.

d. *Other plants or operations.* Sharing of benchmark information from other maintenance operations as to their experiences with certain types of equipment and how they are PMing the equipment is often a good source of information. Most maintenance people will share information if not a competitor. In the green industry for golf course superintendents, the open sharing of information is a very important synergy of lessons learned to help the total population of golf course superintendents.

e. *Arbitrary starting point.* Some PMs may require starting based solely on the best available assessment info as to what level of inspection and intervals are required. The criticality of the equipment must be factored into the decision along with the best available technical knowledge base; both the in-house and out-house available knowledge. As experience and equipment history is gained, the PM instructions and intervals can be revised. PM instructions/checklists should be developed based on priority determined by criticality of the piece of equipment.

Step 2-B: start visual inspections. Step 2-A can be a lengthy documentation period where no results are produced until PM instructions/checklists are developed. Conduct common-sense visual inspections concurrent with Step 2-A. Capitalize on gaining results early and maintaining enthusiasm. These initial inspections will also provide valuable information for developing the PM instructions/checklists and will provide a means to test the effectiveness of instructions/checklists as they are developed in draft form. Launching the PM program as soon as possible with visual inspections will provide immediate benefits.

Step 3: develop estimated times. As instructions/checklists are finalized, develop estimated times for each. Preliminary manpower estimates in Phase II, Step 4 should be finalized based on the specific requirements for each PM inspection.

Step 4: establish inspection intervals. The determination of time between inspections should be based on the best information available in Step 2-A. However, the optimum may not be achieved initially. Adjustments to intervals between inspections will be considered later as actual program results are monitored.

Step 5: finalize manpower requirements. Total manpower can now be estimated based on completed instructions/checklists and the recommended intervals for each inspection. At this point, maintenance can estimate total hours for each craft and determine workload weekly, monthly, quarterly, and so on, for a master PM schedule.

Step 6: initiate inspections. As PM instructions/checklists are completed, initiate these in lieu of the visual inspections being done in Step 2-B. Phase in implementation beginning with the most critical pieces of equipment. Coordinate and schedule equipment shutdowns with operations staff. Insure that optimum routing is considered when planning actual inspections. Ensure that PM inspectors are adequately trained to effectively conduct each inspection.

Step 7: report results. Typically, work orders are used for scheduling and tracking the PM inspections by each asset or group of assets. For a production line or series of assets this might be one work order to cover the entire line. Deficiencies found during the inspection, which require corrective repairs, should then be noted by specific maintainable asset or subsystem, and reported so that a work order can be issued for planned repair for that specific asset or subsystem.

The number of repairs generated from the PM program is one measure of success as we will discuss more when we review the *Maintenance Excellence Index* (MEI) in later chapters. The detection of problems in their early stages and corrective repairs to eliminate catastrophic and emergency breakdown should be measured. During the PM pilot program, it is important to keep top leaders and the PM leadership team informed of progress. Defining major problems avoided to top leaders provides an excellent measurement of PM results during the early stages of implementation. If a downtime baseline is available, you then have an important operational benchmark for comparing CRIs.

Phase IV—Evaluate pilot program

This phase includes a complete evaluation of the overall effectiveness of the pilot program. It provides a review of results achieved, costs, procedures, methods, and the like. Further top leader decisions and commitments are made during this phase for expanding the overall program.

Step 1: evaluate pilot results. Based on measures established by the PM leadership team, evaluate overall results of the pilot program. Measures should include:

a. Percentage PM compliance-percentage of time PMs were actually performed as scheduled
b. Number of corrective repairs detected early and scheduled prior to breakdown
c. Significant problems or savings achieved as a result of early detection of problem
d. Level of reduction in equipment downtime
e. PM craft time required versus estimated
f. Increased level of planned work
g. Craft performance/utilization increases as a result of planned work
h. Product quality improvements and any other results that can be contributed directly to the PM program

Step 2: review PM procedures/methods. It is important to continually review PM instructions/checklist for accuracy, completeness, and format. This review process begins in Phase II and should be continued throughout the program. At this point in the pilot program, all additions/deletions should be included along with other changes to the instructions and checklists.

Inspection intervals/frequencies should be adjusted based on pilot program experience and results. The PM inspection routes should be reviewed and verified that the most optimum routing sequence has been established. Methods for doing actual inspections should be reviewed to ensure that the most effective methods are being used with the proper tools and equipment. Using applied common sense is important here at all times.

Step 3: evaluate manpower requirements. At this point a history of actual times for PM inspections compared to estimated times is available along with the total craft time required for the PM workload for the pilot program. The overall impact of the PM workload can be reviewed against other maintenance work. The key question is "Have breakdown repairs been reduced to offset the PM craft hours being used?"

Step 4: determine scope of total program. The remaining equipment is reviewed to determine scope of total coverage. Priorities should be established based on the same factors used to determine pilot program equipment (criticality, downtime level, costs, and so forth). This step should provide a prioritized listing of the remaining equipment that is recommended to be phased into the PM program over time.

Step 5: determine costs/benefits. Based on experience from pilot program and other available baseline information, the projected costs/benefits for a total PM program should be developed. Initial projections on reduction of downtime, emergency repairs, and the like, should be fine-tuned based on results of the pilot program.

Manpower requirements should be estimated based on the scope of the total program. Consider the fact that phasing in new equipment can be planned and not necessarily added all at one time. OBM and selected PM inspections should also be considered.

Step 6: develop plan of action and recommendations. This step requires consideration of many factors. First, has the pilot program achieved its intended results? Does the pilot program need more time for refinement? If Steps 1 to 3 provide positive results and Steps 4 and 5 support continuation, a plan of action is the next step. Each maintenance operation is different and each plan of action will vary. However, it is critical that top leaders be provided a complete evaluation of the pilot program and a recommendation with a plan of action for further PM expansion.

Step 7: top leaders commitment to total program. Top leaders and PM leadership team are presented the results from a thorough evaluation of the pilot program. A complete review of the plan for expanding the program is presented along with projected costs and benefits. Required resources are clearly defined and firm commitments are pursued.

Phase V—Expand and operate the total PM program

This phase requires the same basic steps as outlined in Phase III—pilot PM program design. The PM team should remain in place as the primary working group to develop PM instructions and checklists for additional equipment. Based on the priority of equipment, visual inspections should be phased in prior to final PM instructions and checklists being developed. Phase V is oftentimes a "trial and error" period in terms of matching available manpower with expansion of PM inspections to all other equipment.

- *Manpower.* The scheduling or phase in of new equipment must match available PM manpower if manpower is a severe limiting factor. Neglecting equipment already under the PM program to expand PM inspections to new equipment will only weaken the entire effort.

- *Continue development of PM instructions/checklist.* The effort to develop the PM instructions/checklists should continue based on priority of equipment even though manpower limits restrict starting actual inspections on some equipment.
- *Decisions on additional manpower.* They should be made or delay phase-in until breakdown repairs are reduced to provide additional craft time for expanded PM inspections. Integrating the PM workload into the normal maintenance schedule without added craftsmen, at best, is a challenge. As we say in the south, it is putting 10 lb of cow manure in a 5-lb bag. However, if the objectives of the ongoing PM program are achieved, planned work replaces inefficient emergency work and time becomes more available to invest in expanded PM work.
- *Continue to evaluate results.* Evaluating progress and results should be a continuing process. Measures established to monitor the PM program should be kept in place. Periodic reports to top leaders and the PM leadership team should be made to review progress of the initial plan and to resolve problem areas such as available manpower to accomplish new PM inspections.

Phase VI—Continuous reliability improvement and evaluation

This phase runs concurrent with Phases IV and V and continues throughout the life of the PM program. It is well defined and focuses on a philosophy of CRI that is based on a continuous evaluation of the actual PM program effectiveness.

Phase VI can include:

1. Focus on eliminating causes of problems rather than detecting problems early before breakdown.
2. Implementation of OBM to include selected PM tasks.
3. Continuous review of instructions, checklists, and inspections intervals for improvements.
4. Continuous monitoring of PM program measures.
5. Application of PdM tools and techniques on selected pieces of equipment.
6. Evaluation of computerized PM as part of overall CMMS (if not currently using CMMS).
7. Clear documentation of benefits projected in previous phases.
8. Develop valid justification for craft labor resources needed to expand or enhance program.

Summary: 23 Key Points to Ensure a Successful PM Program

A successful PM strategy must be implemented to achieve results as Yogi Berra might have said. Consider these key points in your plan of action for PM and CMMS implementation.

1. Develop complete understanding about PM benefits throughout the organization.
2. Create a complete awareness of PM benefits and costs.
3. Develop an understanding that savings through PM take time and that initial maintenance costs will probably go up. You find more, fix it, find more, and fix it at greater cost than doing nothing.
4. Develop a realistic plan of action.
5. Gain a total commitment from top leaders.
6. Involve and gain support/cooperation from "customers" of maintenance.
7. Establish a PM leadership team and a working PM team. Again this is a chartered leadership-driven self-managed team and not a committee or a self-directed group of people.
8. Use *all* available sources of information to develop PM instructions/checklists.
9. Reach agreement with "customers" on scope and frequency of services.
10. Start with the most critical equipment—10 percent of PM could be 50 to 70 percent of major problems. Do the hardest thing first: the second, third, and so forth will be easier.
11. Develop manpower requirements early to determine overall PM workload. As was noted several times in previous chapters, this is very important. Defining total maintenance requirements to top leaders with a clear, honest message from "the maintenance messenger" must be done by the maintenance leader(s).
12. Agree on methods to measure PM compliance and overall success of PM efforts, and continually measure progress.
13. Ensure that the top leader acting as financial officer understands! Also ensure that the top leader acting as financial officer's boss also understands!
14. Launch the program as soon as possible with visual inspections.
15. Achieve early results and publicize them widely to gain credibility and momentum.
16. Fine-tune, formalize, and continuously improve as you go.

17. Keep top leaders and PM leadership team informed of progress and problems with customer acceptance.
18. Celebrate the point in time where the cost of carrying out the PM program is more than offset by the documented savings gained through its use. But celebrate anywhere along the way when you tear down fortifications in the minds of your customer, top leaders, and even naysayer craftspeople.
19. Establish PM as a prominent, highly visible function within the maintenance operation and organization..
20. Continually review manpower requirements and PM workload as PM inspections and coverage increases.
21. Focus on solving and eliminating the root causes of chronic problems and not just providing early detection, and prediction of problems that continue. Use valid value engineering techniques, work simplification, and full strength or 2-percent fat RCM as required.
22. Integrate the PM program with OBM and transfer-selected PM tasks to a trained operator.
23. Develop a philosophy of CRI and evaluation and upgrade with reliability centered maintenance (RCM) technologies anything new that will work for you and your unique maintenance environment.

Chapter

21

Today's Predictive Maintenance Technology: Key to Continuous Reliability Improvement

Predictive maintenance (PdM) generally follows the *preventive maintenance* (PM) implementation in most organizations. PdM has come a long way from a craftsperson listening to bearing noise through a screwdriver's tip placed on the bearing housing. Progressive organizations with new operation start-up include PdM applications (condition-based systems) concurrent with PM (interval-based system) as part of the total system to reduce downtime and emergency repair.

This chapter provides the maintenance leader or team leader with a general guide to today's PdM technology. It is by no means all inclusive and serves only to give an appreciation for the power of technology related to predicting the need for repair and the condition of critical equipment. Teamed with effective PM, the organization that applies PdM technology wisely and effectively is preparing for proactive maintenance excellence and not gambling with maintenance costs with a reactive breakdown maintenance strategy.

The purpose of this chapter is to provide a better understanding of the potential for using PdM technology. It will also provide a basic strategy for putting PdM technology to work as part of a practical maintenance and reliability program.

It is difficult to describe PdM without contrasting it to PM. Both have the objective of extending the total useful life of the equipment and to detect defects and equipment problems before they lead to catastrophic failure and major repairs.

Difference between PM and PdM

The difference between preventive and predictive maintenance methods may be simply stated as the difference between an interval-based system (PM) and a condition-based system (PdM).

PM—An interval-based system

PM is an internal-based surveillance method in which periodic inspections are performed on equipment to determine the progress of wear in its components and subsystems. When wear has advanced to a degree which warrants correction, maintenance is performed on the equipment to rectify the worn condition. Corrective maintenance work can be performed at the time of the inspection or following the inspection as part of planned maintenance. The decision of when to perform the corrective repair related to PM inspections depends on the length of shutdown required for the repair.

Consideration is given to the impact of the shutdown on the operation caused by the repair versus how immediate is the need for repair. If it is judged that the worn component will probably continue to operate until a future repair can be scheduled, major repairs are postponed until they can be planned and scheduled.

Periodically, a PM inspection is made. If the inspection reveals serious wear, some maintenance operation is performed to restore the component or subassembly to a good state of repair and reduce the probability of failure. A PM system increases the probability that the equipment will perform as expected without failure until the next inspection date.

Determining the interval between inspections requires considering the history of maintenance for the equipment in each unique operation. Time between PM inspections (intervals) will ultimately be guided by a number of resources. These include manufacturer's recommendations, feedback information from repair history of breakdowns, and the subjective knowledge of the maintenance crafts and supervisors who daily maintain the asset. Equipment operators may also be a good source of information in some operations.

A central characteristic of PM is that in most major PM applications, the asset must be shut down for inspection. For example, a heat exchanger must be shut down and isolated when a nondestructive eddy current inspection is made on its tubes. The inspection process would require a discreet amount of downtime for the unit, which is typical for PM.

The loss of operational time when significant PM inspections are made is one of the reasons PM programs are often less than successful. This is especially true in applications where there are few redundant units and equipment must operate at 100 percent of capacity. In some situations the loss from shutdown is considered too high a penalty and PM inspections are resisted. An example is the world's leading ladder manufacturer, Werner Company and their Anniston, Alabama operation. Maintenance finally got one of three extruders shut down for 8 h and promised to operations that it would be down for only 8 h! However, because this was needed, shutdown was not allowed for "literally years" by production, the needed repairs and overhaul of subsystem exceeded 8 h. When operations finally understood why more downtime hours were needed, they then saw greater capacity by "paying now with more needed overhaul hours and gaining many more uptime hours of capacity later."

However, the need for downtime during significant PM inspections/overhauls, should not be construed to mean that PM is not a valid and effective reliability tool. PM, when properly applied, unquestionably increases overall equipment availability. You must be aware, however, that in exchange for reducing breakdowns, there is a price that must be paid in addition to the maintenance-related costs. This cost is expressed in terms of lost productive time for the asset during the inspection. It is the *mean time for preventive maintenance* (MTFPM). The total price paid for PM equals MTFPM plus maintenance labor and materials cost involved in the PM inspection process. This may be described as the *total price for PM* (TPFPM).

PdM—A condition-based system

In contrast to PM, PdM is a condition-based system. PdM measures some output from the equipment that is related to the degeneration of the component or subsystem. An example might be metal fatigue on the race of a rolling element bearing. The vibration amplitude produced by the rolling element as it passes over the degenerating surface is an indicator of the degree of severity of wear. As deterioration progresses, the amplitude of vibration increases. At some critical value, the vibration analyst concludes that corrective action should be taken if catastrophic failure is to be avoided. Or, PdM might measure internal pipe corrosion within wastewater treatment facilities and help avoid failure of sewage and grey water pipes along a clean mountain stream avoiding a major fish kill.

PdM usually permits discrete measurements which may be trended and compared to a baseline or defined limit and tracked using statistical control charting. When an anomaly or major variance is observed, warning is provided in sufficient time to analyze the nature of the problem and take corrective action to avoid failure. And if you never knew that extreme corrosion occurs inside piping systems and you never learned about the current PdM detection methods now available, then you could be in serious trouble as a wastewater treatment supervisor or technician. Thus, PdM accomplishes the same central objective as PM.

Confidence in the continued "on specification" operation of the asset is increased. By early detection of wear, you can plan for and take corrective action to retard the rate of wear, or to prevent or minimize the impact of failure. The corrective maintenance work restores the component or subassembly to a good state of repair. Thus, the equipment operates with a greater probability of trouble-free performance.

The enhanced ability to trend and plot "real information" collected from PdM measurements give this method greater sensitivity than traditional PM methods. PdM technologies yields earlier warning of severe wear and thus provides greater lead time for you to react. Corrective actions may be scheduled proactively so that they have minimum reactive impact on operations.

Principal Advantages of PdM

A principal advantage of PdM is the capability it offers the user to perform inspections while the equipment is operating. In particular, in order to reflect routine operating conditions, the technique requires that measurements be taken when the equipment is normally loaded in its production environment. Since the machine does not need to be removed from the operating cycle, there is no shutdown penalty. The ability to conduct asset inspections while running is especially important in continuous operations such as in utilities, chemical, petro-chemical and all continuous processing type of operations.

Another advantage of PdM is that the cost of surveillance labor can be much less than the cost of PM activities. Although the technical knowledge required for PdM inspections is usually higher than those for PM, the inspection time required per asset is much less. With PdM, assets do not have to be disassembled for inspection. For example, with vibration analysis, 50 to 60 assets may be inspected in a single day using modern computer data collectors.

When comparing cost advantages of PdM over PM, consider customer downtime costs, maintenance labor costs, maintenance materials costs, and the cost of holding spare parts in inventory.

Determining When to Use PdM

If PdM methods are superior to PM, why use PM at all? The answer is simple. The nature of your operation will determine which methods are most effective. In actual practice, some combination of PM/PdM is required to ensure maximum reliability. The degree of application of each will vary with the type of equipment and the percent of the time these machines are operating. Pumps, fans, gear reducers, other rotating machines, and machines with large inventories of hydraulic and lubricating oils lend themselves to PdM surveillance methods. On the other hand, assets such as those which might be involved in high-speed packaging and bottling may be better inspected using traditional PM methods. Assets which have critical timing adjustments, which tend to loosen and require precision adjustments, or have many cams and linkages which must be reset over time, lend themselves to PM activities.

The strategy for selecting the appropriate or predictive approach involves the following decision process*:

- Consider the variety of problems (defects) that develop in your equipment.
- Use the predictive method if a predictive tool is adequate for detecting the variety of maintenance problems you normally experience. One or a combination of several PdM methods may be required.
- Use PM if it is apparent that PdM tools do not adequately apply. Inspection tasks must be developed that reveal the defects not adequately covered by preventive maintenance.
- After you have decided the combination of inspection methods, determine the frequency at which the particular inspection tasks must be applied.

Some equipment will be satisfactorily monitored using only PdM. Other equipment will require PM. Ultimately, some combination of methods will provide the required coverage to ensure reliable performance. It is wise to apply a combination of methods to ensure that equipment defects do not go undetected.

Putting Predictive Technology Tools to Work for You

Vibration analysis*

Today, electronic instrumentation is available that goes far beyond the human limitations with which the old time craftsperson had to contend when trying to interpret vibration signals with a screwdriver-handle-to-the-ear method. Today's instruments can detect with accuracy and repeatability, extremely low amplitude vibration signals. They can assign a numerical dimension to the amplitude of vibration and can isolate the frequency at which the vibration is occurring. When measurements of both amplitude and frequency are available, diagnostic methods can be used to determine the magnitude of a problem and its probable cause.

---

*This section is contributed by Remy Kolb of RMK Maintenance Solutions, LLC, Charlotte, NC.

When you use electronic instruments in organized and methodical programs of vibration analysis you are able to:

- Detect asset problems long before the onslaught of failure
- Isolate conditions causing accelerated wear
- Make conclusions concerning the nature of defects causing asset problems
- Execute advance planning and scheduling of corrective repair so that catastrophic failure may be avoided
- Execute repair at a time which has minimum impact on operations

Vibration analysis is probably the most comprehensive and universally applicable of the various PdM methods. Its applications have been commercially developed to sophistication and maturity that permits its successful use almost anywhere rotating assets exists. Vibration analysis can be used in any industrial or commercial environment. If you have rotating assets which can cause financial loss if maintenance failure occurs, vibration surveillance can reduce or prevent that loss. Vibration can reveal problems in assets involving mechanical unbalance, electrical unbalance, misalignment, looseness, and other degenerative problems. It can uncover and track the development of defects in asset components such as those occurring in rolling element and journal bearings, gears, belts, drives, and other power transmission components. Any system deterioration which shows up as an increase in vibration amplitude can be disclosed with the use of vibration analysis.

Various asset problems manifest themselves by generating certain frequencies that are related to the rotational speed of the asset and its physical construction. For example, unbalance will cause a machine to vibrate at high amplitude at a frequency that is equal to the rotational speed of the unit. If a machine rotates at 3600 RPM, and it is out of balance, an unbalance frequency will occur at 3600 CPM. Thus, by noting the amplitude of the vibration, the frequency, the rotational speed of the asset, the type of construction, and the direction at which the vibration pickup is placed on the asset, you can judge the nature of the problem causing high vibration. A collection of diagnostic information exists which permits the analysis of problems. The analysis of problems in a machine using vibration is analogous to a doctor when examining a patient for a disease. The doctor looks for symptoms, and then compares these symptoms with a history of information collected by the profession. A similar approach is taken by the vibration analyst, although with vibration, the complexity is obviously far less.

Another important measurement in vibration is phase. Phase tells us where in the cycle of vibration the vibration measurement is occurring with respect to some fixed reference or some other vibration. Phase measurements are important for two reasons:

- They permit the comparison of the relative motion of two or more parts of a machine. For example, if a machine's vibration is out-of-phase with its foundation, looseness of the foundation bolts or similar problem is suggested.
- They are used to balance rotating equipment. Phase shifts with a change in the center of gravity. By noting the direction and amount of phase shift, when using a trial weight for balancing, the asset may be balanced.

Shock pulse

Shock pulse is a method of surveillance that is specific to rolling element bearings. Because rolling element bearings (sometimes referred to as antifriction bearings) are so common in assets, the method has many applications. The shock pulse method also permits crafts workers to judge the adequacy of the lubrication program applied to this type of bearing.

The shock pulse instrument measures a transient high-frequency wave produced when the moving elements in a bearing strike a defect and release mechanical energy. The device to measure this transient signal, which occurs over an extremely short time (several microseconds), is a piezo-electric accelerometer. The accelerometer measures the initial compression (shock) wave due to the impact before detectable deformation takes place which generates wave impulses normally measured with vibration instruments. The transducer is mechanically and electronically turned to a baseline frequency. Therefore, one can then differentiate the baseline type of signal from that of normal vibration.

The shock pulse instrument electronically filters out other signals and compares the peak amplitude of the "shock" signal with known data. Known data is based on "shock" measurements of similar bearings, which are properly installed, properly lubricated, and in good condition. The difference between the signals produced by a good bearing as compared to the signal from the measured bearing indicates the severity of damage to the measured bearing.

Several models of the shock pulse instrument are available. Recent models of *shock pulse method* (SPM) instruments measures the quality of lubrication in undamaged bearings by analyzing the shock pattern put forth by the bearing. When a measured bearing is adequately lubricated and has a good lubrication film thickness, the moving surfaces of the bearing are separated sufficiently to reduce the incidence of impact. If the lubrication film thickness is inadequate, the surfaces are more likely to strike each other and, thus, the output of shock impact energy increases. The built-in microcomputer in these instruments analyzes the shock pattern and reports two factors pertaining to the quality of lubrication. These are "condition number," which relates to the oil supply to the bearing, and "lubrication number," which is a statement on the thickness of the oil film that carries the load.

Spectrometric oil analysis

Oil in machines carries the products of deterioration resulting from wear and mechanical failure. Analyzing the oil resident in a machine, or the debris the oil carries, predictions can then be made about the state of health of the asset. The critical measurement reflecting the condition of asset wear is the amount of microscopic metal in the oil.

Many laboratories throughout the country perform this analysis for client locations at a nominal fee per sample depending on the number and types of analysis performed. The sampling price usually includes a sample bottle and a mailing package that permits both ease of collection and handling. The analysis results are returned to the client by FedEx/UPS/DHL with comments and recommendations. If an extremely severe wear process is detected which requires immediate action, the lab will telephone or e-mail the client.

The spectrometric oil analysis process is a laboratory technique which uses various instruments to analyze a used oil sample from a machine. A spectrometer is used to show when a significant wear mode is underway. Some varieties of these instruments can isolate up to perhaps 80 different types of metals in the sample at levels as low as one part per million. The spectrometric result is compared to a baseline level of metal found to be typically suspended in the oil under normal operating conditions. When the wear is meaningful, the sample will show high levels (in parts per million) of wear metals compared to the baseline oil sample.

An important feature of the spectroscopic method is that it not only determines the amount of metal in the sample but also the type. Using knowledge of the physical asset's construction, you can match a particular metal type found to be present in the sample to the machine component undergoing excessive wear. The analysis not only permits the discovery of severe wear, but also analysis of the possible location of severe wear in a machine. With this information you can take timely action to prevent further deterioration, either by oil purification, oil replacement, or some other means appropriate to the problem.

It is common for the oil analysis laboratories to maintain a large data base of their findings in computer files. Thus, wear metal levels in machine systems operating with normal wear may be statistically evaluated. These same laboratories will also build a history on assets unique to your facility. After a number of samples that establish a baseline they may make judgments concerning normal versus abnormal wear conditions for distinctive equipment you may have.

The relatively low costs of spectrometric oil analysis make it a very valuable and commonly used PdM method. It is practical for asset systems and subsystems that have a reasonable inventory of oil and are provided with "reuse" methods of oil application such as circulating, bath, splash, flood, and ring-oiling designs.

In addition to providing advance warning when severe wear occurs, spectrometric oil analysis can give important assistance in machine lubrication programs. Conventional oil change frequencies are arrived at according to the traditional PM approach based upon time, hours, or miles operated. A best estimate (or guess) is made of the time or calendar period over which the oil in the system will degenerate and require changing under current operating conditions. Spectrometric oil analysis enables you to change oil only when the actual condition of the oil requires. This approach can save you a great deal of money since you do not have to change oil unless analysis shows it is degenerated beyond safe use. By prolonging the useful life of the oil you obviously reduce your oil change labor, oil replacement, and oil disposal costs.

Standard oil analysis

In addition to the spectrometric analysis, the oil laboratories also check the oil using common oil analysis techniques. For example, the oil is usually checked for viscosity. If the viscosity of the oil has changed 5 percent from new oil of the same type, it is probably time for an oil change due to contamination. Typical of these types of analysis are total acid number, percent moisture, particle count (for hydraulic systems), total solids, and percent silicon (representing dirt from the atmosphere in the form of silicon dioxide or perhaps just from an additive). With prior agreement, special tests are performed at additional cost.

Ferrographic oil analysis

Ferrography provides the maintenance manager with two critical sets of decision support information—condition monitoring that prevents unnecessary maintenance and precise trend information that allows maintenance to initiate repairs before equipment failure. This is accomplished by the analysis of wear particles in lubricants to determine their size, distribution, quantities, composition, and morphology—form and structure.

Ferrography is a technique that provides microscopic examination and analysis of particles separated from fluids. Developed in the early 1970s as a PdM technique, it was initially used to precipitate ferrous wear particles from lubricating oils magnetically. This technique was used successfully to monitor the condition of military aircraft engines. That success has led to the development of other applications, including testing of fluids used in vacuum pumps within the semiconductor industry.

The direct reading ferrograph monitor is a trending tool that permits condition monitoring through examination of fluid samples on a scheduled periodic basis. A compact, portable instrument that is easily operated even by nontechnical personnel, the direct reading ferrograph quantitatively measures the concentration of wear particles in lubricating or hydraulic oil. Additional information about a wear sample can be obtained with the analytical ferrograph system, an instrument that can provide a permanent record of the sample and analytical information. The analytical ferrograph is used to prepare a fixed slide of wear particles for microscopic examination and photographic documentation. This slide is an important predictive tool, since it provides an identification of the characteristic wear patterns of specific equipment.

Ferrography enhancements. Enhancements in Ferrography have been oriented toward the expansion of capabilities and application of proven ferrographic techniques.

Computer-assisted evaluation. New software for personal computers lets the maintenance leader enter ferrographic data into a computer database. The database, containing a complete history of direct reading and an analytical ferrography outputs for each piece of equipment, can generate trend analysis and management reports. In addition, this database may be used to generate "caution" and "failure imminent" warnings automatically.

Expansion of applications. From its first application in jet engines, ferrography has been improved and applied to a diverse range of other materials. Fixers, diluents, and magnetizers enable even organic materials to be analyzed successfully. Another tool is a grease analysis testing kit. This conditioning kit uses a premixed solution that breaks the grease into liquids suitable for use in ferrographic instruments.

Integrated system approach. As the technology of ferrography has advanced, new and better systems and methods have been developed that make effective use of emerging generations of electronic technologies. Today, these individual pieces of equipment are increasingly being integrated into a single system that provides direct reading, in-depth evaluation, photographic recording, and database management. The intent is to provide the maintenance leader with a turnkey system that provides all the information required for PdM decisions.

Online ferrography. A wide range of online ferrography systems already exist, using ultrasonic, capacitance, and other deposition and monitoring systems. The typical system uses a monitor located within a lubricating or hydraulic system while the equipment is in operation. The device avoids the need for regular sampling and the cost of negative sampling while the lubricant is "clean" within acceptable levels. Future developmental trends for online ferrography will be systems to digitize data and send it to a centralized system for trend analysis and *failure warnings.*

Infrared thermography*

Infrared thermography has grown by leaps and bounds in the past 10 years. Equipment is easier than ever to use and more effective. The real power of thermography is that it allows quick location and monitoring of problems. It presents critical decision-making information in *visual* form making it easy for management to understand. Infrared imaging systems, as they are generally called, produce a picture, either black or white or color, of the invisible thermal patterns of a component or process. These thermal patterns, when understood, can be used to monitor actual operating conditions of equipment or processes.

For instance, viewing a thermogram (heat picture) can clearly show the heat of a failing bearing or a pitted contact on a disconnect switch. Today's sophisticated imaging equipment is capable of acquiring a thermal video. This allows us to see *dynamic* thermal patterns—of a casting process or a belt wear pattern in real-time for instance. When teamed with the power of computer-based analysis systems, it can go one step further to compare and trend the critical thermal changes that often precede equipment failure or loss of production quality.

Imaging systems can see temperature differences as small as 0.5°F or less. Even after a few minutes, a handprint on a wall is still visible! Thermography is also noncontact. This makes it safe to view energized electrical systems or moving machinery. It also means the actual temperature measuring process will not change the thermal characteristics of the material or process you are looking at.

Using infrared thermography. Thermography can be used to quickly locate equipment and process problems and in preventing the recurrence of the following problems:

- Catastrophic electrical failures
- Unscheduled electrical outages or shutdowns
- Chronic electrical problems in a piece of equipment or process

*Summary by Remy Kolb, RMK Maintenance Solutions, LLC.

- Excessive steam usage
- Frozen or plugged product transport lines
- An inability to predict failures accurately
- Inefficient use of downtime maintenance opportunities
- Friction failures in rotating equipment
- Poor product quality due to uneven hearing or cooling or moisture content
- A fire in a wall or enclosed space
- Inability to locate or verify a level in a tank
- Replacement of refractory in a boiler, furnace, or kiln
- A leaking flat roof
- Uneven room temperatures affecting product quality or employee productivity
- Trouble locating underground water, steam, or sewer lines

The most common application is the routine and periodic inspections of electrical equipment to locate "hot spots" caused by bad connections. Unscheduled outages due to such electrical failures are often reduced by a factor of 10 or more.

Thermography is also proving to be a valuable ally for vibration analysis. Most mechanical stresses exhibit both thermal and vibration signatures. While vibration analysis is often a more effective diagnostic and monitoring tool, the two together are even more valuable.

An infrared acceptance inspection of a large steam pipe insulation contract nearly guarantees the contractor will do the job right. A steam trap that is "blowing by" can cost $50,000 annually. Failed or not, many companies either routinely replace traps or never replace them! A periodic trap inspection program using thermography, especially in combination with vibration or sonic tools, can quickly identify failed traps for replacement.

Getting started with infrared thermography. Remy Kolb provides some key points on getting started with PdM. There is no single right way to get started. You will want to clarify, and perhaps quantify, the reasons to invest in a program. Investment includes equipment, personnel, and support. Look at the cost of past failures. Examine production processes to see where thermography can help you increase uptime. When someone says "we do not have the money," respect their point of view, but do not accept it. The money you need to start an infrared program is already being spent—either on repairs, reduced product quality, lost uptime, or a combination of the three. Our job as maintenance leaders is simply one of developing ways to begin redirecting the flow of funds through maintenance and back onto the profit and customer-centered side of the organization.

If you are purchasing equipment, invite several vendors to your site. Try the equipment on your applications before you buy it. Do not let the vendor use all the time during the entire sales call! Let them do on-the-job training and show you how to use their equipment. Remember, too, the most expensive is not necessarily the best. Many companies own systems that are too complex to be used by anyone, but a full-time thermographer with an engineering degree. Ask for references of others in your business who are using the equipment and call them. Strong after-the-sale service is critical to keeping equipment, and an effective program, running smoothly. Remy Kolb's interesting professional background includes being an aircraft mechanic literally around the world, a pilot, a corporate level maintenance leader, a copilot/navigator, and now a consulant who truly understands predictive technogolies like thermography.

Ultrasonic detection

Today, a variety of tools using airborne ultrasound technology (commonly called ultrasound) have revolutionized many maintenance programs, allowing inspectors to detect deteriorating components more accurately before they fail. Ultrasound, by definition, is beyond the limits of normal human hearing, so an inspector uses a sophisticated detector to transpose ultrasonic signals to the range of human hearing.

The theory of ultrasonic detection is relative simple. Frequency, the number of times a sound wave cycles from trough to crest is expressed in "cycles per second" and measured in "hertz." One kilohertz (kHz), for example, is 1000 cycles per second. Generally, the best human ears can hear noises in range of 20 Hz to about 20,000 Hz (20 kHz). Many ultrasonic detectors start at approximately 20 kHz and can work upward to sounds as high as 100 kHz. Thus, mechanics using the ultrasound instrument can tune in to and "hear" what is going on in operating machinery. My wife Joyce hears at about 50 kHz to detect me "popping tops" down in my shop!

Fluid and gas systems and other working machinery have constant ultrasound patterns. Changes in the "sonic signatures" can be recognized as wear in components. An ultrasonic detector senses such subtle shifts in the "signature" of a component and pinpoints potential sources of failure before they can cause costly damage.

The longer wavelengths of lower pitched sounds can penetrate solid materials and can be heard without special equipment. Higher frequency sounds cannot penetrate most solids, yet they will slip through the tiniest of openings. Ultrasound detectors are ideal for isolating such sound leaks. Many of these lightweight, hand-held tools are battery-powered; so operators can easily move from machine to machine. They contain circuitry that translates the high-pitched ultrasound to sounds in the human hearing range. Some instruments feature a frequency adjustment dial to provide tuning capability. This lets operators hear the ultrasounds through headphones and gauge their intensity by the definitions registered on an analog meter.

As a rule, operators begin their PdM routine by using the ultrasound instrument to establish the baseline sound patterns of properly operating equipment. Usually they mark a spot on the housing of each component to make sure they are measuring from the same point each time they conduct the test. If the sound changes in subsequent checks, they have an indication that something has physically changed. In bearings, for example, sounds typically get rougher as the balls or raceways wear. Ultrasonic allows you to hear if the bearing is running dry and even detect microscopic cracks in the balls.

Ultrasonic detection has been used successfully in a wide range of application to include the following:

- Steam traps can be audited with ultrasonic tools. Identification and replacement of faulty traps reduces energy losses substantially. This allows steam plants to run more efficiently.
- Ultrasonic instruments can be used to test for corona discharge. Corona discharge is made of destructive flashes of electricity in transformers, cables, and switchgear.
- Ultrasonics have been used to locate switchgear faults that have eaten away about 35 percent of the blade of a switch to the high-voltage fuses. When problems like this are not identified and corrected, damages can run into tens of thousands of dollars.

Today, ultrasound technology in combination with other advanced PdM methods is revolutionizing traditional maintenance programs in virtually every industry.

Tailor the PdM technology to your operation

The methods discussed earlier are the most commonly applied PdM systems. Do not restrict yourself to a commercially available conventional PdM technique. Because the predictive system is based on the measurement of some signal or output, which is indicative of the condition of the machine you are monitoring, many possibilities exist for original applications. Within your organization there may be potential for developing monitoring methods which are distinctive and one of a kind. The particular applications that are best for your facility can vary significantly and will depend on your own inventiveness and initiative. To be effective you merely have to apply the PdM concept in a manner which best fits your site's facility and equipment asset characteristics.

Vendors of the PdM technology, previously mentioned, are excellent sources of information on unique applications and monitoring needs for a wide range of operations. Use all the resources available to determine the right type of PdM surveillance equipment for your operations. PdM technology continues to expand at a very rapid rate as more and more organizations begin to see maintenance as an invaluable part of a business' bottom line. That is what this book is all about!

Implementing a PdM Program

Because vibration analysis is the most universally applicable PdM technique, this section will focus on this method when discussing how to implement a PdM program. If you have rotating equipment in your facility, you have an application for vibration analysis.

The first step in program start-up is to select those items of your facility's equipment which will be included in the monitoring program. This will be discussed in more detail later. Once having selected equipment, the next step is to choose the appropriate surveillance method. When selecting the surveillance method, you must first consider the kinds of defects and problems that develop which must be avoided. Having defined these defects and problems, you are then in a position to decide which of the PdM methods will best do the job for those problems in that particular asset under consideration. Perhaps you will require more than one predictive method for complete surveillance assurance. For example, it may be desirable to conduct vibration analysis and oil analysis inspections on a major, high, horsepower gear reducer in your facility.

This would provide you with high confidence that you are providing full protection to your equipment. You would not only detect early signs of mechanical degeneration in the system through the use of vibration analysis, but perhaps have warnings of severe wear before damage occurs through the use of oil analysis.

With some items of equipment, it may also be necessary to conduct some traditional PM task inspection in conjunction with predictive applications. PM should be used when predictive applications cannot do the entire job. This is a logical approach because PM activities are more costly than PdM. However, because of some unique wear modes experienced by a unit, supplemental PM inspections will also be required.

The Basic Steps for Implementing a Vibration Analysis Program

Step 1: Select equipment to be monitored

The first step in any PdM program is to select the assets to be included in the program. Although benefits can be justified for even moderate size horsepower equipment, a limit must be placed in an initial program. When embarking on a new surveillance method like vibration analysis, take care that you are not overwhelmed with too much data. You should start with a relatively small and manageable group of machines. Limit yourself to critical equipment. When you have resolved the problems and are comfortable with the successes of the prototype program, then expand your efforts to a second phase effort and beyond.

Selecting equipment. Critical equipment for the first phase of a PdM program can be ranked for selection using four evaluating considerations. These are:

- The impact that failure of a candidate unit has on other equipment. If the machine is one of a series of equipment in a continuous operation and its failure means it will shut down an entire system or operation, it must be ranked high for selection in the program. For example, a utility such as plant air compressor can affect an entire facility if it fails because all pneumatic controls in the facility cease to operate.
- The amount of machine capacity used considers redundancy or installed spares. If a machine is one of a kind, and is functioning at full capacity 24 h/day, 7 days/wk, 52 wks/yr; it is a good candidate for selection. Under such conditions, lost production due to mechanical failure cannot be made up. Thus, machines used at full capacity should be ranked higher than those which are not loaded in that manner. A one-of-a-kind machine should be ranked higher than machines which have installed redundancy.
- The profitability of the equipment or machine or costs of downtime. A candidate unit would rank high for selection if a serious profit loss occurs when it fails. Perhaps equal to profit loss would be the impact failure would have on delaying fulfillment of the operation's mission. If downtime costs have been determined for critical equipment, this provides an ideal factor for determining criticality and priority for PdM.
- Other factors such as safety, environmental, government mandated requirements, and repair history should also be considered in this ranking process.

Some factors cannot be evaluated on the basis of cost alone. If a unit's failure increases unsafe exposure or is a hazard to the environment, it should receive a high ranking for selection in the monitoring program. Another consideration when ranking equipment is its frequency of failure. Obviously, if a unit frequently breaks down, it is contributing to high-maintenance costs. Reduction of breakdown through PdM control will provide a significant return on investment for PdM technology application.

Another factor to consider is whether the equipment has a high potential for demonstrating success. If you do not have all the resources you would like to have for vibration analysis activities and have yet to fully convince management on the viability of the method, you need evidence to prove that the effort is practical. You might select a unit because it has a high potential for demonstrating program success. Those machines with high downtime may also be good candidates in this latter case. In this respect, you should try to select equipment for which you have good cost and repair documentation. This data would be essential as baseline reference when you attempt to show cost-reduction improvements and gained value from uptime.

Step 2: Establish vibration limits

A second step in setting up your PdM program would be the selection of reasonable vibration limits which define the boundary between normal operation (baseline) and the need to take corrective action. Every PdM program has three basic steps. These are:

- *Detection.* Under normal surveillance, you collect vibration readings periodically from the equipment. When a unit vibration signal exhibits some anomaly, you have detected a possible machine problem. This constitutes the detection phase.
- *Analysis.* When a problem is noted, the next step is to determine the underlying cause of the abnormal vibration signal. This requires the collection of additional information. Normally, the vibration analyst takes his vibration analyzer to the machine and collects additional data at the source in order to isolate the cause (or causes) of the problem. In many cases, there may be a combination or "stack up" of problems which create the excessive vibration signals. The high-vibration signal may be a machine defect in development or may simply be due to some aberration such as an error in taking the reading. If the cause is a mechanical problem, then the appropriate corrective action must be planned. The process from problem discovery to the corrective plan can be considered the analysis portion of the PdM application.
- *Correction.* Finally, the goal of the whole PdM effort, assuming there is a problem, is correction of the condition producing the problem. It is thus necessary to perform some repair function on the unit if the program is to fulfill its role of avoiding catastrophic failure, reducing, or retarding the impact of failure. The repair may be as simple as an in-place

balancing of the rotating element of the asset, or as complex as a complete dismantling for component replacement. This step constitutes the corrective phase of the PdM process.

The establishment of a vibration limit for each checkpoint on the asset is therefore a critical program element. A vibration reading above this limit is the signal for the start of the three-phase process:

Level of vibration. "How much vibration is too much vibration" is a question often asked by beginners in a vibration analysis program. Unfortunately, the answer to this question is not simple. Several observations must be made when attempting to answer this question.

First, keep in mind the objective of the program. Remember that the program's primary purpose is to provide sufficient advance warning of a problem so that you can schedule and execute corrective action before catastrophic failure. The vibration limit selected therefore does not have to be precisely the point prior to failure. It must only provide sufficient warning of the development of a real problem so that you can appropriately analyze the nature of the problem and plan and execute proper corrective action with minimum impact on the operation.

Second, a conventional rule-of-thumb instructs that when a vibration signal increases to twice the value of the original vibration signal, it is a clear warning of degeneration. Some action must be taken because the significant change in signal gives evidence to the potential of ultimate machine failure. The asset should receive attention.

Beyond these observations you may use a number of different resources for setting up the vibration limits. Prediction of impending failure is based on the comparison of current vibration readings with vibration information of the machine in consideration, or similar machines, when in a condition of satisfactory operation. Comparisons are made several ways:

- By comparison to standards developed by engineering standards organizations, manufacturer's associations, or governmental bodies (e.g., API, American Gear Manufacturer's Association, NEMA, ANSI).
- By comparison to previously measured baseline, this technique—sometimes referred to as "trending"—can be applied graphically. At a selected measurement level, there is a determination that the unit requires repair. The repair is then carried out when measurements reach a designated higher level thereby bringing the measured vibration back to original satisfactory levels.
- Deviation from a statistical control chart. Using statistical quality control charting methods, the baseline vibration information can be plotted over time, using ±3 sigma limits; deviation beyond that limit alerts the technician that machine degeneration is probable.

When automated data collectors are used, the selection of vibration limits may be more complex. This is because many of these instruments can collect several different types of reading from a single check. For example, some data collectors take three readings simultaneously. One of these readings is what is termed "bearing units." This is an acceleration reading for frequencies above 18,000 CPM. Selection of vibration limits for these three readings is obviously somewhat complex than for a simple single velocity reading.

Step 3: Select checkpoints

For the same reason that it is necessary to draw representative oil samples form a lubrication system, it is important to properly locate vibration checkpoints. The resulting data (analysis) should truly reflect the condition of the unit. The checkpoint is the spot on the machine that the vibration transducer is placed in order to collect a vibration reading. Since the dynamic forces of the machine are transmitted through the bearings, bearing caps are a logical point. If this is not possible you should get as close to the bearing caps as possible, for example, the bell ends of a motor where stiffeners are located. The coupling bearings of a driver/driven machine set are more important than the outboard bearings. On the coupling side, the driven bearing is the more important of the two.

To properly compare current readings with past data, you should always take the vibration readings from the same location on the machine. Vibration transmissions are absorbed and will vary as you move over the surface of the machine frame. Therefore, having selected a checkpoint, you should permanently mark the spot so that the transducer can be placed in the same location for each reading.

You should initially take readings in all radical directions. That direction which provides the highest reading should be the checkpoint spot that is selected. It is the plane of most sensitivity for a given load and RPM. Take five readings per machine set including four radical, one for each bearing location, and one axial, to check thrust.

Step 4: Check overall vibration level and correct existing deficiencies in start-up machines

At this point, a basic requirement would be to collect baseline information. This step would first involve collecting overall vibration amplitudes from each of the checkpoints established for each machine. If the overall amplitude exceeds the vibration limit originally selected, it will be necessary to analyze why. If mechanical problems are indicated, you must correct them.

Experience tells us that in an initial program of this type you may find that as many as 30 percent of your machines will have mechanical problems. Conditions such as unbalance and looseness are common. If corrected at this stage you will have already started to receive a return from your effort. At the very beginning of the vibration program, the life of your equipment will be extended by reducing the destructive forces of vibration acting upon them.

Step 5: Determine optimum check intervals

If your staff resources permit, try taking one vibration reading at each checkpoint each week. This frequency will help in detecting a degenerating machine before damage progresses too far. If your staff resources do not permit weekly checks, you should space the readings on some logical determination. Include in your analysis:

- How fast deterioration in a machine can occur
- What the consequences would be of missing a problem and permitting a machine to go to failure
- History of failure

Weigh considerations such as machine speed, service factor, and operating environment when evaluating the speed at which a machine may degenerate. The outside limit for surveillance checks should be at about one-month intervals. Intervals beyond this are not considered good practice.

Step 6: Collect baseline information

Once all machines have been brought into proper operating condition, take baseline spectral plots at each checkpoint. A spectral plot is a graph showing frequency in the x-axis versus amplitude in the y-axis. These plots are usually generated over a wide span of frequencies. The range covered would include output typical for the problems developed at the machine's operating speeds. These baseline "signatures" serve several important purposes.

If you are monitoring your equipment with the use of a meter, the spectral plot will help you determine if a simple unfiltered reading will indicate the problems in the machine under surveillance. The spectral plot will show if vibrations from nearby sources are being transmitted, collected, and disguising the true signals from the machine being monitored. If this is the case, simple meter monitoring will not be adequate.

The signature also provides a baseline which can be used for comparison with future plots. The growth in amplitude at a particular frequency can be compared to the baseline frequency so that a better judgment can be made concerning developing problems.

With automated data collectors, baseline signatures and subsequent spectral plots are stored in electronic files. Comparisons are thus made more quickly and computer displays aid in clarity and interpretation.

Step 7: Design a data-collection process

Data collection should be designed so that it can be as a good manual collection system that can later be computerized for electronic collection. Key elements should include:

- *Schematic section.* A section for showing a simplified sketch of the machine, checkpoint locations, and identified as "A" through "Z."
- *Equipment data section.* Key information including pump name and number, initial vibration readings for each of the checkpoints, and vibration limits established for these checkpoints.
- *Vibration trend chart sections.* A vibration trend chart with the vertical axis showing time.
- *Data section.* The date and results of period vibration checks.

If you are going to use an electronic data collection, the system will provide the data-collection format for you. Design of a data-collection sheet will be unnecessary. The data will be stored in the host computer data files, and data reports generated will be determined by the software blueprint. Electronic systems usually include trend reports both as screen displays and as printed charts.

Step 8: Implement training for vibration analysis

Since a vibration analysis program requires specialized knowledge, you must obviously provide training for your personnel who will be assigned to run the program. There are two levels of training required in most facility vibration analysis programs. These are defined by the tasks data collection and data analysis. In many facilities, the person who collects data may be the same person who does the analysis. In this discussion, however, I will treat these tasks as separate responsibilities involving different people.

Training for data collection. The person collecting the data should have the following minimum abilities:

- Be familiar with facility layout and machine location
- Be able to identify unusual conditions in machines
- Know the objectives of the PdM program and how it works

- Know how to use the data-collection instrument, how to place the transducer for collecting representative readings, and how to take the readings
- Know how to react to indicated problems and what to do with the data after it is recorded

The person doing the data collection does not have to be your most highly technically qualified person. This person, however, should have enough knowledge of the machines from which data is to be collected to be aware of symptoms of malfunction. This person should understand what excessive heat at a bearing cap might indicate or understand that a leaking oil seal is a problem that should be reported. This person should have enough understanding of machine operation to be aware of strange sounds abnormal to the operation.

The person collecting data should have a minimum of training in vibration fundamentals so that he or she is aware of basic vibration measurements and what they mean. The person should certainly have received training in the use of the instruments with which data is collected. If the vibration data in a particular measurement is not consistent with what would be expected, he or she should be able to discern this and seek out the causes.

The type of training in specific vibration subjects required for this person can probably be covered in a two-day course typically provided by the instrument vendor. With several weeks of on-the-job training, this person will soon be fully qualified to perform the tasks of data collection. Good candidates for this responsibility are craftsmen who have worked in the operation and who have a strong desire and are openly enthusiastic about becoming involved in PdM activities.

The task of data collection can often be rotated among crafts where small team problem solving and worker involvement is the norm. In these and in more traditional maintenance cultures, all maintenance personnel should be trained in PdM techniques. It is an excellent addition to the technical knowledge and skills resource area in maintenance. The knowledge of how PdM systems work will bring with it a strong desire for success of these programs. Effective training of maintenance personnel will help make the PdM program successful. Full craft participation is a sign of movement toward maintenance excellence.

Training for data analysis. The person who is assigned the responsibility for analysis of vibration data should as a minimum:

- Have a good understanding of rotating machinery
- Be familiar with the types of troubles these machines experience and how these troubles may be corrected

- Be trained in methods for analyzing vibrating signals and how to interpret these signals and the mechanical problems they suggest (vibration analysis)
- Be computer literate (with today's technology, the analyst must be able to freely interface with computers and be well versed in their application and nomenclature)
- Have enthusiasm for the job, have analytical ability, be thorough, methodical, and patient, and above all have a high degree of personal integrity in the decision process

A good candidate for a job of this type is a graduate with technical training such as from the military: an aircraft mechanic (like Remy Kolb), an electronics technician, or a graduate from a two-year technical college. It would be desirable for this individual to have several years of hands-on experience in the repair of machinery, and specifically, in the type of machinery in your facility. As a minimum, this person should have a reasonable technical knowledge and skill application background.

Many organizations assign degree engineers to this type of activity, especially those larger ones that can support and afford such staff. When an engineer is involved, he or she generally is given broad responsibility for establishing and overseeing the operation of reliability programs. However, a degree is not a necessary requirement for a good vibration analysis program. In some situations, it is a greater advantage to have a hands-on technical person in this position. This is especially true if this person has broad hands-on field experience and understands the machines under his or her area of responsibility.

Training for this level of proficiency involves more advanced vibration subjects. Principal in this area is the need for an in-depth knowledge of analysis of data to form real information. The person doing analysis should have a clear understanding of all the varied types of signals different machine problems project. Because many vibration outputs overlap, rarely are the details of the data collected simple and straightforward. The analyst needs to separate many confusing and sometimes contradictory information.

Vendor training. All manufacturers and vendors of vibration instruments offer training to their customers. In some cases, they charge for their training only if it goes beyond simple instructional information concerning the use of the instrument purchased. Fundamental classroom and hands-on training, involving analysis and balancing, would require probably a minimum of five days of instruction. This training, often provided in separate sessions of basic and advanced short courses, is a starting point of training for the analyst. Beyond this formal instruction, it is important that the analyst make a commitment to a program of self-instruction and professional development. One must continuously read, learn, and apply new topics from the many magazine articles available and from the many texts published on the subject of maintenance, such as this book and many others from McGraw-Hill. There are always a number of short courses and workshops available. One nonprofit organization offering such training is the Vibration Institute, located in Clarendon Hills, Illinois. This organization, dedicated to the advancement of vibration studies, also provides publications, library searches, bibliographies, and surveys on particular topics.

Conclusion

This chapter provides only a general introduction to the area of PdM technology that is available to today's maintenance leaders at all levels in all types of maintenance organizations. It provides for a basic understanding of the differences between PdM and PM. It also provides a review of the major tools and techniques as well as a basic strategy for implementing a vibration analysis program.

PdM technology is constantly changing, being improved, and becoming more affordable to the maintenance operation that must improve physical asset reliability. Effective application of today's PdM technology along with PM can provide a significant return on the investment required. And an investment will always be required *just as maintenance is always forever*!

Chapter

22

Auto Identification Strategies to Support Maintenance Storeroom Excellence*

Bar coding and other autoidentification technologies, to include *radio frequency identification* (RFID) are fast becoming as necessary to maintenance and maintenance repair operations (MRO) materials management operations as vibration analysis equipment is to *predictive maintenance* (PdM) and *continuous reliability improvement* (CRI). As I said in Chap. 21, the days of listening to a noisy bearing through the end of a screwdriver are almost over. Today's automatic identification technology offers progressive maintenance leaders another important tool: a tool that collects accurate and timely maintenance-related information to manage and lead mission-essential maintenance storerooms, to support MRO procurement processes, *preventive maintenance* (PM) data collection, and much more. Carla Reed, President of New Creed LLC provides some important points to help your understanding of RFID and bar coding.

Assume RFID Technology Is Cost-Effective

Begin with the premise that there is bar coding technology available that can cost-effectively serve many needs within maintenance operations. This is true, but because there are so many options that can be pursued, the permanent actions seem endless. First of all, there must be a plan.

*In Collaboration with Carla Reed, President, New Creed LLC, Raleigh, NC.

All too often, the objective of a bar coding project is to reduce labor. This is very often the case, but it is the wrong objective for your plan. Bar coding will increase labor productivity in collecting information related to MRO parts/supplier receipts, issues, stock adjustments, transfers, and the like. Bar-coded work orders, asset tags and equipment IDs for pumps, piping systems, instrument loops, and so forth, all improve labor productivity during the data-collection process. However, the key benefit is having accurate, reliable, and timely maintenance data. That you can now use as true information; one of TMEI's six maintenance resource areas discussed earlier.

What Is RFID?

RFID is a technology solution based on the use of radio waves (RF) to form an electromagnetic communication with another device to identify an objective, event, or condition (sensing) in the environment.

How does RFID work?

With RFID, an allocated frequencies band within the *electromagnetic spectrum* is used to transmit signals. A radio signal or circuit is created by the sending and receiving of signals between devices as shown in Fig. 22.1.

Devices can be a simple integrated circuit or chip, which has a an *antenna* attached (Fig. 22.1), all the way up to fairly powerful "active" technologies that are like small computers or processes that can process a fair amount of instructions, like a cell phone, or personal digital assistance (PDA) devices, which will also have its own battery power source.

Based on the *power* and *frequency* of these devices they can be read at small or quite large *ranges* (inches to hundreds of yards). So, as devices move through facilities such as stores, warehouses, or hospitals,

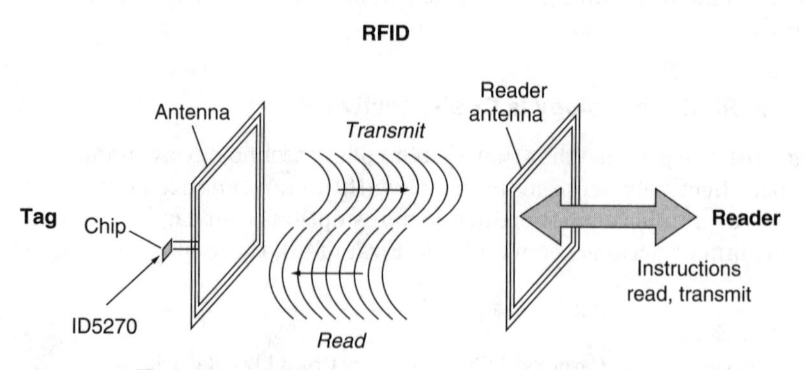

Figure 22.1 The basics of RFID.

as they come into *range* of another device and "activate" they send signals back and forth. These signals can simply be to reject each other (no, you are not the one I am looking for) or transmit a significant amount of data to one another.

RFID devices can be what are called *active* or *passive*. Active devices have their own power sources (batteries) versus passive which is activated when a reader transmitting in the appropriate frequencies comes within range.

Readers can also be quite simplistic, collect data, and send on, or perform functions like read, write to the devices as well as sort, filter, and then determine if it should send data onward.

And simply stated interference, from a variety of sources, can interrupt the transmitting and receiving of signals and data between these devices and potentially many of wave frequencies and energy sources that come into proximity with the RFID device.

Glossary of Bar Code and RFID Terms

Bar code

Bar code is a standard method of identifying the manufacturer and product category of a particular item. The bar code was adopted in the 1970s because the bars were easier for machines to read than optical characters. Bar codes' main drawbacks are they do not identify unique items, and scanners have to have line of sight to read them.

Radio frequency identification (RFID)

RFID is a method of identifying unique items using radio waves. Typically, a reader communicates with a tag, which holds digital information in a microchip.

In addition there are "chipless" forms of RFID tags. This is a type of RFID tag that does not depend on an integrated microchip. Instead, the tag uses materials that reflect back a portion of the radio waves beamed at them. A computer takes a snapshot of the waves beamed back and uses it like a fingerprint to identify the object with the tag. Companies are experimenting with embedding RF reflecting fibers in paper to prevent unauthorized photocopying of certain documents. But chipless tags are not useful in the supply chain, because even though they are inexpensive, they cannot communicate a unique serial number that can be stored in a database. RFID tags comprise the following key elements:

Integrated circuit (IC). A microelectronic semiconductor device comprising many interconnected transistors and other components. Most RFID tags have ICs.

Antenna. The antenna is the conductive element that enables the tag to send and receive data. Passive tags usually have a coiled antenna that couples with the coiled antenna of the reader to form a magnetic field. The tag draws power from this field.

Active tag. An RFID tag comes with a battery that is used to power the microchip's circuitry and transmit a signal to a reader. Active tags can be read from 100 ft or more away, but they are expensive—more than $20 each. They are used for tracking expensive items over long ranges. For instance, the U.S. military uses active tags to track containers of supplies arriving in ports.

Passive tag. Passive tag is an RFID tag without a battery. When radio waves from the reader reach the chip's antenna, it creates a magnetic field. The tag draws power from the field and is able to send back information stored on the chip. Today, simple passive tags cost around 50 cents to several dollars.

Semipassive tag. Similar to active tags, but the battery is used to run the microchip's circuitry and not to communicate with the reader. Some semipassive tags sleep until they are woken up by a signal from the reader, which conserves battery life. Semipassive tags cost a dollar or more.

Smart label. Smart label is a label that contains an RFID tag. It is considered "smart" because it can store information such as a unique serial number, and communicate with a reader.

Frequency

The number of repetitions of a complete wave within 1 second. 1 Hz equals one complete waveform in 1 second. 1 KHz equals 1000 waves in a second. RFID tags use low, high, and ultrahigh, and microwave frequencies. Each frequency has advantages and disadvantages that make them more suitable for some applications than for others.

- *High-frequency tags.* They typically operate at 13.56 MHz. They can be read from about 10-ft away and transmit data faster. But they are consuming more power than low-frequency tags.
- *Microwave tags.* These are radio frequency tags that operate at 5.8 GHz. They have very high transfer rates and can be read from as far as 30-ft away, but they use a lot of power and are expensive.
- *Low-frequency tags.* They typically operate at 125 KHz. The main disadvantages of low-frequency tags are that they have to be read from within 3 ft and the rate of data transfer is slow. But they are less expensive and less subject to interference than high-frequency tags.

- *Ultrahigh frequency (UHF).* Typically, these tags operate between 866 to 930 MHz. They can send information faster and farther than high- and low-frequency tags. But radio waves do not pass through items with high water content, such as fruit, at these frequencies. UHF tags are also more expensive than low-frequency tags, and they use more power.

Interrogation and Reader Definitions and Functional Capabilities

Sensor

It is a device that responds to a physical stimulus and produces an electronic signal. Sensors are increasingly being combined with RFID tags to detect the presence of a stimulus at an identifiable location. In addition to the data capture and transmission tags, an RFID network includes a series of interrogation devices that are used to read, transmit, and distribute information. These include:

Scanner. It is an electronic device that can send and receive radio waves. When combined with a digital signal processor that turns the waves into bits of information, the scanner is called a *reader* or *interrogator*.

Read. This is a process of turning radio waves from a tag into bits of information that can be used by computer systems.

Read rate. It is the maximum rate at which data can be read from a tag expressed in bits or bytes per second.

Reader (also called an interrogator). The reader communicates with the RFID tag via radio waves and passes the information in digital form to a computer system.

Reader field. It is the area of coverage. Tags outside the reader field do not receive radio waves and cannot be read.

Read-only tags. These are tags that contain data which cannot be changed unless the microchip is reprogrammed electronically.

Read range. It is the distance from which a reader can communicate with a tag. Active tags have a longer read range than passive tags because they use a battery to transmit signals to the reader. With passive tags, the read range is influenced by frequency, reader output power, antenna design, and method of powering up the tag. Low-frequency tags use inductive coupling, which requires the tag to be within a few feet of the reader.

Therefore, care in the design of the business process as well as deployment in the physical surroundings is the key to successful RFID use with minimal interference.

Implementation Strategies

A common mistake is applying automatic identification technology, such as bar coding or RFID, directly on top of an existing system. When this occurs, one of two scenarios will occur, both less than optimal. The first scenario occurs when auto ID is put on top of an already bad process (an inaccurate or incomplete equipment numbering system). In this case, the bar-coding methodology will never get the asset management process straightened out.

The next scenario occurs when keying of data into the system is eliminated. Full benefits of the technology are not realized because MRO inventory and related maintenance storeroom labor are not managed. We can get instantaneous data on asset numbers that are lubed and PMed on an irregular, "do it when there's time" type of schedule. This picture is wrong and completely out of focus.

The objective of your auto ID strategy should be to create additional time for analysis, for doing things neglected in the past and for making sure your maintenance management database provides the key information to manage maintenance as a profitable internal business enterprise.

Strategic Planning

To achieve maintenance excellence it is necessary to take a step back and prepare a maintenance strategic master plan, rejecting the short-term, fire-fighting approach which is often taken to maintenance operations. Once you have addressed the long-term total operations requirements for maintenance, over time these requirements can be adapted within the context of the total requirements. A maintenance strategic master plan looks at the current business operations and determines its current maintenance capacities. This effort entails quantifying the existing maintenance process, its current capabilities, and the true maintenance needs of the total operation. Then, forecasted growth is obtained and there is a determination of whether or not the current maintenance process can accommodate this growth. If there is a shortfall, then best practice alternatives are generated to satisfy the future maintenance requirement. Each of these best practice alternatives are quantified in terms of the estimated investment and operating costs, and an economic analysis is performed based on these costs. However, qualitative issues for each alternative are also investigated and factored into the decision process.

Once the maintenance strategic master plan is developed, a time-phased action plan must be established to guide growth and implementation of best practices. This implementation plan could result in the associated conceptual layouts of a storeroom modernization plan to include storage, handling, and automatic identification equipment. Therefore, the objective of a maintenance strategic master plan is to determine the optimal maintenance best practices including automatic identification alternatives, *computerized maintenance management system (CMMS)* strategies, continuous reliability improvement, craft skills development, and overall MRO materials management strategies that best meet the future requirements. Bar coding, in turn, is just another tool we must consider using within our overall plan for maintenance excellence.

The bar-coding segment of the maintenance strategic master plan must identify what level of automatic identification is required and who or what is driving the requirement. Is accounting to be responsible for MRO inventory levels, dollars, and accuracy? Is it engineering that is concerned about life-cycle asset management, mechanical integrity, and reliable management of change? After understanding the requirements, one needs to develop a method to maximize the potential maintenance investment. This is a four-step process that ties in with the strategic master plan. The steps are as follows:

- Determine the alternatives for implementation of automatic identification.
- Evaluate the quantitative and qualitative impact of these alternatives.
- Decide on the best option for the specific application at hand.
- Choose the best way to pursue system implementation.

The driving force may be an external requirement where a supplier/vendor partner mandates an operation support, a quick response system, or *electronic data interchange* (EDI). This can also be an internal requirement where a bar-coded asset tag is necessary to support an asset management system. If it is an external force driving the requirement, then it is best to work with the customer to determine what labeling standards are expected. If the external requirement is industrywide, then one must identify the appropriate industry group and find out what type of labeling standards exist. Internal forces that require automatic identification provide more latitude. In either case, the next step is to determine the type of environment in which the bar-code identifier will be used. For example, roof top air-handling units in hurricane-prone Puerto Rico. Scanning distance, human readability, harshness of environment, and available space to print are all factors that must be considered, and these all impact the bar-code design. Next you must determine how to print the bar code. Many options are available on how to determine whether the bar code should be printed in-house or off-site. The final issue is to determine how to affix the label to the part, the container or asset.

Detailing the System

The next step in the overall process is detail planning. In this phase, you must also write a functional specification for the recommended bar-coding equipment that required purchase. This is a step that is too often overlooked because no one likes to do homework. A functional specification is more than contacting a couple of vendors and telling them your problem, and then they help you solve it. The vendor is interested in your real problem as long as his/her equipment solves it. The danger is the vendor is also interested in a sale. Do some homework and prepare a document which identifies your specific requirements and the maintenance operating environment of the equipment you need. This allows the vendor to understand the problems and also provides him/her sufficient latitude to utilize creativity to provide a solution. Once the bids are submitted, they must be qualitatively reviewed based on a set of prioritized criteria. This results in a numeric rating and ranking for each vendor. Then, site visits need to be arranged to view a similar maintenance operation. During these site visits, many questions need to be asked about implementation schedules, training, support, the system's effectiveness, and the vendor's overall capability. Once the rating forms are filled out and the site visits completed, a vendor may be selected to proceed with implementation of your system.

The Big Sell Not the Big Chill

Throughout the planning process, we are striving toward implementation. However, the planning effort is only half of the total effort. Completing the task requires something most technical and maintenance people do not enjoy—that is excellence in salesmanship. Many are straight-to-the-point shooters who assume top leaders understand. We know top leaders do not understand just as a deer does not understand bright headlights in their eyes on a moonless night on the highway. There will be objections and misunderstandings throughout the justification process, And there will be the inevitable request for funding that may have go to the *chief financial officer* (CFO). Each of these hurdles requires the Zig Ziglar salesmanship traits in all of us to come out.

There are five steps to a successful sales effort for bar code or RFID technology applications:

- Satisfy the accountant
- Educate your audience

- Address what and how things will change
- Incorporate suggestions
- Sell everywhere, up, down, and across the organization

The first step is to satisfy the accountant, CFO, or top leaders and any of the many roadblocks above the maintenance shop level. The accountant's job is to get the maximum return on the company's investment. As the maintenance leader, you must do the same with your maintenance investment. Make sure to follow the accountant's rules for performing an economic analysis. Remember the golden rule! Whoever has the GOLD rules. If you have your own personal questions, they will need answers and most likely be helpful in supporting your justification analysis and acceptance. It is absolutely necessary to perform this analysis and to get the accountant's blessing, otherwise there may be no funding.

The second step is to educate the people who will be using the system. Often the employees have no idea what automatic identification technology is other than what they have seen in grocery stores. The education process must be able to familiarize employees with the system's terminology and concepts. Equipment brochures, photographs, case studies, and articles are helpful; but the most helpful aspect is actually going to see a system similar to that you are proposing. True story: I met a rather short, second-shift storeroom attendant at Broward County (Florida) Public School system's central warehouse. Asked him about bar coding and the antique wooden ladder he was using for getting boxes down from shelves above 8-ft high. He said, "Yes I really like bar coding and RFID. This is my evening job. I work at Lowes during the day and we could not live without it. Yes I can use bar code and RFID equipment." It is a "piece of cake." When some operations are ready for change to bar coding and RFID, some of your hidden assets may just emerge and be your trainers.

The third step is to address change. Just like maintenance, change is a constant forever type of thing. Moving toward a new system with new procedures and new technology will scare some people, but not those who have worked with Lowes. Today's CMMS systems are a prime example. We often like our current routine and believe the current methods are adequate. To overcome this, one needs to justify the change and identify what and how the operations will change.

The fourth step is to be flexible. As you go through the sales process, there will be some who will want to modify the system. Some of these suggestions will be helpful and will meet the requirements better; do not perceive all of these suggestions as an attack. If you incorporate these suggestions it can help build a consensus of support. I like to say "be rigidly flexible!"

The last and most exhaustive step is to sell everywhere. Management is a critical step because it controls the dollars. However, do not underestimate the importance of selling to the craft workforce. If there is not "buy-in" at this level, the solution your team spent so much time and effort on may be wasted.

Support and funding for your recommendations can never be obtained without a sales effort. The effort spent in design is only half the work. If there is belief in the recommendation, you must continue to work and complete the second half of the effort—selling a solution that can be implemented

Because bar-coding technology has become so widespread, implementing a system involves people who are not familiar with the intricacies of data acquisition. The purpose of this chapter is to help create understanding toward this tool for maintenance management and leadership. It should assist in educating those of you who have to embark on the journey to improve maintenance. The successful implementation of bar-coding and RFID technologies can support maintenance excellence now and in the future.

So that is auto identification technology? In the simplest term this relates to the automatic data capture using a combination of bar codes, RF readers, and associated network technologies. Bar codes are universally used in the commercial and industrial environment although there are some restrictions to use related to data reads based on line of sight. This is one of the primary advantages of RFID over bar codes—this brief introduction to RFID and automatic identification with the overview of terms and terminology can lead you to beginning or maybe just renewing the continuous improvement process for better maintenance information plus another important element of CRI, MRO parts and material resources shown in Fig. 4.4.

Chapter

23

Case Study: Planning for Maintenance Excellence in Action at Lucent Technologies*

Planning for maintenance excellence requires planning at the strategic level and at the shop-floor level. Without effective planning and scheduling, maintenance operations continue to operate in a reactive, fire-fighting mode wasting their most valuable resource—craft time. Achieving world-class status for the total operation requires a world-class maintenance operation. The effective planning and scheduling of your most valuable maintenance resource—craft skills and labor—can provide the next step forward in your plan for maintenance excellence.

Taking the Next Important Steps

In late 1995, Lucent's Omaha Works Maintenance department (covering two distinct plants) decided it needed to put together a program that could improve maintenance effectiveness, efficiencies, and quality—basically

*This chapter is in memory of Eric Peterson, Maintenance Planner, Lucent Technologies, Omaha, N.E. In collaboration with David Holley, Maintenance Planning and Scheduling Supervisor, Lucent Technologies in 1996.

overall craft effectiveness (OCE) and customer service. The department had traditionally been a break-and-fix organization. Break-and-fix maintenance is sad because it is a reactive style of maintenance that depends on craftspeople to address maintenance problems or equipment problems that have already occurred. The equipment is run until near death, then dies/breaks with no maintenance in between. When dead, it fails to deliver its designed function(s). George S, the department manager for machine maintenance, realized that there had to be a more efficient way to manage the maintenance operation. He was absolutely right! In January 1996, he brought in David Holley, a supervisor who had a background in developing *planned maintenance* programs into the maintenance organization. Mr. Holley had an extensive background in designing planned maintenance programs for break-and-fix organizations. Before installing a planned maintenance program, Mr. Holley's first step was to educate everyone in the maintenance organization about some of the wasteful characteristics of break-and-fix maintenance: travel time, repairs, wait time, obtaining parts and materials, and idle time. Valuable craft time is wasted by a strategy of "run-to-failure" and "fix it when it breaks."

Maintenance Planners: A Critical Technical Resource

After identifying areas to improve, the next phase of this project was to find out what part of the new planned maintenance program needed to be initiated first. It soon became apparent that the greatest need was creating and staffing maintenance planner positions. As Red Motley once said (a) "You can not sell from an empty wagon" and (b) "everybody sell something!"

When asked by management why there was a need for maintenance planners, Mr. Holley said, "It has been my experience and that of experts in the field of maintenance improvement programs that most of these programs will fail without maintenance planners." The reason for this is that maintenance supervisors, especially break-and-fix supervisors, are so caught up in their operations that they cannot adequately plan maintenance work. The crafts are also so busy being fire fighters "they must look up to look down" they are so far behind. They cannot effectively plan their jobs either. On the other hand, maintenance planners have no people responsibilities and are usually not drawn into the do-it-now environment. It is a simple formula for success: maintenance supervisors are leaders responsible for the do-it-now process and people issues, and the maintenance planners are technical resources responsible for the effective planning, estimating, and developing schedules so that maintenance leaders have a good chance of executing for the customer.

Once selected, the maintenance planners' training included learning shop level planning, estimating, and scheduling skills, *preventive maintenance* (PM), and *predictive maintenance* (PdM) techniques. Each gained enhanced computer skills and knowledge of the existing CMMS, and developed improved communication skills. After their training was well underway, the maintenance planners were given an explicit set of standard operating procedures to guide the new planning for maintenance excellence process. They were to improve the following:

- *Equipment performance* by planning projects and/or repairs that returned the equipment as closely to its original operating conditions as economically possible
- *Equipment reliability* by collecting and using *computerized maintenance management system* (CMMS) data to track recurring equipment failures, to plan repairs, to correct them, and to document each as a future standard job plan as a reusable template in CMMS
- *Equipment availability* by using PdM technologies to predict equipment failures before they happened

Metrics to Measure Results and Total Operations Success

As expected, working together was a challenge for the maintenance supervisors, planners, and production supervisors. However, both parties were open to the changes that were ahead of them. Mr. Holley met these challenges by employing metrics that tracked the progress of the planned maintenance program and set goals for the supervisors and their people. Key areas included:

- *Planned maintenance versus unplanned maintenance.* The amount of planned work was being done on a monthly basis.
- *Trades effectiveness (utilization).* Wrench time by crafts in actually working on equipment versus total paid craft time available.
- *Planning effectiveness (craft performance).* Maintenance department planned workload efficiency versus planner estimates—by craft areas and not by individual craftsperson!
- *Planned work backlog.* How much planned work was in backlog to include reliable estimated times for PM and PdM tasks.

These metrics and others, 12 in all, permitted the Omaha Works Maintenance department to monitor the progress of the planned maintenance program and make adjustments to the program throughout the first year. All from a one-page report, the Lucent Maintenance Excellence Index.

Results translate to your bottom line

The first year of the planned maintenance program was very active and produced some very encouraging results. The ratio of planned work to unplanned work has gone from around 3 to 18 percent. The second year goal has been set at 28 percent. Trade effectiveness (hands-on wrench time) has gone from 37 to 53 percent and planning effectiveness has risen to 85 percent.

Typically for a 30-person craft workforce, the planner position can provide a five to one payback in terms of improved craft productivity. The gained value of the increased wrench time from existing craftspeople more than paid for the Lucent investment in full-time planners. And both old and recent surveys consistently validate and show that only 30 to 40 percent of craft time is hands-on wrench time in a purely reactive, unplanned maintenance process. For some facility maintenance organizations with widespread customers, like UNC-Chapel Hill, the City of Durham, and Broward County Public School System, the wrench time often is much lower than 30 percent. With almost every operation, small, medium, or large, there is tremendous value to be gained with effective shop-level planning and scheduling.

Determining Your Overall Baseline

At the beginning of the planned maintenance program, Mr. Holley used the "The Scoreboard for Maintenance Excellence" (1993 version) to get a benchmark to compare future results against. At the start of the program the score was well below average, but at the end of one year the score had risen to slightly above average. The three to five year goal is to score in the excellent range which is considered world class maintenance.

A maintenance benchmark assessment as discussed in Chaps. 4 and 7 provides for a complete benchmark evaluation of your current maintenance practices. Planning and scheduling is only one of the many best practices that can improve your total maintenance operation. Benchmark your total maintenance operation with the key tools in this book. Determine how you compare in all the other benchmark areas that can improve your total maintenance operation. Determine how you compare in other areas such as PM/PdM, maintenance storerooms, operator-based maintenance, maintenance shop facilities/tools, equipment, CMMS, craft skills training, and so forth. Define your current needs and priorities as the baseline for a maintenance strategic master plan. This plan is essential and is the most important step.

Conclusion: Maintenance Operations can Plan for Maintenance Excellence

The results of the first year of the Omaha Works Planned Maintenance program was even better than expected. What has surprised everyone was the positive impact that the planners have had on the planned maintenance program and every customer connected with the program. Considering the amount of training and the learning curve involved, it is amazing that they have been able to accomplish so much in such a short time. Based on the Omaha Works experience, if you want to begin a maintenance improvement program of any type, make staffing maintenance planner positions the No.1 priority of your maintenance improvement program.

Achieving world-class status for your maintenance operation requires the investment in best practices such as planning and scheduling. Your maintenance strategic master plan will take time. It should focus on continuous maintenance improvement because there are no quick fixes available. The profession and practice of maintenance in all types of organizations continues to grow in its importance. We must look beyond the status quo, invest in maintenance, and document the return on investment that is possible through more effective maintenance planning.

Planning and scheduling also focuses on the customer and provides a greater level of safe, quality repair service. It helps the customer (operations) become accountable for their part of the maintenance process. It will add professionalism, credibility, and accountability to your maintenance operation. Take the time to understand what the benefits of this best practice area can mean to your organization and your customers and then develop your strategic master plan for maintenance with effective planning and scheduling as a top priority area.

Chapter 22 of this book is dedicated to one of the original Lucent planners, Eric Peterson. I was like his shadow for almost one full day while on site at Lucent in Omaha back in 1996. He was a very skilled craftsperson who met the challenge when Dave Holley went looking for planner candidates. I got to know him pretty well while on site. I vividly remember him as we attended the production meeting on Thursday afternoon. Here he reviewed the next week's PM and repair schedule. His plan was "sold" to the customers in production. His positive attitude and pride in work as a planner clearly came through for all to see in this plant. His physical presence is missing at this plant now due to a fatal accident away from work. But his memory and spirit live on with his friends, coworkers, and family. Even those like me who knew him for just a few days in 1996 can never forget him.

Part 5

Validating Best Practice Results with the Maintenance Excellence Index

Chapter

24

Determine and Quantify Benefits and Gained Value

This chapter is the first of the four chapters in Part V, "Validating Best Practice Results with the Maintenance Excellence Index." Part V gets down to the very detailed level of validating results after previous chapters have helped define what your results can be if implementation is successful. This chapter provides a methodical process for you to apply. The devil is in the details of course for getting baseline data/information. Savings (both tangible and intangible) must be estimated and established at their most realistic levels. When we do this in every "Scoreboard for Maintenance Excellence" assessment, we bet our reputation on your outcomes, that is, if you apply our specific recommendations from the major scoreboard assessment categories. As we have said repeatedly in many previous chapters, maintenance leaders and the craft leader and support staff must be given the best practice tools, the people resources, and capital investments to address the improvement opportunities. Then they in turn can be accountable for results that are projected as gained value, increased capacity or whatever term you use for defining your bottom line benefits. We will review the typical general benefits we all have read about before in other books and articles and heard during countless speeches by the well-known "maintenance experts and gurus." Some we see selling snake oil, some peddling the over 200 brands of *computerized maintenance management system* (CMMS), and some doing maintenance excellence consulting like we do. Later we will get into more specific areas of benefits and gained value.

General Benefits

Maintenance operations that have committed to continuous maintenance improvement and are applying today's best maintenance practices, principles, and leadership philosophies can achieve significant improvements. Investments in maintenance that successfully implement the practices listed earlier can achieve results that are comparable to the following:

- 15 to 25 percent increase in equipment uptime
- 20 to 30 percent increase in maintenance productivity
- 25 to 30 percent increase in planned maintenance work
- 10 to 25 percent reduction in emergency repairs
- 20 to 30 percent reduction in excess and obsolete inventory
- 10 to 20 percent reduction in maintenance repair costs

Other improvements that are possible will include:

- Improved product quality
- Improved utilization of equipment operators
- Improved equipment effectiveness and capacity
- Improved equipment life
- Improved productivity of the total operation
- Improved profit and customer service

There are a number of key areas where direct savings, cost avoidances, and gained value can be established and clearly documented.

a. Gained value from increased craft labor utilization/effectiveness due to increase in wrench time
b. Gained value from increased craft labor performance
c. Gained value of clerical time reductions for supervisors, planners, engineering, and administration staff
d. Value of asset/equipment uptime
e. Value of facility availability or cost avoidance from being nonavailable
f. Value of *maintenance repair operation* (MRO) materials and parts inventory reduction
g. Value of overall MRO materials management improvement via better procurement practices
h. Value of overall maintenance costs reductions with equal or greater service levels

i. Value of increased facility and equipment life and net life-cycle cost reduction

j. Value of increased direct labor utilization (production operations)

k. Value of production operations productivity increases

Sources of Benefits and Gained Value from Investments in Maintenance

The purpose of the following material is to present specific sources of benefits and gained value from maintenance process improvements. The results of using this process can then be used to support your organization's methodology for calculating *return on investment* (ROI) and support your personal justification for investments in maintenance improvement. The material presented is broken down to allow for various cost savings elements to be used where they apply. You can omit those that are not applicable to your operation or where good data is not available.

This process will allow your operation to customize and tailor the cost justification to meet your own circumstances and requirements for determining ROI for capital investments. First, we will look at a summary of the four primary sources of benefits and then go through a step-by-step process. The four primary sources include:

a. Craft labor productivity

b. MRO materials management

c. Major project cost savings

d. Asset availability and throughput capacity

Craft labor productivity

The details measuring and defining gained value from craft productivity improvement was included previously in Chap. 13. The following summarizes the three elements of craft productivity and *overall craft effectiveness* (OCE).

Element 1: craft utilization. The first element of the OCE factor is *craft utilization* (CU). This element relates to measuring how effective we are in planning and scheduling craft resources so that our people assets are doing value-added, productive work. CU is about wrench time. Effective planning/scheduling within a proactive maintenance process is key to increased wrench time. It is having the right part, at the right place, in time to do scheduled work with minimal nonproductive time on the part of the craftsperson or crew assigned to the job.

Element 2: craft performance. The second key element affecting OCE is *craft performance* (CP). This element relates to how efficient we are in actually doing hands-on craftwork when compared to an established, planned time, or performance standard. CP is directly related to the level of individual craft skills and overall trade experience, as well as the personal effort and motivation of each craftsperson or crew. Effective craft skills training and continuous technical skill development contribute to a high level of CP.

Element 3: craft service quality (CSQ). The third element affecting OCE relates to the relative level of the methods being used considering personal and shop tools, special shop equipment, shop work areas, repair methods, and so forth as compared to "current state-of-the-art methods." This element can include callbacks where the poor quality of the initial repair requires another trip to fix it right the second time. Typically, the *crafts service quality* (CSQ) element is a more subjective value and is not determined based on actual data like CU and CP. However, the CSQ level does affect overall craft labor productivity.

Increasing OCE provides greater craft capacity. Chapters 13 and 14 have illustrated that increasing OCE provides greater craft capacity and gained value in terms of available wrench time. Doing more with available craft assets does not lead to immediate head count reduction. Typically, operations that gain productive craft hours desperately need to invest the time elsewhere. We cannot automatically and indiscriminately reduce head count when we measure and subsequently improve OCE. Long-term stabilization and reduction of head count can occur. We feel strongly that attrition can absorb valid staff reductions that may result over the long term. All areas related to craft labor productivity can contribute to well-documented ROI for best practices and effective CMMS. Effective CMMS allows for the performance measurement data to be easily collected as well as almost all metrics that we will later cover for developing the *Maintenance Excellence Index* (MEI).

MRO materials costs

The MRO parts and material cost is related to the frequency and size of the repairs made to the equipment and facility assets. The sheer volume of MRO parts and material, in addition to the storeroom's operating policies, methods of storage/retrieval, MRO purchasing, and overall inventory management practices contribute to the overall MRO maintenance materials costs and potential savings.

Very little attention is paid to MRO materials in some operations, and inventories may be higher than necessary by some 30 to 50 percent. This results in increased MRO inventory holding costs and makes materials unnecessarily expensive. The inability of the storeroom to service the customer (the maintenance department) needs or to maintain an accurate inventory, results in "pirate" or "illegal" storage areas for just-in-case spares on the shop floor. This practice also drives up the cost of MRO parts and materials.

Good inventory control enables companies to lower the value of the inventory and still maintain a service level and inventory accuracy of at least 95 percent. This allows the maintenance department to be responsive to the operations customer group, while increasing their own personnel productivity. Successful CMMS users can achieve lower material costs and overall measurable reduction in total inventory value.

Major project cost savings

In many operations, maintenance is involved in major projects: shutdown repairs, major outage and turnaround projects, or overhaul and refurbishing activities. These activities, if not properly controlled, can have a dramatic impact on the operation's capacity and throughput. These activities are usually performed with the critical equipment out of service. This means that any time that can be eliminated from the major project work can be converted back to production time.

Improved planning and coordination can be achieved with effective CMMS and effective planning and scheduling. This will often help to shorten the downtime, even if the operation is currently using a project management system. Successful CMMS users can expect an average 5 to 10 percent reduction in outage time due to better execution and planning of major project work.

Asset availability and throughput cost savings

Improving asset availability and increasing throughput capacity is an important gained value for many operations. Downtime cost for equipment assets may vary from several hundred dollars per hour to virtually hundreds of thousands of dollars per hour of asset production time.

Private and public sector examples. One pharmaceutical company I visited in Virginia has a production asset in its plant, with the value of downtime being worth $1 million of product per 24-hour day. Another operation has critical equipment where a 1 percent gain in uptime equates to $50,000 in profits. In some operations, the levels of downtime can run as high as 30 percent or more. This can result in lost sales opportunities, unnecessary expenditures for capital equipment, poor quality, reactive maintenance, and generally puts the company in a weak and unreliable competitive position. By enforcing good maintenance policies and best practices, and by utilizing the CMMS as a tracking tool, asset uptime can be dramatically improved. We will now look more specifically at the sources of benefits for these four major areas.

Determining craft labor productivity cost savings—two methods

Method one. The following method provides a guide for determining specific areas where craft time is being wasted by estimating five individual items:

1. Time wasted by personnel looking for spare equipment parts. This is based upon varying levels of key best practices being in place.
   - No inventory system = 15 to 25 percent
   - Manual inventory system = 10 to 20 percent
   - Work order (WO) system and inventory system = 5 to 15 percent
   - Computerized inventory and manual WO system = 0 to 5 percent
2. Time spent looking for information about a work order —— percent
   - Manual WO system = 0 to 5 percent
   - No WO system = 5 to 10 percent
3. Time wasted by starting wrong priority WO —— percent
   - Manual WO system = 0 to 5 percent
   - No WO system = 10 to 15 percent
4. Time wasted by equipment not being available —— percent
   - Manual WO system = 0 to 5 percent
   - No WO system = 10 to 15 percent
5. Training time, leave, meeting, and other time —— percent
6. Total of all percentages of nonwrench time
   - Items 1 + 2 + 3 + 4 + 5 = —— percent
7. Total number of craftpeople = —— total crafts
8. Total number of crafts (Line 7) × average hours per person per year —— = —— total hours
9. Total time wasted: Multiply the number of total crafts hours (Line 8) × percentage time wasted (Line 6) = —— hours
10. The average craft hourly rate: $ ______ per hour
11. Total cost of wasted time: multiply time wasted (Line 9) by average craft hourly rate Line 10 = $ —— costs
12. Multiply the $ cost figure in Line 11 by the percentage that best describes your operation: Line 11 $ —— costs × —— percentage estimated here
    - No WO or inventory system = 75 to 100 percent

TABLE 24.1 Example from a Large Multisite Maintenance Operation with a Craft Workforce of 2762 and a Projection of a 5 Percent Improvement in Wrench Time/Craft Utilization

Gained value from improving wrench time (WT) by 5%								
Cost per hour ($)	Paid craft hours per year	Labor cost per craft employee ($)	WT (%)	Total WT hours	Cost per WT hour ($)	Gained value of WT hours per craft employee	Total ACC craft employees	Total potential gained value WT
33	2080	68,640	30	624	110.00		←If baseline @ 30% WT	
33	2080	68,640	35	728	94.29		←Improvement to 35% WT	
		Gained value	+5% increase WT Gain	+104 WT Hours Gained	$15.71 per WT hour decrease	$1634 gained value/craft	2762	Total $4,513,108 gained value potential

- Manual WO system = 50 to 75 percent
- Manual WO and inventory system = 30 to 50 percent
- Computerized inventory and manual WO system = 25 to 40 percent

13. Total savings (Line 11 $ —— cost × Line 11 —— percent = $ —— costs

This represents one method to calculate the projected number of craft labor hour savings and craft labor productivity improvements.

Method two. Method two was shown previously in Chaps. 13 and 14 and is the recommended method for defining craft productivity gains. This method requires that a simple method of tracking nonwrench time be established for both wrench time (CU) and CP. Appendix G provides a very convenient customer service pocket pal that can track nonwrench time as well as provide a very convenient pocket work order and time-keeping method. When using this method, simply substitute your actual number of craftspeople and the average hourly craft labor rate into the calculations. This method requires that you use an estimate of baseline CU and performance levels and measure both or even just one of these areas on a regular basis to document cost savings. An actual example is shown in Table 24.1 of a large multisite maintenance operation with a craft workforce of 2762 and a projection of a 5 percent improvement in wrench time/CU only.

Determining MRO materials management benefits

1. Total dollar value of MRO parts and materials purchased per year $ ——
2. Percentage of time MRO items are in stores when same items are purchased —— %
   - No inventory system = 25 to 30 percent
   - Manual inventory system = 10 to 20 percent
   - Computerized inventory system = 5 to 15 percent
3. Cost total (cost avoidance) Line 1 × Line 2 $ ——
4. Additional savings (inventory overhead cost) $ ——
   - Multiply Line 5 × 30 percent average carrying charge
5. Estimated total inventory valuation = $ ——
6. Estimated inventory reduction
   - No inventory system = 15 to 20 percent
   - Manual system = 5 to 10 percent
     - Estimated one-time inventory reduction (obsolete or unnecessary spares) = $ ——

7. Estimated additional savings Multiply Line 6 by 30 percent average carrying cost = $ ——
8. Number of actual stock-outs that cause downtime, Total = ——
9. Amount of actual downtime (in hours), Total caused by stock-outs = —— hours
10. Actual cost of downtime (per hour) $ ——
11. Total cost of MRO materials related downtime. Multiply Line 10 by Line 11 $ ——
12. Percentage of savings obtainable —— %
    - Current controls poor—75 percent
    - Current controls fair—50 percent
    - Current controls good—25 percent
13. Savings in MRO parts and materials-related downtime $ —— Multiply Line 11 by Line 12
14. Total MRO Materials Management Related Savings
    - Add lines 3, 4, 7, 8, and 13 = $ ——

Determining major project benefits

1. Number of major projects, outages, and overhauls per year Total = ——
2. Average major project length (in days), Days = ——
3. Cost of equipment downtime in lost sales using hourly $ —— Downtime rate times total hours of outages for major projects
4. Total estimated cost per year
    - Multiply lines 1, 2, and 3 = $ ——
5. Estimated savings percentages ——
    - No computerized WO system = 5 to 10 percent
    - Project management system = 3 to 8 percent
    - Project management and inventory control systems = 2 to 5 percent
6. Total cost savings ——
    - Multiply lines 4 and 5

For savings from a project, outage, or an overhaul to be used in this total, there must be equipment downtime, production losses, or delays impacting "sold product." Line 2 is the total number of days for all of the major projects, outages, and/or overhauls for the company. These must be measurable periods of downtime for the equipment resulting in production loss or delays. The average cost of equipment downtime per day or per hour must be established for the equipment impacted by the major projects, outages, and/or overhauls identified in Lines 1 and 2. These costs are usually incurred when the equipment capacity is required or the resulting downtime requires the product to be diverted or delayed.

Line 5 is the percentage of the total equipment downtime related to major projects, outages, and/or overhauls for the company that could be saved if good maintenance controls and project management processes were implemented. The averages are based on the current level of controls presently utilized by your operation. Line 6 provides the total projected savings from better controls on the major projects, outages, and overhauls using an improved CMMS.

Asset availability and throughput capacity

1. Percentage of equipment downtime for one year —— (If not known, use a realistic estimate—average for industry is 5 to 25 percent)
2. Total number of production hours for equipment for one year ——
3. Total of all lost production hours for year —— (multiply Lines 1 and 2)
4. Multiply total lost production hours for one year (from earlier Line 3) by your percentage from below:
   - No WO system = 25 percent
   - WO system = 20 percent
   - WO and stores inventory system = 10 percent
5. Total of increased uptime hours (Line 4) —— hours
6. Cost of downtime per hour $ —— per hour
7. Total value of uptime ——
   - Multiply Lines 5 and 6 potential operational savings
8. Total direct labor wages and benefits times the $ —— total of all lost production hours
9. Lost sales volume for one year $ ——
   - Divide the total sales for year by the total number of yearly production hours and multiply this figure by the total downtime hours saved
10. Increased production costs to make up production loss due to downtime $ ——
    - This would include the extra labor required on weekends or off shifts to operate the equipment, extra energy costs to operate the equipment, and the like

Asset availability and throughput capacity costs savings explanation. Line 1 is the percentage of equipment downtime for the total operation. In order to obtain the total value for the plant equipment downtime, it may be necessary to calculate this section for the major critical pieces of equipment. The values given are industry averages. These should only be used if your actual downtime is not known.

Line 2 is the total number of production hours the equipment was scheduled for the year. Because the downtime is only accumulated when the equipment is scheduled to operate, this number must be derived from the production schedule.

Line 3 is the total number of lost production hours per year, obtained by multiplying Lines 1 and 2.

Line 4 is the percentage of Line 3 that could possibly be saved by implementing the disciplined controls in a CMMS and other best practices. This percentage is based on the current condition of the operation. The industry averages listed provide guidelines for the possible savings.

Line 5 is the result of multiplying the percentage in Line 4 and the lost production hours in Line 3. This is the total number of downtime hours that are projected to be saved by implementing best practices.

Line 6 is the cost of 1 h of downtime for the production equipment used to compile the total in Line 3. This line may vary from equipment to equipment within the plant. You may take an average cost for all of the equipment, or perform the calculations in this section for each individual piece of equipment and total the individual sums.

Line 7 is the result of multiplying Line 5 by Line 6. This total is the downtime savings projected by implementing the controls and disciplines of a CMMS and other best practices.

Line 8 is an optional consideration for downtime cost savings. It is the wages and benefits for the total number of idle operator hours incurred during the equipment downtime. This is typically calculated by multiplying the downtime hours by the number of operators assigned to the equipment. In some cases, this value is already added into the equipment downtime cost figure. If this is the case, this value should not be used.

Line 9 is another optional consideration for downtime cost savings. It is the value of the lost sales for the plant. This value is only usable if the plant is in a sold-out condition and the downtime resulted in a lost sale for the product. The value would be the amount of the product not produced during the downtime.

Line 10 is another optional consideration for the downtime cost savings. It is the increased production costs to make up the product lost during the downtime. These costs are somewhat more difficult to determine, but would include the extra wages paid to the operating and maintenance staffs to cover the new production times scheduled. This is usually one-and-half times their base rate. These costs would also include the additional equipment utility charges for operating the equipment when it should have been down. There are also other paperwork charges incurred for changing material flow, transferring material from the equipment to storage and back to the equipment, and so on. The value of Lines 8 through 10 should be added to the total in Line 7, if they are applicable.

Total projected benefits

1. Total from craft labor productivity $ ——
2. Total from MRO materials management $ ——
3. Total from major projects $ ——
4. Total from asset availability and throughput capacity $ ——
5. Total savings possible from effective CMMS installation $ ——
   - Total of Lines 1 to 4
6. Total projected investment for CMMS $ —— (Include hardware, software, support agreements, training, and implementation cost to include costs of all internal and external personnel resources)
7. Simple rate of ROI = ———
   - Divide Line 6 by Line 5 or use Line 5 for your operations economic evaluation requirements for new investment

Other Important Areas Providing Benefits

The quality improvement impact of maintenance excellence

Because the maintenance department is responsible for the equipment condition, quality costs are impacted by poor maintenance practices. For example, what percentages of all quality problems eventually are solved by a maintenance activity? Even if the activity is performed by the operator, the activity is one of maintaining the equipment condition.

In some companies, 60 percent or more of the quality problems are equipment related. In order to calculate the possible cost savings, the value of the annual production for the plan should be calculated. Next, the first pass quality rate should be determined. The difference between this and 100 percent gives the current reject rate. The next step would be to determine the reasons for the rejects. Usually, a top 10 list will provide the majority of the rejects. After examining the list, determine which causes have a maintenance solution.

This is the percentage amount that could possibly be reduced. An estimate must be made of what percentage of all the maintenance-related losses could be eliminated by a good maintenance program for PM and PdM. This percentage multiplied by the dollar value of the company's annual product, will produce the possible quality-related savings. This number should then be added as a line item to all of the previous savings.

Paperwork/Clerical Savings

This section looks at the decrease in paperwork and clerical support that an improved CMMS requires compared to the current methods. The current time required may be supplied by maintenance clerks, planners, supervisors, or even the maintenance manager. If the responsibilities are spread over multiple individuals, use the average labor cost for the calculation.

Maintenance paperwork/clerical cost justification

Use reliable self-estimation of this time or use pure work sampling data for weekly averages for:

1. Amount of time to plan WOs ——
2. Amount of time to closeout WOs ——
3. Amount of time preparing management reports ——
4. Amount of time to update equipment histories ——
5. Amount of time to generate weekly schedules ——
6. Amount of time spent preparing PMs ——
7. Amount of time spent preparing bill of materials for the —— weekly schedule

Clerical/paperwork savings for purchasing

Use weekly averages for:

1. Amount of time to prepare purchase requisitions ——
2. Amount of time to consolidate purchase requisitions ——
3. Amount of time to prepare purchase orders ——
4. Amount of time to update purchase order history ——
5. Amount of time to generate purchasing reports ——
6. Amount of time spent contacting vendors for pricing ——
7. Time spent on other purchasing-related clerical activities ——
8. Total purchasing-related paperwork/clerical time per week —— (total Lines 1 to 7)

9. Anticipated reduction in clerical/paperwork time required —— (usually 20 to 30 percent)
10. Total hourly savings —— (multiply Line 8 by 9)
11. Average hourly clerical/paperwork cost ——
12. Total dollar savings from clerical/paperwork improvements —— (multiply Line 10 by 11)

Clerical/paperwork savings for engineering

Use self-estimated weekly averages for:

1. Amount of time to find drawings for WOs ——
2. Amount of time to update equipment drawings ——
3. Amount of time updating PM program ——
4. Amount of time performing failure analysis ——
5. Amount of time performing reliability engineering ——
6. Amount of time spent providing maintenance information ——
7. Time spent on other maintenance-related engineering clerical activities ——
8. Total maintenance engineering-related paperwork/clerical time —— per week (total Lines 1 to 7)
9. Anticipated time reductions of engineering staff doing clerical/ paperwork tasks, usually in a range of 15 to 25 percent) ——
10. Total hourly savings (multiply Line 8 by 9) ——
11. Average hourly clerical/paperwork cost ——
12. Total dollar savings from clerical/paperwork improvements —— (multiply Line 10 by 11)

Summary

When maintenance leaders and others question profit-centered maintenance for in-house maintenance, look at the numbers; you can be very surprised. If your estimate is only $100,000 and you are only half right, that is still $50,000. You can be sure today's growing numbers of contract maintenance providers who will know their numbers when they propose your operation as a takeover target to your top leaders. Let us hope they do not do it before you get your plans of action in action and implemented. This material has provided proven methods for defining sources of benefits and justifying the implementation best practices to include a CMMS. And this material will support determining your unique ROI options. The total costs of many improvement projects with a new CMMS can be in access of $1 million with all internal and external costs as shown for SIDERAR in Chap. 12. Management must view maintenance process improvements as ROI and not merely another cost associated with maintenance and physical asset management. This information can be your guideline for determining the ROI for almost any maintenance excellence initiative.

Chapter

25

Developing Your Maintenance Excellence Index to Validate Results

The key to profit- and customer-centered maintenance is measuring results and *return on investment* (ROI). If you were a maintenance contractor, your results would be profit and customer service. *The maintenance excellence index* (MEI) is the third benchmarking tool that takes benchmarking down to actual results at the shop level as shown in Fig. 4.1.

Developing Key Performance Indicators

The process for developing your MEI includes defining and gaining consensus on very specific key performance indicators related to the total maintenance operation as well as metrics that affect the success of the total operation. This chapter covers a recommended set of internal benchmarks or metrics for maintenance leaders to consider, the purpose for each metric, where they traditionally can be found in the *computerized maintenance management system* [CMMS (or financial system)] database, how to calculate each one and how to determine your current baseline. Each element for developing and calculating your MEI is covered step-by-step in this chapter. Most importantly, this chapter recommends an attainable performance goal and how your own uniquely developed MEI will validate results and ROI for maintenance operations improvements.

Key performance indicators to measure the overall effectiveness of a maintenance operation can include:

- *Percent craft utilization (CU).* Evaluates actual wrench time (hands-on time) for craft labor. Provides one of the two key elements for

	Breakdowns
Availability Issues ⇒	Set-up and Adjustments
	Idling and Minor Stoppages
Performance Issues ⇒	Reduced Speed
	Operator Issues
	Process Defects
Quality Issues ⇒	Reduced Yield

Figure 25.1 Six major losses plus operator issues.

measuring the *overall craft effectiveness* (OCE). Measures the overall increase in craft labor wrench time due to a proactive, planned maintenance strategy with effective planning and scheduling, positive impact from the *preventive maintenance/predictive maintenance* (PM/PdM) program, effective *maintenance repair and operation* (MRO) materials management service, and improved CMMS.

- *Percent craft performance (CP).* Evaluates actual CP against a reasonable/reliable planned time for a planned repair job or task such as PM inspections. Where craft labor utilization measures "effectiveness," this measure addresses the "efficiency" factor for *overall craft effectiveness* (OCE). This measure is improved by having effective craft skills to do the job along with the motivation to work efficiently. It is directly impacted by shop working areas, having the right personal and special tools available, and safe working conditions.
- *Overall equipment effectiveness (OEE).* A world–class metric originated from the *total productive maintenance* (TPM) movement that evaluates critical equipment in terms of equipment availability, equipment performance, and the quality of output. Improving OEE focuses on eliminating the six major losses; we feel it should be seven and include operator issues as highlighted in Fig. 25.1.

  The average OEE factor is in the 40 to 50 percent range before an improvement process starts. A world-class OEE factor is around 85 percent, which means that all three elements must be around 95 percent, that is, 0.95 (availability) × 0.95 (performance) × 0.95 (quality) = 0.857 × 100 = 85.7% OEE factor.

  The OEE factor = availability % × performance % × quality %
- *The overall craft effectiveness (OCE) factor.* The OCE factor relates to craft labor assets as compared to the metric for OEE that measures the combination of equipment asset availability, performance, and quality output—basically *asset productivity*. The OCE factor focuses upon measuring and improving the value-added contribution that people assets make to total physical asset management. Chapter 14 included more detailed information about the OCE factor, an important and often overlooked segment for measurement. OCE includes CU, CP, and craft service quality (CSQ) as shown in Fig. 25.2.

OCE =	CU%	×	CP%	×	CSQ
	Craft Utilization	×	Craft Performance	×	Craft Service Quality

Figure 25.2 Overall craft effectiveness.

- *Craft service quality (CSQ).* Measure the number of callbacks or percentage of craft rework, and the quality of maintenance repair work. It provides one key indicator for maintenance customer service and helps focus on "doing the repair right the first time."
- *Schedule compliance.* Percent of jobs completed as scheduled, evaluates the overall effectiveness of executing the planned work on the agreed upon schedule by the craft workforce.
- *Percent of planned work.* Measured either by the percentage of work orders (WOs) planned or percentage of actual craft hours on planned work. This metric evaluates the overall effectiveness of the planning process as well as the impact of all maintenance best practices to promote a proactive maintenance repair strategy, that is, PM, PdM, reliability improvement actions, effective MRO support, and so forth.
- *Percent of WOs with reliable planning times.* Provides a key measure for the maintenance planner position. Evaluates the ability and effectiveness of the maintenance planner/coordinator position to establish reliable planning times using maintenance standard data or other available data for determining planning times. The objective for the planning process is to have as many jobs scoped and planned as possible to the level that reliable estimates are established.
- *Percent of planned WOs generated from PM/PdM program.* Provides valuable feedback that the PM/PdM effort is helping avoid catastrophic failure and is truly providing benefits. Measures how well PM/PdM program is detecting deficiencies before catastrophic failure or downtime occurs. The reliability improvement goal, however, is not to continue to fix before failure but to eliminate root causes of failure and in turn the failure rate. However, this is an important metric as a client's PM/PdM program is reinforced or renewed and in turn begins to provide measurable results.
- *Percent PM/PdM compliance.* Measures the execution and compliance to completing scheduled PMs, PdM data collection, lube services, scheduled structural/process inspections, calibrations, and the like. Typical completion of scheduled PM/PdM actions required within a week's window of time is used as criteria for compliance. This metric can also apply to instrumentation calibration and to any regulatory inspections such as crane inspections and fire protection testing/PM.

We have seen large operations even print out daily operator-based inspection work orders for material handling equipment—a very real requirement from OSHA.

- *Percentage storeroom service level or number of stock-outs.* Evaluates the MRO inventory/materials management process capability to have availability of stock items that are normally stocked in the maintenance storeroom. A 95 percent plus service level should be the goal. Provides an excellent countermeasure to ensure planned inventory reduction goal is not detrimental to storeroom customer service. Does not measure nonstock items that require requisitioning/purchasing as direct purchases or availability of project-related items
- *Percent of inventory accuracy.* Maintains accurate fiscal accountability of stocked items to ensure total confidence in current inventory levels and dollar value of MRO inventories. Measures the effectiveness of the cycle-counting process and storeroom control
- *Dollar value of MRO inventory reductions.* Helps to ensure that proactive MRO inventory management practices and well-planned MRO inventory reductions are given proper credit and recognition within the organization. This reduction may also serve to offset additions to inventory of items that are more useful and critical spares. Again, wherever we recommend this one we always recommend that our client also measure stock-outs, because we can shoot our self in the foot with indiscriminate, on site inventory reductions demanded by financial top leaders.
- *Percent of asset utilization / availability.* A good metric that evaluates how well the overall capacity of an asset is being used in the operation. This metric can often be used alone for critical assets. In addition, it can easily be expanded into the complete measurement of OEE. Smart operations will have this information flow via the shop-floor reporting system. In this case, maintenance must ensure that the downtime call by operations is correct. Manufacturing managers can easily manipulate this call when their production goals are threatened and they need an easy excuse for not making their production target. Therefore, maintenance and operations must both work with a cooperative spirit on this one because everyone loses when production misses promised shipment of sold products.
- *Maintenance cost per unit of output.* Measures bottom-line maintenance cost per unit of output and evaluates net improvements related to maintenance improvements on total operation costs. Excellent metric for process-type industries and/or for discrete manufacturers with major single product output and a strong standard cost and production reporting system. BigLots headquartered in Columbus, Ohio has a measured day-work plan for distribution center employees. As

a result, they have a very close count on cartons shipped each day for operations. Therefore, we were able to use maintenance cost per carton very easily by using operations counts of output, which are "equivalent cartons being shipped to BigLots' stores." In addition, each of their four major distributions centers had significantly different average maintenance cost per carton. This provided a very good internal benchmarking tool.

- *Maintenance cost as percent of total operations cost.* Provides an overall comparison of how maintenance cost impacts the total operations cost. For very similar operations, this metric may be useful. It can be harmful in the hands of financial people who do not realize that it is extremely difficult to get true comparisons between different organizations, even those operations like universities. Comparing, centuries-old Oxford University facilities maintenance cost to a newer modern facilities complex, such as UNC-Wilmington, requires clear definition of total maintenance requirements.
- *Other metrics to consider:*
  - Percent of overtime.
  - Percent of actual craft time charged to WOs.
  - Percent of overdue WOs based upon a clearly defined priority system.
  - Percent of craft hours for pure emergency work (or percent of total WOs).
  - Percent of PM/PdM coverage, that is, the equivalent craft hours invested in PM/PdM tasks which can also be shown as equivalent craftspeople.
  - Response time to priority jobs, such a true life safety issues, and based upon a priority system with a response time focus (e.g., our local natural gas service company responded in 6 min to an emergency call by the author in reference to a gas leak in my home furnace).
  - Percent of closed WOs without reliable planning times (excludes pure emergency jobs).
  - Percent of craft hours charged back to customers: We highly recommend charging back all maintenance cost to users of maintenance services whether in a plant, a facilities complex, as well as fleet of healthcare maintenance operations.
  - Percent of jobs by one-person crew (when safety does not require a two-person team).
  - Percent of craft hours by WO type.
  - Percent of craft hours by cost center/profit center.
  - Customer or maintenance satisfaction with planner support: (from results from periodic surveys with customers of maintenance and with maintenance supervisors and crafts. Appendix D provides a craft

survey used for over 3000 crafts throughout the Boeing Commercial Airplant Group's five manufacturing regions in late 1990s.

- Percent of work orders covered with planned job packages: Measures volume not quality of the planned jobs where a complete job package was required includes more detailed list of repair tasks by craft type, parts bill of material, permits, drawings, and the like.

Maximizing maintenance for profit optimization and improved customer satisfaction is the goal. It is not a report with evaluation details from "The Scoreboard for Maintenance Excellence" neatly bound in a report and a Powerpoint presentation to top leaders and maintenance leaders. The goal is successful implementation of prioritized improvement opportunities from the benchmark evaluation and to help improve all internal resources to do a better job for the tenant/customer. Doing a better job for the internal customer is also direct support to profit optimization. This section introduces the third benchmarking tool from this book—"The Maintenance Excellence Index." By applying all benchmarking tools in this proven approach from The Maintenance Excellence Institute, the journey toward maintenance excellence is well underway with these three essential tools to measure results and long-term contribution to real profit or "imagined" profit and validated gained value in not for profit operations.:

1. "The Scoreboard for Maintenance Excellence" for maximizing overall best practices
2. "The CMMS Benchmarking System" for optimizing your IT investment
3. "The Maintenance Excellence Index" (MEI) for validating bottom line shop-level results

This section covers the process of defining and gaining consensus on very specific key performance indicators related to the total maintenance operation. It covers a recommended set of internal benchmarks or metrics for today's maintenance leader, the purpose for each, where they traditionally can be found in the CMMS (or financial system), how to calculate each one, and how to determine your current baseline. Each step and element for developing and calculating your own MEI is covered. Most importantly this chapter recommends an attainable performance goal and how your own uniquely developed MEI will validate results and ROI for maintenance operations.

Develop Method to Measure and Validate Results

This topic is last, but it must be foremost in our minds as we begin the benchmark evaluation of physical asset management and the maintenance process. Each of the 300 evaluation items on "The Scoreboard for Maintenance Excellence" must be viewed in terms of whether or not there are tangible or intangible benefits possible. If we are able to make improvements that generate benefits, can we measure them? Often performance measurement is something new to the in-house maintenance operation, but we highly recommend that a performance measurement system be put in place. Contract maintenance providers understand the value of measurement so that their customers clearly see value-added services received. Justification for investments in maintenance best practices for in-house maintenance operations must be validated. If your maintenance operation was a third-party contract maintenance provider you would expect a profit. Therefore, we too must measure and validate results from internal maintenance improvement.

Initiate an maintenance excellence index

Our approach has been to help clients focus on results and create an important deliverable: the MEI that includes 10 to 15 key performance indicators with agreed upon weighted values. These metrics are then used to provide a one-page Excel spreadsheet that brings all client metrics together into a composite total MEI performance value as shown in Fig. 25.3—an example with 13 key performance indicators.

The MEI process gives a composite internal benchmark as to how all six maintenance resources are contributing to profit optimization and customer service. The metrics selected should be applicable to the specific type of maintenance organization. For example, a pure facilities maintenance operation without critical production or operations equipment to maintain would not use OEE as part of its MEI. OEE measurement is an excellent metric best suited to a small number of mission-essential critical assets within a production operation. The following Table 25.1 provides a review of 21 key metrics that also can be considered with purpose, data source, and how to calculate. These 21 metrics encompass the measurement of all key resources necessary for effective physical asset management:

- People resources and craft labor
- Dollar resources related to total operations and overall budget dollars from both maintenance and the customer
- MRO material resources
- Planning resources and customer service
- The physical asset as a key resource, its uptime, availability, and reliability to perform its primary function
- People resources such as a skills-training hour conducted and paid by the employer.

Maintenance Excellence Index Example: 13 Performance Measures															
A. Performance Measures	1. Actual Maintenance Cost Per Unit of Production	2. % Major Work Completed within 5% of Cost Estimate	3. % Overall Maintenance Budget Compliance	4. % Overall Schedule Compliance	5. % Overall PM Compliance	6. % Planned Work	7. % Craft time for Customer Charge Back	8. % Work Orders with Reliable Planned Time	9. % Critical Asset Availability	10. % Wrench Time (Craft Utilization)	11. % Craft Performance	12. % Inventory Accuracy	13. Number of Stock-Outs of Inventoried Stock Items	F. Performance Level Scores	
B. Current Month	1.30	90	94	90	94	68	75	50	90	34	90	90	15	Perf Level	
C. Performance Goal	1.00	95	98	95	100	80	85	60	95	50	95	98	10	10	
D. Baseline Performance Levels	1.05	94	96	94	98	78	83	58	94	48	94	97	11	9	
	1.10	93	94	93	96	76	81	56	93	46	93	96	12	8	
	1.15	92	92	92	94	74	78	54	92	44	92	95	13	7	
	1.20	91	90	91	92	72	75	52	91	42	91	94	14	6	
	1.25	90	88	90	90	70	72	50	90	40	90	93	15	5	
	1.30	89	86	89	88	68	69	48	89	38	89	92	16	4	
	1.35	88	84	88	86	66	66	46	88	36	88	91	17	3	
	1.40	87	82	87	84	64	63	44	87	34	87	90	18	2	
	1.45	86	80	86	82	62	60	42	86	32	86	88	19	1	
	1.50	85	78	85	80	60	57	40	85	30	85	87	20	0	
E. Performance Level Score	4	5	8	5	7	4	6	5	5	2	5	2	5	Perf Level Scores × Weight	I. Total MEI Value
G. Weighted Value of Metric	10	6	6	7	11	7	6	7	13	8	8	6	5		
H. Performance Level Score (E) × Weight (G)	40	30	48	35	77	28	36	35	45	16	40	12	25	467	
J. Total MEI Value Over Time	Date	2/05	3/05	5/05	6/05	7/05	8/05	9/05	10/05	11/05	12/05	2/06	3/06		
	MEI Value	467													

Figure 25.3 Maintenance excellence index example: 13 performance measures.

TABLE 25.1 Potential Performance Metrics for Using on a Maintenance Excellence Index

No.	Performance metric, purpose, and data source	Goal	How to calculate metric
1.	**Percentage of overall maintenance budget compliance:** To evaluate management of dollar assets: *obtained from monthly financials*	98%	Traditional budget variance =/– % Variance to actual planned budget
2.	**Actual maintenance cost per unit of production:** To evaluate/ benchmark actual costs against stated goals/baselines or against industry standards: *obtained from asset records and monthly CMMS WO file of completed WOs for the month. Obtained from production results and financial report. Provides ideal support to ABC costing practices*	TBD	Total maintenance materials and labor per reporting period ÷ total units produced Note: Production units could also be expressed in equivalent standard hours if traditional standard cost system or activity-based costing (ABC) is being used
3.	**Percentage of customer or capital-funded jobs completed as scheduled and within +/– 5% of cost estimate:** To measure customer service and dollar assets plus planning effectiveness: *obtained from customer or capital funded WO types from the CMMS WO files, comparing date promised to date completed and estimated cost to actual cost*	98%	Total number of customer or capital-funded jobs completed as scheduled within budget variance goal ÷ total number of customer or capital-funded jobs completed
4.	**Percentage of other planned work orders completed as scheduled:** To measure customer service and planning effectiveness: *obtained from a query of all planned WO types in CMMS WO files and comparing date promised to date completed.* Could be expressed in percentage based on craft hours.	95%	Total number of planned jobs completed as scheduled within time variance goal ÷ total number of planned jobs completed (Typically higher priority jobs where customer was given a promised completed date, not PM work which is planned which is measured via per cent PM Compliance)
5.	**Schedule compliance:** To evaluate how effective scheduling was for executing planned work to meet scheduled dates/time: *obtained from query of CMMS completed WO file where all scheduled jobs coded and their actual completion compared to actual planned completion date/time*	95%	Total scheduled jobs completed as per the schedule ÷ total jobs scheduled

(Continued)

TABLE 25.1 Potential Performance Metrics for Using on a Maintenance Excellence Index (*Continued*)

No.	Performance metric, purpose, and data source	Goal	How to calculate metric
6.	**Percentage of planned WOs versus percentage of true emergency WOs:** To evaluate positive impact of PM, planning processes, and other proactive improvement initiatives (CRI,/RCM/VE,etc): *Obtained from a query of all true emergency WO types in CMMS WO files and comparing to total WOs completed.* Could be expressed in percenatge based on craft hours	80 to 85% planned	Total emergency type WOs completed ÷ total WOs completed per reporting period (could be expressed as percentage also using craft hours)
7.	**Percentage of craft time to WO for customer charge back:** To monitor craft resource accountability for internal revenue generation (or external): *obtained from a query of all WO types in CMMS WO files that are charged back comparing these craft hours to total craft hours paid*	85%	Total Craft Hours for all work charged back to customer ÷ total craft hours paid for reporting period
8.	**Percentage of craft time to WOs:** To monitor overall craft resource accountability and to support internal revenue generation: *obtained from a query of all WO types in CMMS WO files and summation of actual craft hours reported to WOs.*	100%	Total craft hours charged to all WO types (including standing WOs) ÷ total craft hours paid for reporting period
9.	**Percentage of craft utilization (actual wrench time):** To maximize craft resources for productive, value-adding work, and to evaluate effectiveness of planning process: *obtained from a query of all craft hours reported to noncraft work from CMMS time-keeping WO files and summation of actual craft hours paid from financials*	60 to 70%	Total craft hours of pure wrench time charged to all workorder types minus total time to nonwrench time standing WO types) ÷ total craft hours paid for reporting period
10.	**Percentage of craft performance (against reliable estimates for PM and planned work):** To maximize craft resources, to evaluate planning effectiveness, and also to determine skills training ROI: *obtained from completed WO file in CMMS*	95%	Total actual craft hours charged to completed WOs with planned times ÷ the total planned time from the WOs having reliable planned times

TABLE 25.1 Potential Performance Metrics for Using on a Maintenance Excellence Index (*Continued*)

No.	Performance metric, purpose, and data source	Goal	How to calculate metric
11.	**Craft quality and service level:** To evaluate quality and service level of repair work as defined by customer: *obtained from WO file in CMMS where all call-backs are tracked and monitored via work control, WOs and the planning processes*	95%	100%—total number Call-Backs/etc ÷ total number of WOs completed per reporting period
12.	**Overall craft effectiveness (OCE):** To evaluate cumulative positive impact of overall improvements to craft utilization (CU), craft performance (CP), and craft service quality (CSQ) in combination: *obtained from using results of measuring all three OCE Factors: (a) craft utilization, (b) craft performance, (c) craft service quality*	65%	**OCE = % CU × % CP × % CSQ** **Where CU = 70% .... a realistic maximum** **CP = 95% .... plus is achievable** **CQSE = 95% .... plus is achievable** **Therefore:** **OCE = 0.70 × .95 × .95 = 0.632 ≅ 65%**
13.	**Percentage of WOs with reliable planned times:** To measure planner's effectiveness at developing reliable planning times: *obtained from completed WO file in CMMS where planning times are being established for as many jobs as possible by planner/supervisor*	60–70 %	Number of WOs with reliable planned times ÷ total number of WOs completed per reporting period
14.	**Percentage of overall preventive maintenance compliance (could be by type asset, production department/location, or by supervisor/craft area):** To evaluate compliance to actual PM requirements as established for assets under scope of responsibilities: *obtained from completed WO file in CMMS for PM or PdM work types*	100%	Total number of PM WOs completed as scheduled ÷ total number of PM WOs due and scheduled per reporting period (Note: PMs to be completed within reasonable window of time from date they are generated for scheduling)
15.	**Gained dollar value from craft utilization/performance:** To determine actual gained dollar value of craft productivity gains as compared to original estimate and/or the initial baseline: *obtained only from using results of measuring two of the OCE factors—(a) Craft utilization, (b) Craft performance*	TBD	[Total current craft hours of wrench time minus baseline average wrench time hours] × baseline cost / wrench time hour (or actual cost per hour)

(*Continued*)

TABLE 25.1 Potential Performance Metrics for Using on a Maintenance Excellence index (*Continued*)

No.	Performance metric, purpose, and data source	Goal	How to calculate metric
16.	**Percentage of inventory accuracy:** To evaluate one element of MRO material management and inventory control policies: *Obtained from cycle count results and could be based on item count variances or on cost variance*	98%	Item count variance: Total stock items cycle Counted as correct ÷ total stock items cycle counted Cost variance: Actual inventory cost of total stock Items counted as correct ÷ total actual inventory cost of stock items counted
17.	**Percentage of or dollar value of actual MRO inventory reduction:** To evaluate another element of MRO material management against original estimates and the initial baseline MRO inventory value: *obtained from inventory valuation summation at end of each reporting period*	10%	Actual dollar value of inventory reduction ÷ baseline inventory value
18.	**Number of stock-outs of inventoried stock items:** To monitor actual stock item availability per demand plus to monitor any negative impact of MRO inventory reduction goals: *Obtained from tracking stock item demand and recording stock outs manually or by coding requisition/purchase orders for stock items not available per demand*		Actual stock-outs recorded as they occur. A next step can be the percent of downtime hours as a result of a true stock out
19.	**Dollar value of direct purchasing cost savings:** To track direct cost savings from progressive procurement practices as another element of MRO materials management. Could apply to contracted services, valid benefits received from performance contracting, contracted storerooms, vendor managed inventory: *obtained via best method per a standard procedure that defines how direct purchasing savings are to be accounted for*	TBD	Tracked via best method per a standard procedure that defines how direct purchasing savings are to be accounted for. NOTE. Not the high cost of low-bid buying seen in many places

TABLE 25.1 Potential Performance Metrics for Using on a Maintenance Excellence Index (*Continued*)

No.	Performance metric, purpose, and data source	Goal	How to calculate metric
20.	**Possible metrics for critical operational or production assets** Overall equipment effectiveness (OEE): World–class metric to evaluate cumulative positive impact of overall continuous reliability improvements to asset availability (A), asset performance (P), and quality (Q) of output all in combination. (Similar to OCE above but for the most critical production assets): *obtained via downtime reporting process, operations performance on critical assets and the resulting quality of output.* Important Note: We always look closely at the downtime(DT) reporting process during our benchmark assessments. Three distinctively different numbers are possible as follows: (1) When operations only makes the call as to reason for, (2) When maintenance only makes the call as to reason for DT, and (3) When operations and maintenance use a clearly defined, easy DT reporting process that both agree to using. We always recommend Number 3	85%	Where OEE = % availability × % performance × % quality An OEE factor of 85% is recognized as world-class, which therefore is: 95% for all three factors. **OEE = A × P × Q** **OEE = 0.95 × 0.95 × 95 ≅ 85%**
21	Percentage of asset availability/uptime: To evaluate trends in downtime due to maintenance and the positive impact of actions to increase uptime: *obtained via downtime reporting process*		Total hours asset performs its primary function ÷ total hours asset scheduled to perform its primary function

- Information resources and how data becomes true information via effective CMMS (Your CMMS Benchmarking System results and the CMMS Scoreboard category)

Figure 25.4 is another actual example that includes 14 metrics on a one-page summary spreadsheet that is in its Excel spreadsheet format that calculates automatically the total MEI.

Maintenance Excellence Index Example: 14 Categories

A. Performance Metric	1. Direct Purchasing Cost Savings	2. % Overall Budget Compliance	3. % Estmd Jobs Complete as Scheduled (w/in Compliance)	4. % Other Planned WO's Completed as Scheduled	5. OEE	6. % Craft Time to Maintenance WOs	7. % Craft Utilization ("Wrench Time")	8. % Craft Performance (PM and Planned Work)	9. % Work Orders with Planned Time	10. % PM Compliance Overall	11. Gained Value of Capacity Increases	12. % Inventory Accuracy	13. % Value of MRO Inventory Reduction	14. # of Stock-Outs	Perf level
B. Current Month Performance	200	94	94	91	60	36	56	85	56	74	300	91	2	19	
C. Performance Goal	500	98	98	95	85	40	60	95	60	90	1700	98	10	10	10
	450	96	96	93	80	38	58	90	58	88	1500	97	9	11	9
	400	94	94	91	75	36	56	85	56	86	1300	96	8	12	8
	350	92	92	89	70	34	54	80	54	84	1100	95	7	13	7
	300	90	90	87	65	32	52	75	52	82	900	94	6	14	6
	250	88	88	85	60	30	50	70	50	80	700	93	5	15	5
	200	86	86	83	55	28	48	65	48	78	500	92	4	16	4
D. Baseline Performance in BOLD	150	84	84	81	50	26	46	60	46	76	300	91	3	17	3
	100	82	82	79	45	24	44	55	44	74	100	90	2	18	2
	50	80	80	77	40	22	42	50	42	72	50	89	1	19	1
	0	78	78	75	35	20	40	45	40	70	0	88	0	20	0

F. Performance Level Scores

	1	2	3	4	5	6	7	8	9	10	11	12	13	14	
E. Current Performance Level Scores	4	8	8	8	5	8	8	8	8	2	3	3	2	1	Scores × Weight
G. Weighted Value of Performance Metric	11	8	6	8	13	6	10	5	7	2	9	5	5	5	
H. Performance Level score (F) × weight (G)	44	64	48	64	65	48	80	40	56	4	27	15	10	5	= 565

I. Total Maintenance Excellence Index Value

J. Total MEI Values Over Time	2/02	3/02	4/02	5/02	6/02	7/02	8/02	9/02	10/02	11/02	12/02	1/03	2/03	4/03
	565													

Figure 25.4 Maintenance excellence index with 14 performance measures.

The step-by-step format for MEI calculations. The total MEI value is the composite score of all metrics, considering current performance of each metric as compared to the goal and the weighted value of each individual metric. The total possible score for the total MEI value is 1000. The 10 key steps from A to J for developing and using your MEI are as follows (Table 25.2):

Your MEI Validates Your Results

The MEI provides a composite index that integrates a number of key metrics into a composite value: the *total MEI value*. Each metric can be monitored and trended individually comparing their baseline value to an established performance goal. These metrics are then used to provide the one-page Excel spreadsheet that brings them all together into a composite total MEI performance value. This is your index as to how all resources are contributing toward maintenance's contribution to profit optimization. Also this does not limit you to 10 to 15 metrics for your MEI. Most operations we work with do use other reports generated from their CMMS in addition to their MEI results.

The MEI helps keep in focus the fact that the success of physical asset management and the execution of maintenance processes depend on many factors; therefore, one or two metrics cannot provide the total performance picture. The MEI gives us a broad-based approach to performance measurement. If we have justified the project on craft productivity increases, parts inventory reduction or a decrease in maintenance cost per unit of output, we now can validate the results of our initial projections. Your MEI is developed specifically to validate your results and your projected ROI.

Get started now

Take action and get started now with an evaluation of your total maintenance. Conduct a self-assessment like the one we discussed in Chap. 7 or get help with a well-qualified maintenance consultant for this important first step. Take the time to develop a plan of action, as we will discuss in Chap. 26. Integrate your strategic maintenance plan with your company's business plan. Plan for maintenance excellence. Commit the necessary internal and external resources to the hard task of implementation. Get the most from your CMMS, measure your results, and ROI.

The new millennium view toward maintenance and physical asset management must see maintenance helping directly with profit optimization. The four benchmarking tools and the strategy defined in this book have been proven for application within a multitude of different

TABLE 25.2 Key Steps for Developing and Using MEI

Step	Description	Comments
A	Performance metrics	From 10 to 15 metrics are selected and agreed upon by the organization. If a maintenance excellence strategy team is being used, this is a key item they will consider and approve. See App. F.
B	Current month performance	This is the actual performance level for each metric during the reporting month. This value can also be noted in one of the incremental values blocks below the performance goal. This current month value will correspond to a value for Step F, the performance-level scores which go from 10 down to 1.
C	Performance goal	This is the preestablished performance goal for each of the MEI metrics. For example, if the current month's performance is at the performance goal level, the performance-level score for that goal in Step F will be a 10, the maximum performance level score of 10.
D	Baseline performance	The baseline performance level prior to start of your MEI performance measurement process. This may not always be available before measurement starts. In this case, the baseline can be a 4 or 8-week period of actual measurements using your MEI.
E	Current performance score	Depending on the current month's performance, a performance level score (F) will be obtained. This value then goes to the current performance score (Row E) and serves as the multiplier for the (G) weighted value of the performance metric.
F	Performance-level score	Values from 10 down to 1, which denote the level of current performance, compared to the goal. If current performance achieves the predetermined goal, a performance value of 10 is given. Each metric is broken down into incremental value from the baseline to the goal. Each incremental value in the column corresponds to a performance-level value. This value becomes the current performance score in Row E.
G	Weighted value of the performance metric	The values along this row are the weighted value or relative importance of each of the metrics. These values are obtained via a team process where a consensus on the relative importance of each metric selected for the MEI is established before measurement begins. All of the weighted values sum to 100 along Row G.
H	Performance value score	The weighted values (Row G) are multiplied by (Row E) the current performance scores to get the performance value score (Row H).
I	Total MEI values	The sum of the performance value scores for each of the metrics and based upon the composite value of monthly maintenance performance on all MEI metrics. Maximum value is 1000. This is when all performance goals are achieved.
J	Total MEI performance values over time	Location for tracking total MEI performance values over a number of months.

types of maintenance and physical asset management operations. This includes both the public and private sectors and recently one of our military services—the U. S. Air Force. The approach is simple but powerful in terms of achieving results and validating ROI. Organizations which clearly understand that "maintenance is forever" and find the key to balancing all resources toward optimum total operations success will succeed in the twenty-first century.

For organizations now evolving into today's profit-optimization trend, profit- and customer-centered maintenance can help maximize your profit-optimization efforts. A true profit-centered approach must include maximizing your physical assets, the production assets, the facilities, and related business processes within an organization.

Chapter

26

Nontraditional Return on Investment for Improving Your Maintenance Return on Investment

The top leadership of many organizations, both large and small, must gain a greater awareness and personal understanding as to the importance of maintenance. We have mentioned this many, many times in the previous chapters. Top leaders must always be aware of their total maintenance requirements, the real needs to maintain the primary physical asset function. They must be aware of best practice needs and take advantage of the investment opportunities within maintenance. And top leaders must never, never forget the regulatory life safety and environmental requirements that apply to the maintenance process. Top leaders that have experienced the results of a "Scoreboard for Maintenance Excellence" benchmark assessment along with projected benefits and gained value will take notice. In addition, when they see that you have implemented *The Maintenance Excellence Index* (MEI) that will validate those projected results, all top leaders will take even a closer look. Unfortunately, many organizations simply continue to gamble with maintenance costs and do nothing of significant value. Maintenance requirements always continue to grow from the four maintenance challenges we discussed very early in Chap. 3.

The core requirements for maintenance never, never goes away because maintenance is forever! Top leaders might exclaim in *Business Week* that "Our core competency is assembling cars and not maintenance. It is not even the manufacturing of subassemblies. We assemble cars and trucks!"

The truth is that maintenance remains forever as a core requirement for auto assembly success as well as a core requirement for subassembly manufacturing success througout the total auto industry's supply chain. So when I read or hear about a top leader saying that "we are outsourcing maintenance because maintenance is not a core competency" I cringe and think to myself: here is one more top leader that just does not get it. Maintenance is a core requirement in everything: body, soul, mind, car, house, factory, and the roads to the factories. The core competency for doing maintenance and managing the business of maintenance is strictly another issue. The core competencies to perform in-house maintenance can slip away gradually or very abruptly by key craftspersons retiring or changing companies. And some top leaders maybe are correct if the core competency has vanished. Profit- and customer-centered service providers are everywhere and they are on shore not overseas.

Sound investments that affect the bottom line are still available in almost all existing maintenance operations. You can bet heavily on the fact that profit-centered contract maintenance providers are continuously looking to improve their profit picture and customer service as providers of service. I personally know many from this growing business sector and some are relentless in their pursuit of profit optimization and quality customer service. Tim Viox up in Cinncincinnati,Ohio, I know for a fact is selling customer-centered service. From what I know, Vioxx Services is making a fair profit for providing core competencies in maintenance as part of its parent company, which is now Emcor, Inc. Takeover targets emerge as plant or facilities maintenance operation continue to maintain the status quo for way too long. Then it becomes too late to turn around the in-house operation and a takeover of maintenance occurs. Many times this results from a new top leader who comes in wanting to make an immediate impact. It may be a new top leader who is impatient and a reader of all the literature on outsourcing and says: Let us just do it! Sometimes that top leaders are absolutely right when they pick organizations like Viox Services!

Traditional ROI

We have focused on creating a greater awareness of the traditional *return on investment* (ROI). In the previous chapters, very specific methodologies for you to apply and to use immediately were presented. Much more is possible by applying today's best maintenance practices, principles, and leadership philosophies. Technology of the future will open up even greater maintenance improvement options. Real maintenance ROI is available to the organization that successfully converts the thinking of top leaders into a complete awareness about maintenance improvement as a profitable action they must take. There are some nontraditional ROIs that I would like for you to consider as we reach the end of this book. The following nontraditional ROIs will be important contributors to your success in gaining commitment to action by top leaders to maintenance leaders and to your craft leaders. In addition, if your organization has not really, really begun that critical first step of the journey toward maintenance excellence, these nontraditional ROIs will help on that important first step.

General benefits

Maintenance operations that have committed to *continuous reliability improvement* (CRI) and are applying today's best maintenance practices, principles, and leadership philosophies, have achieved significant improvements. Investments in maintenance that successfully implement the practices from this book and others on the market can achieve results that are comparable to the following summarized again here and in Chap. 24.

- 15 to 25 percent increase in equipment uptime
- 20 to 30 percent increase in maintenance productivity
- 25 to 30 percent increase in planned maintenance work
- 10 to 25 percent reduction in emergency repairs
- 20 to 30 percent reduction in excess and obsolete inventory
- 10 to 20 percent reduction in maintenance repair costs

Other improvements that are possible will include:

- Improved product quality
- Improved utilization of equipment operators; greater production productivity
- Improved equipment effectiveness and capacity
- Improved equipment life
- Improved productivity of the total operation

Specific benefits

The results listed earlier are achieved by maintenance organizations who have committed to CRI. Every maintenance organizations must realize that there are no easy answers and "no quick fixes." Organizations that have invested in their total maintenance process over the long term, will realize a tangible return on that investment and measure results very clearly. Consider what would happen if your numbers were used in the following example:

- Maintenance craft productivity increased by 20 percent
  - OCE improvement in craft utilization, craft performance percentage, craft service quality by 20 percent.

    20% OCE 0.20 × 40 craftsmen × $35,000/year = $280,000/year
- Equipment uptime increased by 25 percent
  - 25%: downtime reduction from 8 to 6% – Savings from improved utilization of operators 0.25 × $800,000 total value of downtime = $200,000/year
- Inventory reduction in maintenance storeroom decreased by 25 percent
  - 25% reduction from $1,000,000 to $800,000 = $200,000 × 0.30 inventory carrying costs = $60,000/year
- Improved pricing from suppliers of 1 percent
  - 1% 0.01 × $1,000,000 purchase volume/year = $10,000/year
- Reduction in maintenance repair costs of 10 percent
  - 10% 0.10 × $750,000 annual repair cost = $75,000/year
- Improved product quality
  - 1% reduction through equipment-related scrap, rework returns, waste, and better yields 0.01 × $2 million value of production = $20,000/year
- Improved equipment life
  - 1/2 year 0.5 year × $10 million capital investment × 0.10 interest rate – $200,000 additional maintenance cost = $300,000/year

Impact on sales

Let us look again at the potential of maintenance to influence the bottom line in another way as shown in previous chapters. If your organization's net profit ratio is 5 percent, then each dollar in net profit requires $20 of equivalent sales. The potential for maintenance to contribute to the bottom line, as shown in Table 26.1, is significant.

Investing in a strategy of CRI is an excellent investment when $50,000 savings by maintenance is equal to $1,000,000 in equivalent sales. These examples illustrate that tangible ROI can be significant, depending on the size of the maintenance operation and the type of organization being supported. Maintenance leaders must be able to gain support for CRI by developing valid economic justifications. Take the time to evaluate the potential savings and benefits that are possible within your own organization. Gain valuable support and develop a partnership for profit

TABLE 26.1 The Maintenance Savings Impact if Profit Ratio is 5 Percent

Maintenance direct savings ($)	Equivalent sales required ($)
1	20
1,000	20,000
10,000	200,000
20,000	400,000
50,000	1,000,000
100,000	2,000,000
150,000	3,000,000
200,000	4,000,000

with operations. Include all other key departments that will receive benefits from improved maintenance. The application of today's best maintenance practices will provide the opportunity for maintenance to contribute directly to the bottom line. However, the pursuit of maintenance excellence requires leadership. There are *no easy buttons*.

Leadership and the Return on Inspirational Leadership (ROIL)

The term "maintenance leader(s)" is used throughout this book for several reasons. First, it is used as an all-inclusive term to recognize and refer to the maintenance leader at all levels. It can include a corporate official with direct maintenance responsibilities, the director of maintenance, a plant engineer, a maintenance supervisor, or a maintenance crew leader. The return on inspirational leadership (ROIL) has profound importance everywhere, especially within maintenance.

Second, the term "leader" (or leadership) is used instead of "manager" (or management) most of the time because there are many distinct and important differences between these two terms that apply throughout an organization. Table 26.2 summarizes important

TABLE 26.2 Maintenance Leaders versus Maintenance Managers

Maintenance leader	Maintenance manager
■ Proactive	■ Reactive
■ Makes things happen	■ Wonders what happened
■ Has vision	■ May wear blinders
■ Promotes growth	■ Maintains control
■ Promotes continuous improvement	■ Maintains status quo
■ Serves in the lighthouse	■ Manages the firehouse
■ Has leadership ability	■ Has management ability

differences between the maintenance manager and the maintenance leader:

Good leaders and good managers are key resources in all successful organizations. Both will continue to exist and be needed. However, to face the maintenance challenges of today and in the future, real maintenance leaders must emerge at all levels. Likewise, top management/managers of organizations who have previously stifled investments in maintenance must provide leadership and develop (or renew) long-term commitments to maintenance excellence.

Beyond the Bottom Line

Investments in maintenance improvement will provide significant tangible results based on the traditional concept of ROI as we discussed previously. Successful improvements in maintenance will impact the bottom line directly while providing both tangible and intangible benefits throughout the organization. Today's leaders must also adopt some nontraditional interpretations of ROI that will also provide significant additional benefits. The ROIs included in Fig. 26.1 are all interrelated and support the traditional concept of ROI. Consider the value of these ROIs and then go beyond the bottom line with your maintenance investment.

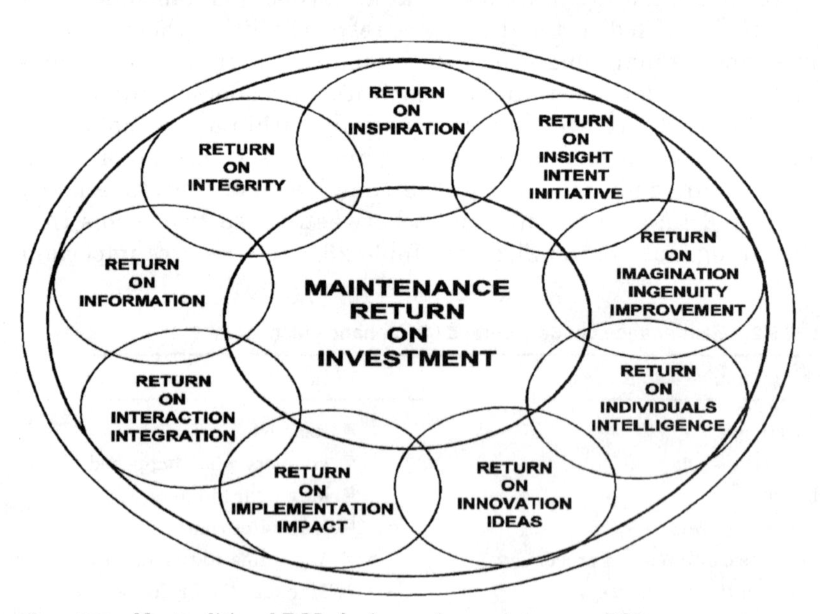

Figure 26.1 Nontraditional ROIs for improving maintenance ROI.

Return on inspiration (ROInspire)

Inspiration begins with a vision and a positive expectation for the future. The maintenance leader must completely understand the organization vision. The leader must be a part of and become aligned with the organization vision, mission, and requirements of success. Maintenance must be established as a top priority within the entire organization and clearly be seen as a requirement for success. The maintenance leader must translate the organization vision into a vision for maintenance excellence and develop a shared commitment from the entire maintenance operation. Each maintenance employee must understand that a shared commitment to improving the maintenance operation is a positive factor. They must have positive expectations about change and overcome the negatives normally associated with change.

The maintenance leader is a critical and valuable maintenance resource. Leaders are not born, they are continuously developed. They must develop a clear understanding of the requirements of success in maintenance. The investment in maintenance leadership development and technical skills development will provide return. The maintenance leaders that have a clear communicated vision of maintenance excellence will provide inspiration to all. The return on inspiration and good maintenance leadership will improve employee morale, attitudes, cooperation, communication, performance, and bottom-line results. I want to share here a personal example of fine leadership. It is about General Kenneth Newbold, Commander of the 30th Separate Infrantry Brigade (a minidivision) headquartered in Clinton, North Carolina, back in the late twentieth century.

I worked for BG Newbold two times; first as his engineer staff officer and his commander of a separate combat engineer company (a mini-battalion) headquartered in Rockingham, North Carolina, as a major in the NC Army National Guard. Following this two-year tour, I then served as General Newbold's *inspector general* (IG) for almost two years before moving back to the 30th Engineer Brigade (Construction). Very briefly, BG Newbold was "the commander" of the best separate infantry brigade in the U.S. Army. This unit just returned from the Iraqi freedom conflict. BG Newbold was also the CMO (chief maintenance officer in my opinion) of this war-fighting unit of North Carolina citizen soldiers. If you have ever read about General Stonewall Jackson from Virginia, BG Newbold was the reincarnation of this famous military leader during the U.S. Civil War. He believed in three things: attack, attack, attack!

1. Attack incompetence and replace it with leadership
2. Attack maintenance issues, do it by the book, and expect maintenance excellence
3. Attack the staus quo and replace it with proactive and positive change

Leaders like BG Newbold do not always know the impact they have had on people like me. This fine Southern gentleman was surprised when I told him what his personal example meant to me back then. Like Stonewall Jackson from the south, Kenneth Newbold was an educator and also like Colonel Joshua Chamberlain from Maine another Civil War leader from the north. All three were called to military duty as leaders. The main point here is that they all were top leaders who listened to the maintenance leader and their soldiers. Do not waste potential. "Do not play defense forever because you will not score." Do what these three leaders did very well—Attack! Attack! Attack!

Return on insight, intent, and initiative (ROIII)

Maintenance leaders with insight understand that the requirements of success in maintenance begin with *hard work from a good team committed to a common goal.* These types of leaders have the insight and understanding that there are no easy answers and quick fixes. They communicate their insights and intent for improving maintenance openly and confidently throughout the organization. They are intent on making a difference and can make positive things happen with the proper investments of time and money. Their vision and goals are not just good intentions. They put goals into action through initiative and determination.

Maintenance leaders, know *where* they are, with what I call 20/20 insight, because they have taken the initiative to totally evaluate their current maintenance operation. They readily accept the good news and the not-so-good news. They also have many lessons learned from 20/20 hindsight. True maintenance leaders know exactly where the next investment of time and money should be. They think like an owner of a maintenance business. They have accepted the challenge to re-engineer and improve their maintenance operation for the future and have established prioritized plans for action. Specific short-term operational goals have been established that the maintenance team can also focus on with intent and immediacy.

Maintenance leaders who have 20/20 insight know their operation as well as the good points of other maintenance organizations. They seek out new ideas, products, and services from all sources. They benchmark continuously against their own high standards of excellence while pursuing a strategy of CRI. They also benchmark against other like organizations, yet do not get trapped into management fads. Maintenance leaders' continued insight reinforces their faith in the American maintenance workforce as the world's finest and guarantees quality for the future by providing supreme leadership.

Return on integrity (ROInteg)

There must be a maintenance champion. The real maintenance leader readily accepts the role as champion for maintenance excellence. Likewise, integrity of purpose and the integrity of the maintenance champions must set an example for others in the organization to follow. Ralph W. Emerson said it very well when he remarked, "What you are thunders so loudly, I cannot hear a word you say to the contrary." Leadership by example and "walking your talk" is essential for the maintenance champion and all top leaders.

An organization with true integrity of purpose does not chase fads. It understands the business it is in and has a consistent vision of success with maintenance as a top priority. Such an organization wants to know the true costs of maintenance—"if we did it right" or "if we achieve regulatory compliance that the law requires." The maintenance champion with integrity is honest, no matter what the situation or cost, and will determine what is really needed to do the job right.

The maintenance champion understands the true cost of deferred maintenance as well as inadequate *preventive/predictive maintenance* (PM/PdM). The champion is prepared to provide proactive leadership and support to the organization's compliance to local and federal regulatory issues. The real maintenance champion is prepared to take bad news to top leaders with courage and fearless commitment.

The return on integrity may be difficult to see and measure when it is present. When integrity is not present within the organization or the maintenance champion, it is obvious. For example, one can see where not buying new tires or changing oil in your car prior to trading it is okay, while turning back the speedometer is against the law. Neglecting plant or facility maintenance due to lack of knowledge can be reversed through education, time, money, and quality people. Neglect through lack of integrity may never be counteracted without a change in leadership. The United States has witnessed this vividly with the Enron scandal, a case of pure and simple corporate robbery.

Return on integrity is easy to see when it permeates the entire culture and is part of the organization's personality. It takes the form of an organization built on trust, mutual support, and mutual respect at all levels. This return on integrity utilizes maintenance, operations, and operators working together to detect, solve, and prevent maintenance problems. It encourages pride in ownership with operators and maintenance as they do their part to fix and prevent maintenance problems through a cooperative team effort of operation-based maintenance. The return on integrity inspires individual integrity and is obvious when all employees do their jobs as if they owned the company. Individual integrity includes pride in work no matter what the task.

Return on integrity goes well beyond the bottom line to ensure long-term survival of an operation and its maintenance function. It is measured by the evidence of success unique to each operation. If allowed by current accounting practices, return on integrity should show up on the left side of the balance sheet under intangibles. When integrity is not present, it represents a current or long-term liability that will eventually show up on the balance sheet and appear clearly on the bottom line.

Return on information (ROInfo)

Traditionally, all organizations have more available data than usable valuable information. In most cases, maintenance is in this same dilemma. Because maintenance may be viewed as a cost center, maintenance information systems are lacking or even nonexisting. Let us now look at some very interesting and thought-provoking "what ifs?" that we discussed in previous chapters.

What if the maintenance operation in your organization or plant was a business? What if your maintenance was a third-party maintenance service that worked out of the same shop area, used the same storeroom, and maintained the same equipment? As the maintenance champion, what if your only job was to determine the scope of services needed, develop the plan for maintenance contract service, plan, estimate, schedule, and monitor the services received, and approve payment based on quality of service per the contract? What if this scenario came to pass?

Would you have the right information to do your new job? Conversely, what if you owned this in-plant, third-party maintenance service? Could you define and measure your level of service in order to make a profit as a business? Would you get the maintenance contract in your own plant as the third-party contractor with your existing workforce and processes? Always remember that there are thousands of profit-centered maintenance service companies that can come in and ensure effective customer service for your top leaders without you.

If your current maintenance practices and maintenance information system does not allow you to manage maintenance like a profitable business, then your organization will continue to view you as a "cost center." Your company/organization must invest to the point that you can manage and monitor your performance and level of service with real maintenance information.

This scenario is not a scare tactic which advocates third-party maintenance in total for an organization. However, third-party maintenance in specialty areas or areas where current maintenance skills or competencies are lacking is a cost-effective practice that will continue to grow. Greater third-party maintenance will occur in operations where maintenance is not treated as a business and where physical assets have deteriorated to the point that a third-party service is more effective and less costly than in-house maintenance staff. Creeping outsourcing often occurs. For manufacturing plants, like the one I supported back in 2002, it may start by outsourcing the facilities maintenance piece. The contractor will then wait patiently for the plant maintenance piece to become available. Trends in government maintenance and service operations are rapidly progressing toward privatization with greater performance, service, and reduced total costs. Third-party maintenance will be a common practice in those organizations that have continually gambled with maintenance costs and have lost. There are also many pitfalls to third-party maintenance and I have seen many of them personally in South Vietnam and via hundreds of scoreboard assessments. And, I am personally pulling for the home teams, the in-house maintenance team, and the total operations team for many reasons.

Maintenance leaders must demand and welcome adequate systems support to ensure that existing maintenance data becomes real maintenance information for managing maintenance as a "profit center." *Computerized maintenance management systems* (CMMS) should be viewed as an important tool to assist in planning, management, and administrative procedures required for effective maintenance. Maintenance data is of little value without being in the form to support decision making.

The maintenance database on equipment repair history, work order status/backlog, PM schedules, repair parts inventory, job estimates, performance measures, repair costs, life-cycle costs, and so forth, must be current, accurate, and capable of providing quality information for timely decisions. The integration of actual maintenance labor costs with planned labor costs must be the basis for labor performance measurements. Customer service criteria must be established to evaluate and measure the level of maintenance customer service. The maintenance customer must also be involved in determining quality indicators and be a part of the flow of information. Information to evaluate overall equipment effectiveness should be readily available and used to identify:

- Causes for breakdown
- Problems created by set-up/adjustments
- Problems created by idling/minor stoppages
- Reasons for running below design speed/feed
- Causes for process defects
- Reasons for reduced yield during changeover

Information about overall equipment effectiveness—a key *total preventive maintenance* (TPM) concept—provides immediate opportunities for improvement by operators, set-up personnel, and maintenance. This type of information gets at the root cause of recurring breakdowns which continuously plague production with uncertainty and maintenance with unplanned breakdown repairs.

It is a good investment to provide maintenance leaders with timely and accurate information to manage and measure maintenance as a business. In turn, the maintenance leader must treat maintenance as a profitable business by providing information to top leaders that clearly shows an ROI. Craft utilization and performance, PM compliance, work backlogs, downtime levels, the effectiveness of planning and scheduling, and the like, should be evaluated as part of a broad-based maintenance performance measurement system. Figure 25.4 illustrates another example of a MEI that uses multiple performance metrics as an overall performance measurement indicator.

Return on interaction and integration (ROII)

Almost all organizations today must think globally in terms of new markets, products, and competition. From the maintenance perspective, we must think globally in terms of the best practices we need to consider, but we must start local at our own unique shop level. The maintenance leader must "think global and act local" to achieve positive interaction with top leaders and staff at all levels. Maintenance cannot succeed without positive interaction and support from others in the organization. The process of CRI does not stand alone within an organization. It must be well integrated with the customers of maintenance as well as all other staff sections. Positive interactions promote effective integration of solutions.

Maintenance and operations/operators must interact and work closely to monitor, service, and prevent maintenance problems. All must take an integrated approach to improving overall equipment effectiveness as a cooperative team of operations and maintenance. Think of maintenance and operations as members of a racing team with the operator as the driver and maintenance as the pit crew as shown in Fig. 26.2. The pit crew cheers the loudest when their driver wins for the team! All leaders at all levels must interact to instill this spirit of teamwork and pride in ownership into the culture of the total organization. Achieving maintenance excellence requires positive interaction with the players internal and external to the maintenance function.

Return on imagination, ingenuity, and improvement (ROImIngImp)

Maintenance craftspeople by nature are ingenious, creative, and normally able to do more with less than your average person. This is from my personal observation over 35 years of exposure to plant, facilities, fleet, healthcare, and golf course maintenance people. They also take

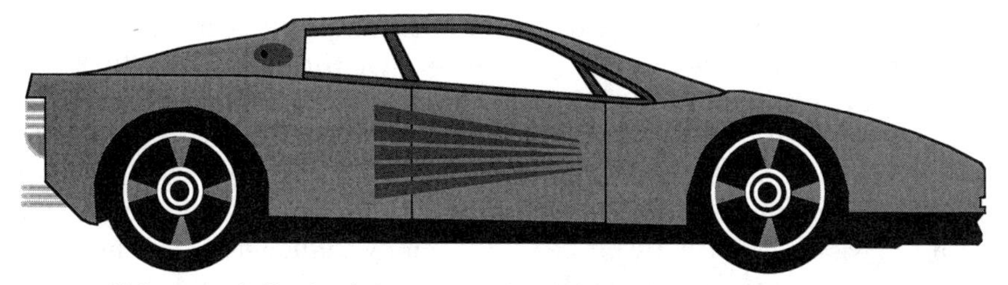

Maintenance is like the pit crew to operations. Perfect, fast maintenance, effective storeroom and MRO procurement performance is the difference between winning and unscheduled downtime

Figure 26.2 Maintenance as the pit crew.

pride in their ability to solve problems and come to the rescue to fix something. However, they are impatient with equipment abuse and continuing problems that could be eliminated if enough time was available to find the real cause. They know that PM/PdM/RCM/VE will work if given the chance. I think that crafts really care about making things better. They are smart and want to get smarter and they want to be involved. They know that improvements can be made and they need to be part of the process. For example, the ACE Team Benchmarking Process reviewed in Chap. 15 provided a good way to involve crafts directly in establishing reliable planning times. They are proud of their profession for the most part but can become less positive about it when a reactive, fire-fighting strategy puts all the blame on maintenance. I do believe that all want to use more of their imagination and ingenuity and really become involved in maintenance improvement. Our support to the UNC-Chapel Hill group of over 300 crafts and support staff from their facilities service division proved this very clearly to me. Maintenance people know they are capable and that they are capable of doing better. They are the most critical and most important of the six key resources we have in the maintenance profession. Why not take advantage of your hidden assets, ideas for improvement from the people doing the work?

These positive statements about maintenance people should apply to your maintenance operation. The maintenance leader that involves maintenance people in the process of team-based continuous improvement becomes a maintenance leader. A leader with vision, insight, and confidence knows that maintenance working together as a team can make a difference. Teamwork works and is a hidden asset. True leaders know that interaction and integration with operations staff, engineering, and operators all will provide a return on time and resources. Become the leader of maintenance who wants to know all about the current best practices in maintenance and try to use them. Be a maintenance leader who knows the advantages of good planning, estimating, and scheduling and serves in the lighthouse, not the firehouse. Just like most craftspeople, the real maintenance leader is ingenious and sees the future with a vision for maintenance excellence. This type of leader gains inspiration from the challenges ahead and will make things happen with a team-based approach to maintenance improvement. Our creator gives all of us many blessings and many challenges over a lifetime. Read the words in red in the Bible for inspiration as a leader.

The return on imagination, ingenuity, and improvement is unlimited. With a real maintenance information system in place, it can be measured in terms of tangible dollars and intangible benefits. Unleashing the power of maintenance people is not a fad. It is profitable, practical, and a proactively positive approach for going way beyond the bottom line. It has been used in the past, it is essential to success now and for success in the future.

Return on individuals and intelligence (ROIndIntel)

Individuals with varying levels of talent and intelligence make up huge armies, large corporations, small companies, maintenance crews, and teams. CEOs and CFOs ponder their next investment opportunity in new equipment, facilities, processes, or new merger possibilities with a view of how it will improve the bottom line. An investment in people, the most valuable asset, is often the most neglected investment opportunity. The maintenance operation that is not investing in continuous maintenance education and development of its people is in danger.

Today's maintenance leader must have a current assessment of the skill level of their maintenance workforce. They must have a plan and the resources to provide craft skills development that is needed in their specific type of maintenance operations. With a strategy for multicraft maintenance, it is necessary to be able to develop multiskilled craftspeople in order to achieve benefits from this approach. Training provides intangible benefits on top of the direct benefits. Maintenance employees will know that the organization cares enough to invest the time and money. Employees will have greater confidence in the long-term future of the organization as well as the importance of maintenance. With new skills, the craftspeople will be more confident and perform at a higher level that is safe and more productive. Improved customer service and quality will result.

Your CFO may have a hard time with the previously discussed return on integrity. The return on information that quantifies tangible benefits will be a little easier.

Your case for an investment to provide return on individuals and intelligence will be hard fought, but remember, if they think education and training is expensive, try ignorance! It has also been stated this way via two questions: (1) What if we invest all this money in craft skills training and they leave us? (2) What is the true cost if we do not train them and they stay? Also for those who also think education and good maintenance is expensive, they should try ignoring today's best maintenance technology and practices! Intelligent investment decisions are made when key leaders at all levels invest in people, their most valuable asset. Maintenance excellence will not be achieved by ignoring today's best maintenance practices. Continuous maintenance education in areas of need is a requirement for success in maintenance. The maintenance leader must not let craft skill training become the weak link in developing maintenance excellence.

Return on innovation and ideas (ROInoIde)

The maintenance leader is innovative and makes the best use of resources available, yet keeps an eye out for innovations wherever they might be. Real maintenance leaders seek out new equipment processes, technologies, techniques, and ideas. They do not chase fads for a quick fix and they realize that maintenance people are a valuable source of good ideas waiting to be unleashed.

Maintenance excellence for the real maintenance leader is more than just a vision. It has taken shape in the form of an action plan that looks first at priority areas. During the process of achieving maintenance excellence, the maintenance leader knows that ideas from within maintenance, as well as from operations, will be a requirement for success. The maintenance leader has developed a partnership for profit with operations that works. The universal team-based approach of using leadership driven self-managed teams becomes the critical process to ensure that all good ideas are given a chance.

Cooperation, coordination, communication, and commitments are developed and strengthened. Teamwork changes involvement to commitment. The true maintenance leader is innovative in personal leadership and is able to gain commitment rather than just involvement and consensus. The maintenance leader going along the path forward to maintenance excellence will take a "no holds barred approach" to the evaluation and audit of the maintenance business. The resulting plan will re-engineer maintenance by looking at a whole new way of doing business. Innovations in new equipment, new technology, and new procedures will be included. Real maintenance information will be available to provide valid economic justifications on required purchases and to measure progress.

The investment in innovations will require capital but provides a positive ROI as shown in previous chapters with step-by-step guide for you to use or modify for your use. The investment in creating an environment which cultivates ideas from the workforce is essentially free, pro bono, and without accounting cost. Team-based CRI in maintenance takes time and leadership. Remember that your maintenance people are the best in the world! Believe that they are ingenious, have imagination, want to be involved. Believe that they want to make the commitment to improve maintenance and they will. With quality leadership, the world's best maintenance people will get better as a team and as individuals. It has been said many times, but first in the Bible's red letters: "Whatever the mind can conceive, can be achieved!" Your vision for maintenance excellence must be stated in a detailed plan that is reasonable, understandable, measurable, believable, and achievable. And if there is a shared commitment to your vision, then it will be achieved.

Return on implementation and impact (ROImplmpac)

Ideas, information, innovation, and all the other ROIs do nothing for the bottom line without implementation. The bottom line impact of a plan for maintenance excellence is but an imprint of numbers and letters on the printed page unless the plan is implemented. Successful implementation is the key to making an impact on the bottom line. Maintenance managers and most people, feel that 90 percent of the work is done when the plan is complete. Some even think the purchase of CMMS is a magical solution. The well-informed maintenance leader knows that 90 percent of the work is just ahead. They see implementation as the way to give value to a good plan and realize that implementation is a very challenging task. The maintenance leader must become the champion for implementing maintenance improvements.

The maintenance leader faces both technical challenges and people challenges in getting the remaining 90 percent of the implementation challenge completed. The key to implementation will be turning people challenges into achievers/helpers and getting the right type of support from within the organization at the right time. Three kinds of support are vital to the success of implementation: formal approval, buy-in, and ownership.

The maintenance leader must now become a good communicator and salesperson. Implementation of a maintenance excellence plan requires selling up, across and down the organization. Top leaders must first give the formal approval for making the investment. The maintenance leader must sell the quantitative benefits and related costs along with the qualitative benefits of the plan. Priorities must be clearly defined.

Top leaders must also clearly understand the requirements of success in maintenance that were discussed in Chap. 2 are the key requirements for profit- and customer-centered maintenance. They must trust the technical knowledge and personal integrity of the maintenance champion who is bringing the message that maintenance needs support. They also expect a traditional ROI that can be identified, measured, and validated.

The implementation plan must then be sold across the organization. Almost all staff organizations will be impacted by a broad-based strategic maintenance plan. Each must understand its role, the purpose of the change, and how the new system or program will work. Each staff group must "buy in" to the plan and understand the importance of maintenance improvement. If the organization has embraced an overall strategy of team-based continuous improvement, this "buy in" will be obvious to the other staff groups and will be readily accepted.

The third and final type of support needed for successful implementation also depends on somebody stepping forward to assume ownership. First, the maintenance leader, along with the maintenance operation, must readily accept ownership. With maintenance people who have been involved, provided ideas, and developed a team-based commitment to maintenance excellence, this will not be difficult. For operation managers and supervisors who have had a part in developing the plan for greater maintenance service, fewer breakdowns, and greater equipment effectiveness, being part-owners will not be a problem. For operators who now are trained to do selected operator-based maintenance services and know how to detect and help prevent maintenance problems, being part-owners in the plan for maintenance excellence will not be difficult. The impact from successful implementation will be apparent all across the organizations from top down to the shop floor and the bottom line. Leadership and the acceptance of ownership at all levels are critical to successful implementation. Successful implementation is the key to creating impact.

Conclusion

We have looked at some new interpretations of ROI in this chapter as well as the traditional concept of ROI. The Maintenance Excellence Institute believes that maintenance operations have a tremendous opportunity to contribute directly to the bottom line with a strategy of CRI across all six maintenance resource areas. Top leaders of today's public and private sector operations, who want to be a part of the future, must look beyond the bottom line with respect to maintenance. Maintenance must be a top priority for success. A near-sighted organization vision focused on short-term results is fatal and will fail as did the Enron corporation. Organizations of all types need long-term commitments from top leaders and from true maintenance leaders at all levels in maintenance.

We also believe that an even greater awareness must be developed toward the investment opportunities that are available in maintenance.

Sound investments in maintenance can impact the bottom line directly. Maintenance leaders must sell their ideas up, across, and down the organization. Top leaders must listen, act, and do the right things in terms of maintenance. Companies with a vision of long-term survival cannot afford to gamble with maintenance costs.

Achieving maintenance excellence requires an investment in both the traditional and nontraditional ROIs that we have discussed in this book. It requires a strategic maintenance plan for applying today's best maintenance practices, principles, and leadership philosophies to your operation. It requires plans of action at both the tactical and operational level. And it, for sure, requires "Do it now" plans that the maintenance leader can do right now. The essence of achieving maintenance excellence goes well beyond the bottom line to a simple, positive affirmation statement . . . *pride* in maintenance. The real bottom line is *pride*—people really interested in developing excellence—in maintenance. This kind of *pride* is needed at all levels from bottom to top. If your organization has this kind of *pride*, then make an investment and achieve a real return from your maintenance operation.

Part 6

The Journey toward Maintenance Excellence

Chapter

27

Developing and Implementing a Profit-Centered Action Plan

When all the different benchmarking has been completed and all the specific assessment recommendations have been clarified and prioritized for action, the real work begins. Now is when our return on implementation and impact begins. All the good ideas, information, innovation, and all the well-defined benefits and gained value do nothing for the bottom line without implementation. The final three chapters in Part VI, "The Journey toward Maintenance Excellence," are to help you put in action what we covered in the first 26 chapters of this book.

The bottom line impact of a "Scoreboard for Maintenance Excellence" assessment will be zero (0.0) when it remains as numbers and letters on the printed page within a binder in your bookcase. Successful implementation is the key to making an impact on the bottom line. Maintenance managers and many people, feel that 90 percent of the work is done when the plan is complete. Some even think the purchase of *computerized maintenance management system* (CMMS) is a magical solution. The well-informed maintenance leader knows that 90 percent of the work is just ahead. Implementation gives value to a good plan of action. Realizing that implementation is a challenging task, the maintenance leader must become the champion for implementing maintenance improvements.

The maintenance leader faces both technical challenges and people challenges in getting the remaining 90 percent of the implementation challenge completed. The key to implementation will be people challenges and getting the right type of support from within the organization at the right time. Three kinds of support are vital to the success of implementation: formal approval, buy-in, and ownership. The maintenance leader must now become a good communicator and salesperson. Implementation of a maintenance excellence plan requires selling up, across, and down the organization. Top leaders must first give the formal approval for making the investment. The maintenance leader must sell the quantitative benefits and related costs along with the qualitative benefits of the plan. Priorities must be clearly defined.

Top leaders must also clearly recognize the requirements of success in maintenance that were discussed in Chap. 2 as the key requirements for profit- and customer-centered maintenance. They must trust the technical knowledge and personal integrity of the maintenance champion who is bringing the message that maintenance needs support. They also expect a traditional *return on investment* (ROI) that are identified, measured, and validated. Secondly, the implementation must be sold across the organization. Almost all staff organizations will be impacted by a broad-based strategic maintenance plan. Each must understand its role, the purpose of the change and how the new system or program will work. Each staff group must "buy in" to the plan and understand the importance of maintenance improvement. If the organization has embraced an overall strategy of team-based *continuous reliability improvement* (CRI), this "buy in" will be obvious to the other staff groups and be readily accepted.

The final type of support needed for successful implementation also depends on somebody stepping forward to assume ownership. First, the maintenance leader, along with the maintenance operation, must readily accept ownership. If your maintenance people have been involved in providing ideas and technical input and have developed a team-based commitment to maintenance excellence, this will not be difficult. For operation managers and supervisors who have had a part in developing the plan for greater maintenance service, fewer breakdowns and greater equipment effectiveness, being part owners will not be a problem. For operators who now are trained to do selected operator-based maintenance services and know how to detect and help prevent maintenance problems, being part owners in the plan for maintenance excellence will not be difficult. The impact from successful implementation will be apparent all across the organizations from the top down to the shop floor and the bottom line. Leadership and the acceptance of ownership at all levels are critical to successful implementation. Successful implementation is the key to creating impact.

Successful implementation of maintenance best practices is the objective and represents the culmination of doing an effective overall "Scoreboard for Maintenance Excellence" assessment and having a well-defined plan of action developed from assessment recommendations. The results from the assessment, sets the stage for a prioritized plan of action. Project teams can then be defined to create detailed project plans with tactical and operational tasks/plans to support strategic plan implementation. As we discussed previously, all assessment will yield some "*Do it now*" actions that should be accomplished immediately. The major phases of a strategic plan of action may look something like the following plan summary:

Phase I: Establish strategic plan of action

- Conduct Scoreboard for Maintenance Excellence benchmark assessment
- Define CMMS functionality requirements
- Evaluate and select CMMS
- CMMS implementation in Phase III
- Best practice development and implementation planning per scoreboard assessment results
  - Initiate effective planning and scheduling (Phase II)
  - MRO materials management and procurement improvements (Phase IV)
  - Storeroom modernization (Phase IV)
  - Establish effective *preventive and predictive maintenance technology (start in Phase I and complete in Phase V)*

Phase II: Develop and establish maintenance planning, estimating, and scheduling function

Phase III: Implement CMMS (concurrent with Phase II)

Phase IV: Implement storeroom modernization, maintenance repair and operations (MRO) improvements plan (concurrent with Phase II)

Phase V: Implement other scoreboard recommended maintenance best practices with CMMS implementation (Continue and complete PM/PdM program started in Phase I)

Phase VI: Measure maintenance performance and validate benefits with the Maintenance Excellence Index (begin in Phase II)

Phase VII: Continuous reliability improvement (Phase I to Phase VII)

The Real Work Is Implementation

Successful implementation is the ultimate objective that provides maximum value to your maintenance operation. It is very important to remember the 90–10 rule for projects. Ninety percent of the real project work comes during the implementation and normally the easiest part is planning, which is only 10 percent of the work. Never underestimate the resources needed for implementation. Begin implementation with clearly defined plans for each area of the strategic plan as well as tactical and operational level tasks and action plans. Never, never begin implementation without methods in place to validate results, especially if you use external consulting resources like TMEI and others. External resources at times can "borrow your watch to tell you the time and then walk off with your watch before implementation." Use external resources/consultants who insist that you measure the benefits that are agreed upon before implementation begins. Involve consultants in the implementation of their recommendations. Who better to shift the blame to than a consultant?

Implementation results of your profit-centered action plans will require new processes for the business of maintenance. These new processes should be documented in easy to understand written procedures or visually via a flowcharting format. Back in my early industrial engineering (IE) days, flowcharting was a classical IE tool. It is now called *business process mapping* or *blueprinting*. It is still a powerful tool and Mick Windsor, Founder of Windsor Business Solutions in Australia, provides a very good summary and key points to consider as you develop your various plans of action for implementation. As an accredited SAP PM consultant, with years of maintenance and production experience, Mick's key points that follow can help map your journey toward maintenance excellence.

What Is Business Process Blueprinting?

Maintenance is a business within a business, which in some of the plants I have worked in, is only measured on cost and equipment failures. Sounds a bit familiar? Breaking out of the rut is hard work and a map is needed to get there. That map is the blueprinting of the business processes, documenting the way you want to run maintenance.

The end result of *business process blueprinting* (BPB) is the documentation of all your business processes and the configuration of the CMMS to support them. The processes are then performed by your staff, all of whom have been well trained in the transactions of your CMMS and the business processes and compliance is measured. Fairy tales? No. This subject is a book in itself, but here are some pointers to help you get there.

To achieve this, business requirements need to be identified, analyzed to determine the most effective method of running your business, and then documented. The documents are then used to develop templates for configuring the CMMS. It is also the foundation of your CMMS training material, and the pertinent points must be included in position descriptions. It is important to see how CMMS flows through all aspects of your maintenance business.

Business processes affect all aspects of maintenance, safety, craft competencies, asset history, life-cycle costing, and procurement—nothing escapes. Best though, it provides a common framework to work in, standardizing the maintenance management function. Whether you use SAP PM or any other CMMS, the same methodology should be applied. An auditable, standardized approach to business management that builds corporate memory and optimizes your investment in the CMMS is then achievable.

Return on investment for effort (ROIEff)

Corporate memory. How often do people leave a firm after many years taking a wealth of knowledge with them? All the time unfortunately! Business mapping keeps knowledge in-house by identifying where data is stored. This is done using a data matrix developed from the blueprint.

Easier training of new starters/employees. What about new starters? Their predecessor has moved on, others in the team are trying to cope with their own workloads, so where do they get their training? Flowcharts and support documents on line, based on the blueprints provide a strong knowledge base for all personnel.

Not just transactional training. I recall talking to a maintenance supervisor with many years of experience. The company had implemented SAP about five years earlier, he was among those who got the initial two solid days SAP training and as such was considered good at SAP. (Two days in five years? Not much.) We were having a chat about various things while he kept creating new work orders and closing ones that were done. All of a sudden, he said, "I hate creating these orders and closing finished ones off. What a waste of time. Why can't it be like before SAP when I just got my guys to do the work and when they finished I'd get them to do something else?" For five years, he had done it and never knew why. He was smart and a good supervisor and he had asked others, but they did not know either. "It was the way SAP wanted it," they all thought. The *dominant failure mode* here is that only transactional training was given, and no training in business processes was given.

Maximize your return on investment

- Develop and document sensible business processes with input from each business group that contribute to maintenance.
- Use the *audit method* to critique the blueprint. When you review each process, use the "Scoreboard for Maintenance Excellence" to audit it. This can be done in a workshop to involve staff and achieve better buy in from staff.

- Use a skilled business analyst who has consultant-level skills with your CMMS to do the blueprinting, so they can configure the system as well, giving a quicker and more seamless result.
- Provide combined transactional/process training to your people.

How to Do Business Process Blueprinting?

Project management. Good project management is necessary. It is usually a large and sometimes an expensive project with a lot of scrutiny from the executive level. Do not have the skills? Obtain them or hire someone who has them. This is the foundation of success and the rewards that come with success.

Skills required for BPB. The analyst needs training and experience in business analysis and must be an accredited *original equipment manufacturer* (OEM) consultant in the CMMS (does not matter if they are internal or external consultant). Business representatives need training in business analysis techniques and in-depth training in the CMMS functionality.

Requirements gathering. Use workshops with representatives of all business groups to determine requirements. This will include safety and environment, warehouse, procurement, production, finance, HR, training, and even administrative people as they all are a part of an effective maintenance effort. Too often administration is overlooked, causing a link to weaken in the chain. Giving *post it* notes to participants to write their requirements makes requirements grouping much easier. Try to have participants from different roles in each workgroup, which helps to reduce the pinpoint focus that can occur if it has all maintainers from one team in the group. Doing it in short sessions over several days with positive feedback from their managers helps keep the interest up; and providing a nice but simple lunch certainly helps attendances. Keep it all positive.

Documentation. When developing your documentation use flowcharts with cross-functional flowcharts or "swim lanes" which delineate roles or responsibilities. This approach makes it much easier for people to follow and clearly shows who needs to achieve each step. An extra "lane" can be used to specify transactions in the CMMS so an integrated approach of both process and transaction steps are presented. Would that make training easier to deliver, give quicker comprehension, and better compliance to the way your business is run? Yes it does.

Rigor. Why spend the money in developing this if you are not going to ensure that people actually use it. This begins with knowledge and adherence being included in the position description than role-specific training and ends with process auditing to monitor compliance. Listen to your people if they say the process is difficult!

Rolling it out. Planning, more planning, and strong scheduling is the key. Use a dedicated planner and scheduler if possible.

Musts:

1. Ensure all associated business groups are included.
2. Sufficiently resourced and budgeted for, with contingency.
3. Top management gives regular, visible support to "enhance" commitment by staff, and encourage attendance to training sessions.
4. Having an executive manager as a sponsor really smoothes the way.
5. Each stage of the BPB and each business group are measured with all the results published.
6. Do not forget the training effort.

Long-term success requires long-term support. All too often, an implementation is overtaken by other priorities that have a huge impact on the ROI. Too often failure is blamed on "that software" or "that consulting company," and the root cause is ignored. Often they are not solely to blame. Well is it not easier to blame someone else? *It can be...*, but if you end up with an inferior result, you have to live with it and life is usually harder. In addition, you cannot go back to the old system either! Doing it right once means doing it once!

Ongoing Support

Monitor compliance

Maintenance leaders do not always hear when there is regular noncompliance to processes. It is often swept under the carpet and only aired when asked directly and there is no perceived retribution. There is usually one person though, who will tell you what is wrong no matter what—yes, they may be a pain but do not ignore them, delve a little deeper. You may find a nugget of gold in there somewhere. As a part of normal conversation with any of your staff, ask what the processes are like: Are they user friendly? Do they achieve our goals? What makes your job hard? Listen, learn, and act—and feed back information to your staff.

Reasons for noncompliance:

- Does not make sense
- Too hard to do
- Do not want to do it
- Do not know how to do it

Try to cover these issues at each step during the analysis and implementation to negate them before they begin. A road map, in documented business processes with management commitment, training, and ongoing support will provide a firm foundation for continually improving the maintenance function—with the return showing in the bottom line. Yes, you can do it. Thanks to Mick Windsor at Windsor Business Solutions Ltd. (Mick@WindsorBusiness Solutions.com).

Profit optimization is about total operations success, profit, and greater customer service. There are many direct and indirect savings opportunities and we illustrated how to capture one of those opportunities: increased craft productivity as gained value. Just one of the many best practice areas from "The Scoreboard for Maintenance Excellence" such as effective shop-level planning and scheduling by a single planner can provide more than a five to one ROI for a 20-person craft workforce.

When maintenance becomes truly profit centered with new attitudes and positive actions, maintenance support to profit optimization is another very real contributor to the bottom line for the total operation. You can maximize your total maintenance operation for profit optimization by successful implementation of strategic, tactical, and operation-level actions. And remember that the "*Do it now*" tasks are now yours to do now.

Chapter

28

Achieving PRIDE in Maintenance

I would like to begin this chapter with something I wrote while serving as plant manager of manufacturing operations for the Cresent-Xcelite tool plant in Sumter, South Carolina. It talks about *pride* in excellence, which parallels all we will talk about for achieving *pride* in maintenance. Late in 1983, when this was written to plant employees, little did we know that over 100 employees from the total plant would face lay offs in late 1984 and 1985 due to dumping of foreign hand tools onto the U.S. marketplace. The following is my message to over 500 very dedicated plant and maintenance employees at this plant.

PRIDE in Excellence

A very famous German sports car manufacturer (Porsche) has a slogan that says: "Excellence Is Expected." To me this is what our customers, distributors, and Cooper Tool Group sales/marketing people expect. "Excellence is Expected" from the world famous Crescent/Xcelite trade name.

In our employee communication meetings, the topic of foreign competition and the real threat that it presents to the American tool industry is an issue that all employees must keep in mind. Crescent/Xcelite is continually addressing both foreign and domestic competition through an aggressive capital investment effort to upgrade old equipment, secure new equipment, and to improve manufacturing methods and processes by whatever means available.

However, while money buys the new equipment and our raw materials; our *people* are the most important assets toward developing continued excellence in our products. Personalized human skills ranging from die making to final inspection are the assets that can determine the outcome of the "big game" with foreign and domestic competitors. To win the "big game," each employee at Crescent/Xcelite must consider taking "pride in excellence" in a renewed effort for 1984.

As we prepare for 1984, think about a few of the thoughts below and add one or two to your list of New Year's resolutions. They apply to your work here at the plant as well as to your personal family life.

1. *Do every job as if you owned the plant, the department, or the piece of equipment you operate or maintain.* Every employee in this plant is a manager and in a real sense a manager of a small business regardless of the operation. Crescent/Xcelite as a plant is made up of many small plants or teams. Be a proud competitor on your team in 1984.
2. *Develop a commitment to excellence in everything you do.* Have fun and seek justice against poor quality where it is due. If work were thought of as a hobby such as golf or fishing, think of the fun it would be to meet that 7 a.m. starting time at the plant.
3. *Develop pride in work regardless of the task.* Give 105 percent performance whenever possible to make up for the times you were at only 95 percent.
4. *Maintain a sincere belief in your capabilities as well as the potential of those you meet each day.* Practice positive reinforcement on yourself and others.
5. *Practice the golden rule.* If it does not work the first time then . . . practice! practice! practice!
6. *Practice good maintenance in all areas: physical, spiritual, family, equipment, mental, financial, and so forth.* Plan to wear out rather than rust out!
7. *Try to "make things happen" rather than "watch things happen."* Try not to be in that group of people that "wonders what happened." Reduce *work-in-process* (WIP) and stamp out *rework-in-process* (RIP)!
8. *Develop pride.* Try to develop pride in yourself, your company, and your country.
9. *Establish specific written personal goals.* Try to establish goals in all areas of your life.
10. *If none of the above items works for you, just keep smiling.* Do this because people will really wonder what you are up to!

Pride in excellence is a very personal matter that we as individual Americans must address. The late Congressman Larry McDonald (who died in the India Airline plane shot down by "a so called accident" in the early 1980s) made a closing statement to the documentary film "No Place to Hide" with these words: "Freedom is not free." Excellence, quality, and success likewise are not free; all take hard work and commitment from *people*.

Pride on the other hand is free when the work is done, or the game is over and you look back and say "I did the very best I could do" or as a departmental team, "We did the best we could !" Let us enter 1984 with the goal of looking back a year later and being able to honestly say: "We did our best!"

The nine key items from this 1984 letter to plant employees apply today, well over 20 years later. Think about leading your maintenance operation forward with these key items as part of your personal philosophy for achieving *pride* in maintenance:

1. Do every job as if you owned the plant or the maintenance department.
2. Develop a commitment to excellence in everything you do.
3. Develop *pride in work* regardless of the task.
4. Maintain a sincere belief in your capabilities as well as the potential of those you meet each day.
5. Practice the *golden rule*!
6. Practice good maintenance in all areas: physical, spiritual, family, equipment, mental, financial, and so forth.
7. Try to "make things happen" rather than "watch things happen."
8. Develop *pride* in yourself, your company, and your country.
9. Establish written and specific written personal goals in all areas of your life.

PRIDE in Maintenance

The term *pride* in maintenance was first used in 1981 while I was on the Cooper Tools Group staff as group manager of Industrial Engineering and striving to improve maintenance processes across seven plant sites. One first step other than creating the original "Scoreboard for Maintenance Excellence" was to sponsor a group-wide training session on maintenance best practices. This was a weeklong session held in Greenville, Mississippi, on the Mississippi River, way down south at the Nicholson saw plant! We had all plant managers attend for one day and then engineering managers, maintenance supervisors/foreman, planners, and storeroom managers for the rest of the week. We even had two *predictive maintenance* (PdM) equipment vendors come in to discuss this new technology. They took turns demonstrating the benefits of vibration analysis and other technologies. To our good fortune during an in-plant demo, one vendor actually found a bad bearing on a major piece of equipment. This was in the steel mill for rolling specialty steel for hacksaw and band saw blades. The training event was a success in creating a better understanding about the importance of maintenance for top leaders and maintenance leaders. It also helped avoid a catastrophic bearing failure at the Nicholson steel mill next door to the main saw plant. So what is the point so far you might ask?

The main point is that we forgot to include any craftspeople in this weeklong event. The consultant that I used for the training was Mr. George Smith, from Orange, New Jersey. He had just recently founded Marshall Institute, now owned by Dale Blann here in Raleigh, North Carolina. George discussed many best practice topics related to the crafts workforce throughout his presentations during the week-long event. This was before RCM and TPM were in the maintenance glossary of terms. Therefore, I said to George, "We forgot a very important group. We did not get your message down to the crafts level." So I told George "I will talk to the plant manager here at the Greenville plant. We will see if we can get the crafts together for at least a 1-h session and you can talk to them too before you leave." Gerry Grieve, the plant manager, quickly agreed and George talked to 30 or more craftpeople on the topic that I called *pride* in maintenance. George was a Navy pilot in World War II and he had some real maintenance stories to tell about why *pride* in maintenance was important. We videotaped George's presentation, and his words apply today just as much as they did way back in the 1980s.

The craft workforce *must not be forgotten* in today's world of fancy new terms and technological advances. Your crafts can be *people really interested in developing excellence* in maintenance when they feel that they are a true member of the team. Our domain name, www.Pride-in-Maintenance.com was due to George Smith's presence in Greenville, Mississippi. It reflects our belief about the importance of maintenance, the value of maintenance people and the work they do, and how we must change attitudes about the profession of maintenance.

I would now like to reinforce some key points made in the previous chapters. Your improvement strategy must include all maintenance resources, equipment, and facility assets as well as the craftspeople and equipment operators. It must also include MRO parts and material assets, maintenance informational assets, and the added-value resource of synergistic team-based processes. Maintenance leaders and top leaders must support their most important maintenance resource of all, the crafts workforce that I consider as your MVPs—most valuable people. Figure 28.1 illustrates the support role that must start with top leaders realizing that the foundation for maintenance excellence begins with *pride* in maintenance.

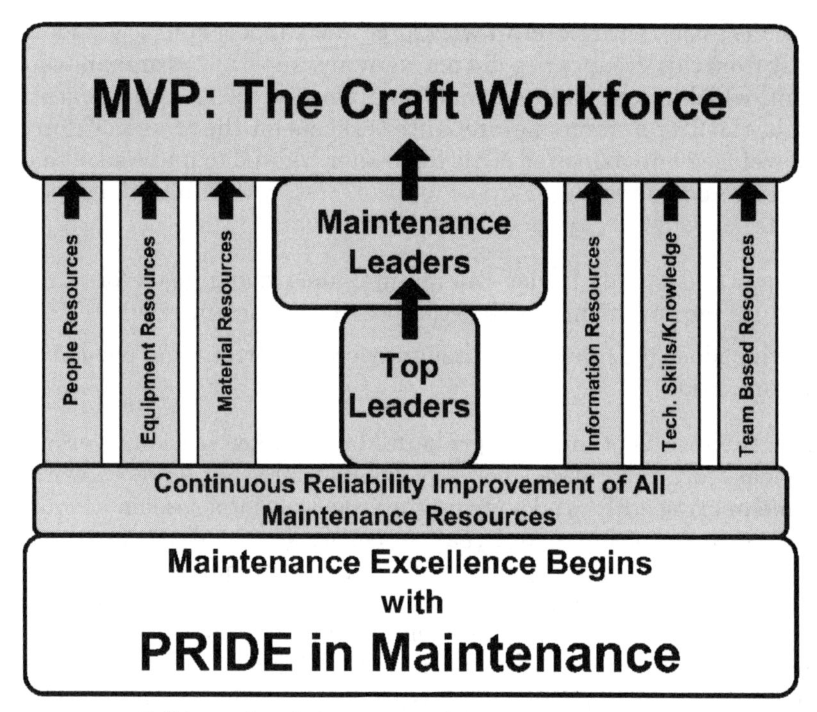

Figure 28.1 Build your foundation upon *pride* in maintenance.

Our vision is to help achieve *pride in maintenance* from within the craft workforce and their maintenance leaders. In addition, it is to have top leaders realize the true value of their total maintenance operation and then take positive action to support the maintenance leaders and their craft workforce. The Maintenance Excellence Institute provides a wide range of consulting with our maintenance excellence services, temporary operational services public training for maintenance excellence and customized in-house offerings. We support all types of maintenance operations. However, the bottom line is that *pride in maintenance* within your craft workforce is the foundation for your success for building long-term maintenance excellence. Your craft workforce can be a valuable source of new ideas and positive reinforcement during your journey toward maintenance excellence.

During each "Scoreboard for Maintenance Excellence" assessment, we always interview a number of craftspeople. We get very candid input and many comments that support the improvement needs of the overall maintenance operation. However, we lacked a focused method that could bring out more ideas and concerns from the craft workforce. Following an assessment at UNC-Chapel Hill, the top leader requested that we conduct a session with the craft workforce and support staff to get ideas directly from this group. They did not want any supervisors or managers present, which might hinder open discussion of ideas and concerns. As a result, the first *pride* in maintenance sessions for the craft workforce was developed and delivered. This top leader wanted to make sure that their craft and support employees (approximately 350) had the opportunity to:

1. Understand the top leader and maintenance leader goals for maintenance improvement per the scoreboard assessment results
2. Provide ideas that the crafts thought were important and needed by the operation

From the results of this first pride-in-Maintenance session, over 300 good ideas were generated. Therefore, we feel sure that it is well worth the time to bring crafts and support staff together for a session devoted to open sharing of ideas and real concerns. We firmly believe that "maintenance excellence begins with *pride* in maintenance." It is important to have people at all levels with PRIDE in maintenance—*people really interested in developing excellence in maintenance. Pride* in maintenance sessions can help you gain people with greater *pride* in maintenance. They will be your own craftspeople who can then add greater value to your maintenance operation by sharing their ideas and being a vital part of helping you implement today's proven best practices for maintenance excellence.

A positive and proven approach

Pride in maintenance sessions with your craft workforce should begin only after the consultant has a clear understanding of your current improvement goals, your current challenges, and your past successes. To do this, a "Scoreboard for Maintenance Excellence" assessment is recommended. The following key steps (illustrated in Fig. 28.2) should be taken to help your craft workforce become a valuable source of new ideas and attitudes.

1. Develop your results from the "Scoreboard for Maintenance Excellence" assessment.
2. Develop your *pride* in maintenance session materials. Gain your approval of materials and your commitment to begin.
3. *Pride* in maintenance sessions for the craft workforce sharing goals/key challenges (1 h).
4. *Pride* in maintenance team exercises with craft workforce focused on your key challenges (1 h).

Figure 28.2 Your craft workforce is a valuable source of ideas.

5. Teams present recommendations. Presentations from each team are videotaped and a summary of all team recommendations prepared.
6. Client reviews assessment results, crafts team recommendations and determines their strategic, tactical, operation, and "Do it now" commitments.
7. Implementation support provided from The Maintenance Excellence Institute only as needed.
8. Continue with *continuous reliability improvement* (CRI) and chartered cross-functional teams as a possible next step.
9. Measure and validate results.

Gain support of craft workforce

Without support from the craft workforce achieving maintenance excellence can be extremely difficult. *Pride* in maintenance sessions developed specifically for the crafts workforce and other maintenance support staff can add much to your maintenance improvement efforts. They serve as a means to gain craft-level support, to achieve better understanding and greater cooperation for current and future maintenance improvements. It is important that maintenance leaders provide very positive reinforcement to the crafts worker that their job is important and that their ideas are needed and are welcomed. Because they perform such a mission-essential role to success, the need for their positive input, ideas, and active participation is critical.

Pride in maintenance sessions also help instill a philosophy of profit-centered maintenance into the thinking and attitudes of each participant. For public service operations, it is about maximizing customer service. *Pride* in maintenance sessions help support internal teamwork as well as eliminate the fear of changing the status quo. It is important for your craftspeople to understand their contribution to greater profit and service. And, we must challenge and support them "to do each job as if they owned the company." Pride in maintenance materials should be customized to the operation whether the goal is for maximizing either profit or service or both. The client should be able to review and approve all client-specific materials that are developed for the presentation. Session materials including participant handouts, case studies, and additional maintenance excellence references are provided to each attendee. Each session should include up to a maximum of 25 participants to allow for three teams of eight craftspeople across typical craft functions. We like to videotape all team presentations and give each client to receive reproduction rights for future use of their *pride* in maintenance video and all custom materials prepared for their session. Determining the true value of training often seems difficult. But when you can incorporate ideas from your crafts workforce into your various plans of action, the value of this type of training can be readily demonstrated. Scarce craft resources are a terrible thing to waste. So take action now and consider investing in your most valuable maintenance resources to get their ideas. And "as you choose so let it be."

The following is a case study example from the client we discussed previously, the Facilities Services Division for the University of North Carolina at Chapel Hill, North Carolina. It shows a section from their newsletter, one of the communications tools they used to spread the word about their workforce's involvement in maintenance improvement. Each of the over 300 recommendations from their initial *pride* in maintenance sessions were reviewed by the maintenance leadership team shown in Fig. 28.3. This team looked at each recommendation and their

Figure 28.3 Maintenance leadership team at UNC-Chapel Hill.

goal while reviewing each employee's idea to anwer one of these three questions:

1. What can I (we) do now to put this idea into action?
2. How can this idea be put into action based upon *additional internal* review and help?
3. How can this idea be put into action with support and cooperation from departments outside the UNC-Chapel Hill's Facilities Services Division and *additional external* review?

And it came down to Question number 3 for some recommendations that needed support and cooperation from other departments. The following is from their newsletter and describes how they created a cross-functional team to address a major improvement opportunity. This newsletter column started out as *pride* in maintenance, but was quickly changed to *pride* in maintenance and construction because constuction renovation work was an important part of the division's overall responsibility. This segment now competes head-to-head with the private contractors for internal work. It is now a true profit and customer-centered service provider within this key division at UNC-Chapel Hill.

Pride in Maintenance and Construction*

PRIDE in Maintenance and Construction has continued with follow-up on each of the over 300 recommendations received from employees during the initial PRIDE in Maintenance sessions. The Team headed by Steve Copeland with Stanley Young, Bob Woods, Bob Humphreys, Mark McIntyre, and Joe Emory completed the review of *each and every recommendation received* back in November 2003.

Their goal while reviewing each employee's idea was to answer one of these three questions:

1. What can I (we) do now to put this idea into action?
2. How can this idea be put into action based upon *additional internal* review and help?
3. How can this idea be put into action with support and cooperation from departments outside the UNC-Chapel Hill's Facilities Services Division and *additional external* review?

Systems Performance Team Chartered

The review of all employee recommendations has now led to a number of recommendations where support and cooperation from departments outside the Facilities Services Division is necessary. Therefore, the systems performance team was chartered.

The objective of this team is to recommend methods to improve overall building systems performance. The scope includes design and construction phases and the commissioning process of new facilities. The team started in January 2004 and has been meeting biweekly. It has been looking at ways that building services can be more closely involved with design and construction processes and how to eliminate additional work that typically must be performed on new facilities. This team has addressed specific quality control improvement opportunities during the design/construction phases to improve contractor accountability. It has considered how to support the construction administration process better and how to support the final user's requirement for the facility.

Team leader Joe Wall from construction is supported by Ricky Robinson—maintenance/plumbing, Rodney Davis—life safety and access, Robbie Everhart—HVAC, Terry Bowers—housing support, Eddie Short—construction administrator, Joel Carlin—engineering information services, Ed Willis, PE—manager of construction management, Cindy Shea—sustainability coordinator, Julie Thurston—PE A/E services,

*Excerpt from *The Resource* newsletter UNC-CH Facilities Services Division with Carolyn Elfland, Associate Vice Chancellor for Campus Services as their leader.

Pete Peters—The Maintenance Excellence Institute, Carole Aquesta and Keith Snead from facilities planning/design.

The systems performance team is now preparing its final recommendations, and a presentation by team members was planned during April 2004. Their work has been very positive actions on important employee recommendations from our *pride* in maintenance session held there.

Results from the systems performance team helped to ensure that maintenance was involved during the design, preconstruction, construction, and commissioning phases of new facilities on the UNC-CH campus. Facilities maintenance staff from each craft area became an important part of the total team. Significant costs were avoided by ensuring that contractors did the job right the first time and by not having to call maintenance in to correct contractor mistakes after commissioning. Tracking of warrantied items was also improved with vendors and contractors being held more accountable.

Summary

Achieving *pride* in maintenance requires many things within an organization. It requires a top leader who understands the value of maintennance and its challenges. It requires maintenance managers, supervisors, and foremen who are true maintenance leaders within the important profession of maintenance. It requires crafts leaders and a craft workforce trained and dedicated to profit- and customer-centered service. Effective storeroom and support staff all combine to perform the business of maintenance. Across all these people resources there must be dedication to the maintenance profession and *pride* in maintenance that comes from teamwork, personal motivation, good leadership, and good maintenance practices. And as you choose so let it be that your maintenance operation continues to improve its progress on the journey toward maintenance excellence. Maintenance excellence is truly not a final destination, but rather a continuous journey.

Chapter

29

The Journey toward Maintenance Excellence: Where Will You Go Now?

The next steps you take after reading this last chapter will be very important. The tools covered in this book will help improve profit and/or customer service, but it is all up to you. Where will you go now on the journey toward maintenance excellence? Will you stay with the status quo and continue to wait before you take action to improve your maintenance operation? Will you wait until there is a takeover from a contract maintenance provider? I hope you will have positive answers to all these questions and that your future *return on maintenance investment* (ROMI) is positive. Accept the improvement challenges you currently face and "Just do it" as we talked about in Chap. 7. If you have imagined these challenges being overcome or have seen others without these challenges then you know for sure *you* can do it.

The last 10 years of the twentieth century saw extraordinary technology advances related to physical assets, to the use of *computerized maintenance management system/enterprise asset management* (CMMS/EAM), and to the use of the computer and the Internet. The remainder of the new millennium will see even more of these extraordinary advances. A number of positive revolutions will occur within maintenance operations of the future. Begin a positive revolution within your maintenance organization. Consider these four key principles from the book *Revolution: Take Charge Strategies for Business Success* by Jim Tompkins, Founder and President of Tompkins Associates Inc.

1. *"Revolution is never a spur-of-the-moment decision. It is a process, and in a true scenario, a continuous process."* We have seen the importance

of maintenance and new technologies progress well beyond what was seen 20 or 30 years ago. The maintenance revolution has definitely been a continuous process. Likewise, the maintenance revolution you might bring to your organization will not be a spur-of-the-moment decision. Your next steps must the based upon facts that are developed from a "Scoreboard for Maintenance Excellence" assessment via an external consultant or from a self-assessment. Continuous professional development will allow the maintenance leader to stay abreast of new technology. *Continuous reliability improvement* (CRI) and leadership-driven, self-managed teams will enhance the application of emerging new technologies for all of your maintenance team members.

2. *"Although Revolution is a grassroots effort, it is characterized by its leaders."* Maintenance leaders will emerge to support the continuous improvement process at the grassroots, shop-floor level. There will be a progression from maintenance manager to true maintenance leadership. The *chief maintenance officer* (CMO) will emerge as a recognized corporate leadership position. The CMO (or an equivalent) will be essential for long-term profitability in the new millennium and will eventually evolve into a recognized corporate position, either" real or imagined." The CEO/COO of a multisite operation that does not have a CMO accountable for physical asset management will be gambling with stockholder's equity. Just as the CEOs who rob shareholders with board-approved golden parachutes, this type of CEO you can be sure will rob your physical assets with continuous bad maintenance (CBM) during his short term stay with short range visions.

   The small single-site operation without a CMO equivalent will realize the high cost of bad maintenance. The CEO must understand the "state of maintenance" in their operation and the physical asset management process. The emerging CMO with "profit ability" and with effective leadership and technical skills will facilitate this process in larger multisite operations. The future capable company will require a proactive, capable CMO just as they need a CFO, CIO, CEO, and the like.

3. *"A Revolution cannot be managed, it must be led."* An individual who is profit- and customer-centered yet understands the fact that people assets are the most important assets will lead new millennium maintenance. The evolution of a CMO will occur in both large and small operations. Successful maintenance leaders will be "profit-centered" and will have established strategic maintenance plans that are integrated with the business plan for the total operation. They along with the Chief Financial Officer (CFO) will lead the effort to validate ROMI with effective measurement processes. There will always be a need to maintain because *maintenance is forever*. Maintenance of our physical bodies, minds, souls, cars, computers, and all physical

assets providing products or services in today's global economy will always be required. Do not let your organization lose their core competencies in maintenance to the point that they have lost complete control. The core requirement for good maintenance remains (forever), but the core competency to do good maintenance may be missing. In some cases, the best and often the only solution is value-added outsourcing. Think about this; if a maintenance leader can bring a contract maintenance team to replace your team . . . why can't you do the same thing with the players you have (or do not have) right now?

4. *"A revolution cannot be carried out by individuals; it must be a collaborative effort."* The maintenance leader of the new millennium will recognize needs of the customer, serve the customer, and bring together all maintenance resources to maximize the value of maintenance. The synergistic effect of teams and collaborative interactions among individuals will be an additional resource and provide measurable gains in productivity. However, contract maintenance will continue to grow. The core requirement for maintenance will be even more important in the next millennium due to technology advances. Organizations will continue to focus on core competencies and outsourcing. Some will forget the core requirement for maintenance and some may lose their core competencies to do effective maintenance. Profit-centered contract maintenance providers will consume internal maintenance operations that continue a cost-centered approach. In-house maintenance operations will continue to lose when they are unable to replenish and/or maintain their core competencies in maintenance. Likewise, retaining the status quo can lead to a quick outsourcing revolution. The neglect of the past and the future will be overcome with services from a growing number of profit-centered maintenance providers who clearly understand how to provide value-added maintenance service at a profit. Neglect of the past can also be overcome internally by the emergence of an internal CMO and true maintenance leaders who can lead maintenance forward to profitability as if they owned the internal maintenance business.

Your next steps on your journey must collectively mesh with other staff groups in the organization to bring about productive change. Overall maintenance productivity and improvement will emerge from well-prepared plans of action that systematically mesh the right solution to the right problem. Do not forget to ensure that the basics are in place. Do not spend valuable time chasing fads or a better CMMS when basic best practices and the discipline for accountability are not in place with your current CMMS. So you can imagine my body language and facial contortions when a maintenance manager blames the CMMS!

You must stop to see where you are and clearly define where you want to go. Somewhere I heard the statement, "It's tough to know when you're 'there'... if you don't know where 'there' is before you start." This applies both to one's personal and professional life. There are hundreds of ideas and concepts for personal and business development that have been successfully tested and passed along to each new generation. All are in the Bible and in red letters. Regardless of the original source, many good ideas never get put into practice.

My objective for this book was to help you maximize the value of your current maintenance operation immediately, not next year. It is to support improving either profit or the customer service in your organization and very often both. It has been about benchmarking maintenance from four key levels that can lead you toward business process improvement for the business of maintenance. We strived to have this book apply to maintenance leaders at all levels and in both large and small operations. Another objective is to give you proven tools for benchmarking that you can use right now! My desire for this material was also to help the small maintenance operations of less than 5 or 10 crafts in, say, a small manufacturing plant or a small facilities complex. Small shops face the same technical and maintenance management challenges as the 100 to 200 total craft employee operations. The science and basic technologies of physical asset maintenance is common all across the world. Only the scope and location provides the difference for the profession of maintenance. This book has universal application to all size operations and to almost all types of maintenance operations.

And as we close this final chapter, I would like to bring it down to the personal level of goal setting and how personal goal setting can help during your journey toward maintenance excellence. Maintenance strategic, tactical, and operational plans are examples of goal setting from a business standpoint. Personal goal setting is often never applied. One important key to professional development is based on the very basic but powerful principle of goal setting. Through goal setting a person can start making an investment in the most important development of all—human resource development. It will be an exciting and rewarding development because it will be focused on your own unique personal and professional resources and success.

One definition of success is that it is "the progressive realization of worthwhile, predetermined personal goals." In this context, success requires a well-rounded goals program to include personal goals in areas such as financial, educational, physical, spiritual, social, and family. Also, success should be viewed as a progressive journey rather than a final destination. Success should be looked upon as a progressive journey dedicated to achieve your own personalized definition of success. The true bottom line measure of one's success is what you are doing compared to your true potential. Personal success is also related to "where you started from" on the socioeconomic scale and "where you finally ended up" on that same baseline scale. This is exactly like using The Scoreboard for Maintenance Excellence and its rating summary from Chap. 4. Goal setting should never be confused with wishful thinking and daydreaming. Goal setting is the most important aspect of all the personal development concepts and is the strongest human force for self-motivation.

Written goals are important for many reasons. First, written goals help us keep track of where we are going and how we are planning to get there. We are born to shape a final destiny for ourselves. The depth and extent of that final destiny is measured by the personal goals we set and achieve. Written goals with clearly defined objectives and target dates help us to measure personal progress. The achievement of predetermined worthwhile goals provides valuable feedback to us in the form of personal motivation. David McClellan's research on motivation revealing that achievement leads to motivation validated what was included in the Bible. In other words, the achievement of a personal goal, whether large or small, helps rekindle the motivational spirit within us.

Written goals help us make good decisions. It is much easier to decide on a course of action, if you know exactly what you are trying to accomplish. A person with positive moral and ethical goals can easily make the right decisions when facing the multitude of negative temptations in our modern world. Written goals add to our self-respect and confidence. There is an intense satisfaction and self-confirmation in reaching a goal that you have committed yourself to achieve. A positive mental attitude develops from clear, calm, and honest self-confidence. Confidence also reinforces abilities, doubles energy, and expands our mental facilities.

Written goals help us to concentrate and to hold firm to a steady course in full view of life's distractions. Once you really make a commitment to a personal goal, you see, hear, and think of more possibilities of reaching it than you ever dreamed possible. Written goals also help reduce conflict. There is real security in knowing what you want to accomplish and having a valid plan of action to accomplish it. Just as maintenance planning and scheduling support the total maintenance operation. Written goals add a sense of value by requiring you to reflect on your values and to look at yourself in relation to your expectations. Written goals can help us save time. With clearly defined goals, you are less likely to become preoccupied with irrelevant activities.

Written goals can also create a renewed sense of purpose and positive anticipation in life, both personally and professionally. What is your purpose in life? What is your purpose as a maintenance leader or a craft leader? It is much easier to stay enthusiastic about something that is important to you. With enthusiasm, you have one of the prime methods for "persuasion without pressure." Enthusiasm helps us to compete with ourselves, to match today with yesterday. The compulsive drive of enthusiasm also helps us to find encouragement and lessons learned from our past mistakes.

At this point, a valid question may be arising that there must be more to goal setting than just writing down goals and waiting for something good to happen. The achievement of personal goals requires many of the same ingredients that are required for maintenance excellence. Maintenance excellence requires the application of some very important four-lettered words: *Hard work from a good team committed to a common goal.*

A plan of action resulting from a well-rounded goal-setting program requires *hard work* for successful implementation. Allan Mogenson, the father of Work Simplification used a term "work smarter—not harder." For maintenance operations, we may need a slight revision of this term to "work smarter and harder" in today's globally competitive world. Goal setting is therefore not a panacea for achieving success without sweat equity. The common denominator for success will always still be this important four letter word: "work."

Written goals should be a part of your plan of action for living. Most people have a written and often very specific legal plan at death. In other words, they have a personal will, which outlines a plan for the final distribution of their possessions. The average family spends days planning family outings and vacations in detail, but neglects developing a written plan for living. Goal setting is something many people talk about and endorse, but they often neglect to really practice it and apply it to a personal plan for living.

Nothing Happens Until Someone Sells Something

You will be required to sell something as you present your plans of action for maintenance improvement to top leaders. You may be selling your ideas to a top leader who does not really understand maintenance. Red Motley, a successful salesman and lecturer, put it quite well when he said, "Nothing happens until someone sells something." P.T Barnum said it a little differently: "Without publicity a terrible thing happens; nothing! The very best way to be sold on the benefits of goal setting (maintenance planning/scheduling) is to put the basic principles into action through practice. If they do not work perfectly the first time, reevaluate your application techniques, reapply them and continue to *practice, practice, practice*! Personal goal setting can provide added meaning and emphasis to your overall organizational efforts to promote change within maintenance and for changing the total operation culture about maintenance.

You will agree that goal setting is not a dynamic new concept. Personal goal setting is simply another back-to-the-basics principle so often neglected in full view of the many "new and exciting concepts." Consider the dynamic potential that personal goal setting can offer and mesh it fully with your organizational goals for maintenance improvement that you may have. Use goal setting as a self-motivating factor to provide that slight competitive edge so often needed in the business world. Human potential is a vast and often untapped resource that can be more fully developed by goal-directed action. Remember that the most important leader of all is you, the leader of *you, yourself incorporated.*

In his 1970 book *Future Shock,* Alvin Toffler states that, "each individual must become infinitely more adaptable and capable than ever before to deal with change." Many people who understand intellectually that change is accelerating, have not internalized that knowledge. They have not taken the critical issue of change into account during the planning of their own personal and professional lives. Maintenance leaders must be prepared to deal with the "Future Shock" that change will bring to both our personal and professional lives. We must have a personal plan of action that develops more of our full potential as productive human beings to deal with change that we will face from the profession of maintenance in the future.

Take your next steps toward maintenance improvement with the knowledge that things will improve and that you are the owner of the maintenance business within your organization. Plan, do, and act as if you were in fact the CMO. Have the courage to take the true message about the state of maintenance up to the top leaders and share candidly with them the facts about the high cost of gambling with maintenance. Give them the complete message with facts on their "total maintenance requirements." Do not take a piecemeal approach to improvement, but rather strive to improve maintenance in total. Start your efforts by knowing clearly, where you are with a tool such as "The Scoreboard for Maintenance Excellence" and set priorities for action. Make sure that top leaders understand that you will be measuring shop-level results and that you will validate the benefits and ROI that you promised. Establish personal goals that mesh with your goals for maintenance improvement. Be an example of someone with PRIDE-in-Maintenance. If you do not take something from this book and apply it now or later then I have failed in this book for you. So just as our Training for Maintenance Excellence is not over when it is over, it is also not over for you as you finish this book. Just call me or Bob Gaskins at 919-270-1173 and we will talk to you personally about application of contents within this book. And one last and the most important comment, "Never, never, never give up on the belief that you can make a positive contribution toward the success of your total operation and the profession of maintenance."

Appendix A

The Scoreboard for Maintenance Excellence

A.	The Organizational Culture and PRIDE in Maintenance	
Item No.	Benchmark Item Rating: Excellent—10, Very good—9, Good—8, Average—7, Below average—6, Poor—5 or less	Rating
1.	The organization's vision, mission, and requirements for success includes physical asset management and maintenance as a top priority	
2.	Senior management is visible and actively involved in continuous maintenance improvement and is obviously committed to achieving maintenance excellence.	
3.	The organization's strategy and plan for total operations success is known to all in maintenance and includes a strategy for maintenance improvement.	
4.	Maintenance is kept well informed of changing business conditions, strategies, and long-range plans.	
5.	The organization's culture and the maintenance environment results in innovation, pride in maintenance, trust, and an obvious spirit of continuous improvement.	
6.	Open communication exists within maintenance and the overall organization to ensure interdepartmental cooperation, idea sharing, and basic teamwork.	
	A. Maintenance organizational culture and pride in maintenance category: Subtotal score	

B.	Maintenance Organization, Administration, and Human Resources	
Item No.	Benchmark Item Rating: Excellent—10, Very good—9, Good—8, Average—7, Below average—6, Poor—5 or less	Rating
1.	The maintenance organization chart is current and complete with fully defined areas of responsibility.	
2.	Clear-cut craft job descriptions have been developed that completely define job responsibilities and skill levels required for each craft.	
3.	Craft personnel are provided copies of their job descriptions and counseled periodically on job performance, job responsibilities, and craft skills development needs.	
4.	One single head of maintenance operations is supported by adequate clerical and technical staff of planners, first-line supervisors, stores personnel, maintenance engineering, and training support.	
5.	The maintenance department head has high visibility within the organization and reports to a level such as the plant manager.	
6.	The first-line supervisors are responsible for the performance of 12 to 15 craftspeople. Responsible for 12 to 15 = 10, 8 to 11 = 8/9, 16 to 20 = 7/6, less than 8 = 5, and over 20 = 5.	
7.	A time-keeping system is in place to charge craft time to each job.	
8.	Monthly or weekly reports are available to show distribution of maintenance labor in critical categories: breakdown repairs, corrective work, preventive maintenance (PM) work, and the like.	
9.	Monthly or weekly reports are available to monitor backlog status and priority of planned or project work.	
10.	Backlog trend data is available to highlight the need for craft increases, scheduled overtime, or subcontracting.	
11.	Guidelines on the level of accepted backlog are established to determine the need for overtime or subcontracting as well as to identify potential problem areas.	
12.	Sufficient man-hour data is available that allows valid decisions as to which jobs must be delayed if new jobs or projects are added to the schedule.	
	B. Maintenance organization, administration, and human resources category: Subtotal score	

C.	Craft Skills Development and PRIDE in Maintenance	
Item No.	Benchmark Item Rating: Excellent—10, Very good—9, Good—8, Average—7, Below average—6, Poor—5 or less	Rating
1.	The types and levels of craft skills required for an effective maintenance operation have been identified.	
2.	Job descriptions include well-defined standards for job knowledge and skill levels required for each craft area.	
3.	An assessment of the current job knowledge and skill level of each craftsperson has been made to determine individual training needs.	
4.	The overall training needs for the maintenance staff have been developed with a plan of action and cost.	
5.	The organization has committed to providing the necessary resources for maintenance training and skills development.	
6.	A program for craft skills development has been designed to address priority training needs and is being implemented.	
7.	Results of training are determined by a competency-based approach which ensures demonstrated capability to perform on newly trained craft tasks.	
8.	A policy to pay-for-skills gained is available or is being developed as part of the craft skills development program.	
9.	The benefits of developing multicraft capabilities within maintenance have been evaluated and incorporated into the craft skills training program as applicable.	
10.	Individual training plans for each crafts person are being used to document compliance-type training as well as training that the crafts person and supervisor see as being needed	
11.	Actual hands-on job competency is being documented (for example by the work order system) to help validate overall craft skill levels.	
12.	The overall craft workforce has shown the initiative for continuous craft skills development.	
	C. Craft skills development and pride in maintenance category: Subtotal score	

D.	Operator-Based Maintenance and PRIDE in Ownership	
Item No.	Benchmark Item Rating: Excellent—10, Very good—9, Good—8, Average—7, Below average—6, Poor—5 or less	Rating
1.	Operators are responsible for cleaning their equipment and trained to perform selected levels of operator-based maintenance.	
2.	An initial cleaning to bring equipment to an optimal or "as new" status has been planned or has been completed for critical equipment.	
3.	Operators have been trained to perform periodic inspections on their equipment and report problems.	
4.	Operators have been trained and have proper tools and equipment to do selected lubrication, to tighten bolts and fasteners, and to detect symptoms of deterioration.	
5.	The process of transferring maintenance tasks and skills to operators has been well coordinated between maintenance, operations, engineering, and human resources staff.	
6.	Operators have developed greater "pride in ownership" and understand their expanded role in detecting and preventing maintenance problems.	
	D. Operator-based maintenance and pride in ownership category: Subtotal score	

E.	Maintenance Supervision/Leadership	
Item No.	Benchmark Item Rating: Excellent—10, Very good—9, Good—8, Average—7, Below average—6, Poor—5 or less	Rating
1.	Nonsupervisory work is minimized as a result of supervisors having adequate clerical support, storeroom support, planner support, and so forth.	
2.	Supervisors perform primarily direct supervision of maintenance to include scheduling work assignments, verifying quality of completed work, evaluating performance, and the like.	
3.	Supervisors actively support good housekeeping and the safety program by conducting/attending meetings, providing ideas, and having an attitude that creates greater safety awareness.	
4.	An effective supervisory development program is available to increase supervisory leadership and technical skills.	
5.	Maintenance supervisors are team players and are able to gain cooperation and support from operations management staff for the improvement of overall customer service.	
6.	Supervisors actively support continuous maintenance improvement with ideas and suggestions.	
7.	Supervisors promote PRIDE in maintenance and encourage ideas from craft employees.	
8.	Supervisors have the technical background to identify training needs of their craft workforce and create positive support to craft skills development.	
9.	Supervisors are managing and leading their workgroups as if "they owned their maintenance operation as a business."	
	E. Maintenance supervision/leadership category: Subtotal score	

F.	Maintenance Business Operations, Budget, and Cost Control	
Item No.	Benchmark Item Rating: Excellent—10, Very good—9, Good—8, Average—7, Below average—6, Poor—5 or less	Rating
1.	The maintenance budget is based on a realistic projection of actual needs rather than past budget levels.	
2.	Maintenance expenditures are charged to work centers or operating departments and budget variances monitored to highlight problem areas.	
3.	Deferred maintenance repairs to operating equipment and facilities-related assets are identified and presented to management during budgeting process with an evaluation as to the negative future impact of deferring maintenance.	
4.	Maintenance provides key input and support to long-range budget planning for new equipment, equipment overhaul and retrofit, facility expansions, rearrangements, and repairs.	
5.	Labor and material costs are established for all work orders accumulated to the equipment history file along with problem, causes, and action taken.	
6.	An equipment history file is maintained for major pieces of equipment to track life-cycle cost, types of repairs, and repair trends.	
7.	The equipment history file is reviewed periodically to analyze repair trends and define root causes on critical equipment as means to evaluate recurring problem and to improve reliability.	
8.	Labor and material costs are estimated prior to the start of major planned repair work and projects.	
9.	Major work order and project-related cost variances are investigated and explained to person authorizing the work.	
10.	Cost-approval guidelines are established for large or special repair jobs as compared to normal repair.	
11.	The cost of downtime is known and published for major pieces of equipment or work centers and is used in determining priorities for repair.	
12.	Maintenance operations can operate as an internal business with current financial, budgeting, and cost-accounting systems.	
	F. Maintenance business operations, budget, and cost control category: Subtotal score	

G.	Work Management and Control: Maintenance and Repair (M/R)	
Item No.	Benchmark Item Rating: Excellent—10, Very good—9, Good—8, Average—7, Below average—6, Poor—5 or less	Rating
1.	A work management function is established within the maintenance operation.	
2.	Written work management procedures which governs work management and control per the current CMMS is available.	
3.	A printed or electronic work order form is used to capture key planning, cost, performance, and job priority information.	
4.	A written procedure which governs the origination, authorization, and processing of all work orders is available and understood by all in maintenance and operations.	
5.	The responsibility for screening and processing of work orders is assigned and clearly defined.	
6.	Workorders are classified by type, for example, emergency, planned equipment repairs, building systems, PM, and project work.	
7.	Reasonable "date-required" is included on each work order with restrictions against "ASAP".	
8.	The originating departments are required to indicate equipment location and number, work center number, and other applicable information on the work orders.	
9.	A well-defined procedure for determining the priority of repair work is established based on the criticality of the work and the criticality of equipment/facility, safety factors, cost of downtime, and the like.	
10.	Work orders are given a priority classification based on an established priority system.	
11.	Work orders provide complete description of repairs performed, type of labor, and parts used and coded to track causes of failure.	
12.	Work management system provides means to provide information to customer; backlogs, work orders in progress, work completed, work schedules, and actual cost charge backs to customer.	
	G. Work management and control—maintenance and repair category: Subtotal score	

H.	Work Management and Control: Shutdowns and Major Overhauls	
Item No.	Benchmark Item Rating: Excellent—10, Very good—9, Good—8, Average—7, Below average—6, Poor—5 or less	Rating
1.	Work management and control is established for major repairs, shutdowns, and overhauls, and includes work by in-house staff and contracted services.	
2.	Work management and control of major projects provide means for monitoring project costs, schedule compliance, and performance of both in-house and contracted resources.	
3.	Workorders are used to provide key planning information, labor/material costs, and performance information for major shutdown and overhaul work.	
4.	Equipment history is updated with information from work orders generated from major repairs, shutdown, and overhauls.	
5.	The responsibility for screening and processing of work orders for major repairs is assigned to one person or unit.	
6.	Work orders for major repairs, shutdown, and overhauls are monitored for schedule compliance, overall costs and performance information including both in-house staff and contracted services.	
	H. Work management and control—shutdowns and overhauls category: Subtotal score	

I.	Shop-Level Maintenance Planning and Scheduling	
Item No.	Benchmark Item Rating: Excellent—10, Very good—9, Good—8, Average—7, Below average—6, Poor—5 or less	Rating
1.	A formal maintenance planning function has been established and staffed with qualified planners in an approximate ratio of 1 planner to 20–25 craftspeople. (Ratio based upon equivalent crafts)	
2.	The screening of work orders, estimating of repair times, coordinating of repair parts, and planning of repair work is performed as a support service to the supervisor.	
3.	The planner uses the priority system in combination with parts and craft labor availability to develop a start date for each planned job.	
4.	A daily or weekly maintenance work schedule is available to the supervisor who schedules and assigns work to crafts personnel.	
5.	The maintenance planner develops estimated times for planned repair work and includes on work order for each craft.	
6.	A day's planned work is available for each craftsperson with at least a half of a working day known in advance.	
7.	A master plan for all repairs and PM is available indicating planned start date, estimated time, duration, completion date, and type crafts required.	
8.	The master plan is highly visible and is reviewed and updated by maintenance, operations, and engineering as required.	
9.	Scheduling/progress meetings are held periodically with operations to ensure understanding, agreement, and coordination of planned work, backlogs, and problem areas.	
10.	Operations cooperate with and support maintenance to accomplish planned repairs and PM schedules.	
11.	Set-ups and changeovers are coordinated with maintenance to allow scheduling of selected maintenance repairs, PM inspections, and lubrication services during scheduled downtime.	
12.	Planned repairs are completed on time and in line with completion dates promised to operations.	
13.	Deferred maintenance is clearly defined on the master plan and increased costs are identified to management as to the impact of deferring critical repairs, overhauls, and so forth.	
14.	Maintenance planners and operations planners work closely to support planned repairs, to adjust schedules, and to ensure schedule compliance in a mutual goal.	
15.	The planning process directly supports the supervisor and provides means for effective scheduling of work, direct assignment of crafts, and monitoring of work in progress by the supervisor and customers.	
16.	Planners training have included formal training in planning/scheduling techniques, training on the CMMS, and on-the-job training to include developing realistic planning times for craft work being planned.	
17.	Benefits of planning/scheduling investments are being validated by various metrics that document areas such as reduced emergency work, improved craft productivity, improved schedule compliance, increased uptime, reduced total operations cost/more profit, and improved customer service.	
18.	Planning and scheduling procedures have been established defining work management and control procedures, the planning/scheduling process, the priority system, and the like.	
	I. Shop-level maintenance planning and scheduling category: Subtotal score	

J.	Shutdown and Major Maintenance Planning/Scheduling and Project Management	
Item No.	Benchmark Item Rating: Excellent—10, Very good—9, Good—8, Average—7, Below average—6, Poor—5 or less	Rating
1.	The planning and scheduling function includes major repairs, overhauls, and project type work not considered as part of day-to-day maintenance work.	
2.	The use of work orders, estimating of repair times, coordinating and staging of repair parts/materials and planning/scheduling of internal resources, and contractor support is also included for major work not considered day-to-day maintenance and repair.	
3.	A project work schedule or formal project management system is used to manage major work.	
4.	Estimated labor and materials are established prior to project start using work orders with effective labor and material reporting to track overall cost, work progress, schedule compliance, and so forth.	
5.	The master plan for all major repairs is available indicating planned start date, duration, completion date, and type crafts required.	
6.	Resources required for day-to-day maintenance work are not compromised by having to perform major repair-type work, installation, modifications, and the like, consuming in-house resources required for PM's and other day-to-day type work.	
7.	Scheduling/progress meetings are held periodically with operations/customers to ensure understanding, agreement, and coordination of major work and problem areas.	
8.	Major work performed by contractors is preplanned, scheduled, monitored and includes measuring performance of contracted services.	
9.	Planning and scheduling procedures have been established for project-type work.	
	J. Shutdown and major maintenance planning/scheduling and project management category: Subtotal score	

K.	Manufacturing Facilities Planning and Property Management	
Item No.	Benchmark Item Rating: Excellent—10, Very good—9, Good—8, Average—7, Below average—6, Poor—5 or less	Rating
1.	The equipment asset inventory system provides an accurate and complete record of asset information for both operational assets and facility assets.	
2.	New facilities planning, equipment additions, and renovations are well coordinated with both operations/plant and facility maintenance staff.	
3.	Maintenance staff provides input into the engineering planning process for new facilities, major facility additions, and new equipment additions.	
4.	An effective procedure for adding new facility and new equipment information to the asset inventory is used as well as the deletion of equipment being removed from the facility.	
5.	Maintenance requirements are clearly defined and designated as to responsibilities for both operational asset maintenance and facility maintenance. The facility is effectively organized and staffed to accomplish both.	
6.	The overall property management function within the organization provides close coordination with the site's maintenance operation when planning new facilities, renovations, or major production equipment additions.	
7.	Facilities planning includes adequate planning for future maintenance requirements.	
8.	Consideration is given to life-cycle costing of systems and/or subsystems when designing new facilities, renovations, or major operational equipment additions.	
9.	Maximum standardization of facility systems and subsystems is planned for within new facilities, major renovations, as well as standardization of operational asset additions and subsystems.	
	K. Manufacturing facilities planning and property management category: Subtotal score	

L.	Production Asset and Facility Condition Evaluation Program	
Item No.	Benchmark Item Rating: Excellent—10, Very good—9, Good—8, Average—7, Below average—6, Poor—5 or less	Rating
1.	A process is in place to periodically evaluate the current operational status of production equipment and in turn defines repair, overhaul, or replacement recommendations for life-cycle cost reduction opportunities. Supports defining total maintenance requirements.	
2.	A facility condition assessment process is in place to evaluate current operational status of facility-type equipment and in turn defines repair, overhaul, or replacement recommendations for life-cycle cost reduction opportunities. Supports defining total maintenance requirements.	
3.	The PM and predictive maintenance (PdM) programs provide key information as to overall state of maintenance for operational assets.	
4.	The PM and PdM programs provide key information as to overall state of maintenance for facility-type assets.	
5.	The current evaluation processes for production and facility assets supports defining the true maintenance requirements of the operation.	
6.	The current evaluation processes for production and facility assets provides an effective baseline for replacement planning and capital justification of new assets: both production and facilities related.	
	L. Production asset and facility condition evaluation program: Category: Subtotal score	

M.	Storeroom Operations and Internal MRO Customer Service	
Item No.	Benchmark Item Rating: Excellent—10, Very good—9, Good—8, Average—7, Below average—6, Poor—5 or less	Rating
1.	The parts inventory system provides an accurate and complete record of information for each stock item. Parts "where used" is included in the master database along with usage, vendor information, warranty information, and the like.	
2.	The "ABC" classification of stock items is known and proper storage methods and accountability is established for each class.	
3.	"A" and "B" items have valid reorder points, EOQ, and safety stock levels established.	
4.	"C" items (50% to 70% of stock items with 5% to 10% of total inventory value) are identified and use two-bin system, floor issue, or self service.	
5.	Inventory accuracy is determined by an effective cycle counting program. Cycle counting used = 10; Count once per year = 7, Count occasionally = 6, Do no inventory counts = 5 or below.	
6.	Inventory accuracy is regularly measured and is 95% or above. 95% or above = 10; 90%–95% = 9; 80%–89% = 8; 70%–79% = 7; <70% = 5	
7.	An up-to-date storeroom catalogue is readily available to crafts (hard copy or electronic) and includes all stock items, storage area and locations, stock numbers, and the like.	
8.	Parts usage history is continually reviewed to determine proper stock levels, excess inventory items, and obsolete items.	
9.	A critical spares listing is available for critical assets and critical spares (insurance items) are denoted in the parts inventory database	
10.	Spare parts for critical instrumentation system components are listed with inventory tracking numbers. and stored in a separate static-free area. Remotely stored items (for safety or quick access) must be listed in main inventory as to storage area and location.	
11.	Storeroom procedures are in place that define issues, receipts, inventory control, access control, parts accountability, reserving of parts, staging, quality control of incoming parts, rebuilt items, and so forth.	
12.	A storeroom operations assessment has been conducted for possible modernization by evaluating current facilities, location, storage and materials handling equipment, staffing levels, inventory levels, procurement systems, and operating procedures.	
	M. Storeroom operations and internal MRO customer service category: Subtotal score	

N.	MRO Materials Management and Procurement	
Item No.	Benchmark Item Rating: Excellent—10, Very good—9, Good—8, Average—7, Below average—6, Poor—5 or less	Rating
1.	Procedures and evaluation criteria for adding new maintenance parts and materials to stores are in place and used.	
2.	Stores requisitions and issues are tied to the maintenance work order and changed directly to the repair job either before or after work completion.	
3.	Maintenance planners and the storeroom personnel coordinate to reserve repair parts and material for planned work. "Kitting" and direct delivery to the job site is done whenever possible.	
4.	Purchasing has an effective program to evaluate vendor performance and quality.	
5.	Purchasing has developed partnerships with selected vendors and suppliers and is committed to purchasing based on fast delivery, quality parts, and service.	
6.	Maintenance storeroom staff is well trained, customer-oriented, and provide a high level of customer service to maintenance.	
7.	Maintenance storeroom and MRO procurement performance indicators have been established and are evaluated and reported on a monthly basis.	
8.	An effective control method is in place for emergency purchases.	
9.	Purchasing and maintenance work together to ensure procurement of quality parts and material and the "high cost of low-bid buying is avoided."	
10.	Stockouts are being monitored as a performance measure of customer service from the storeroom operation and the MRO procurement process.	
11.	Standardization of parts and components is being pursued by engineering, maintenance, and procurement staff.	
	N. MRO materials management and procurement category: Subtotal score	

O.	Preventive Maintenance and Lubrication	
Item No.	Benchmark Item Rating: Excellent—10, Very good—9, Good—8, Average—7, Below average—6, Poor—5 or less	Rating
1.	The scope and frequency of preventive maintenance (PM) and lubrication services have been established on applicable equipment.	
2.	Operations staff supports and agrees with the frequency and scope of the PM and lubrication program.	
3.	Optimum routes for PM inspections and lubrication services are established.	
4.	PM checklists with clear, concise instructions have been developed for each piece of equipment. Lubrication checklists and charts are available for each asset included under the lubrication program.	
5.	Inspection intervals and procedures are periodically reviewed for changes/improvements and updated as required as part of well-defined PM change process.	
6.	Planned times are established for all PM and lubrication tasks.	
7.	The total craft labor requirements by craft type to accomplish the overall PM program has been established to validate staffing needs for effective PM.	
8.	The required level of manpower is being committed to achieve the total scope of PM services.	
9.	Actual craft time devoted to PM is known and evaluated as a percentage of total craft time available.	
10.	Goals for PM compliance are established and overall compliance and results are measured against the operation's benchmark.	
11.	All noncompliance to scheduled PM and lube services is aggressively evaluated and corrected. Lubrication services are viewed as a key part of PM and are not neglected or overlooked.	
12.	Maintenance and operations work with close communication, coordination, and cooperation to schedule PM and lube services.	
13.	The success of PM is measured based on multiple factors: reduced emergency repairs, increased planned maintenance work, reduced downtime costs, the elimination of the root cause of problems, and improved throughput/quality and quality, and so on.	
14.	PM is a highly visible function within maintenance, is well received as an operational strategy.	
15.	The PM staff are well-qualified craftspeople and serves as good maintenance ambassadors and "customer service representatives."	
16.	A PM master schedule is developed to evaluate the weekly or monthly plan and to level load tasks when required.	
17.	Corrective repair work orders are generated and become planned work as a result of PM inspections and provide one measure of PM success	
18.	PM craft labor needs are adjusted to satisfy changing PM inspection requirements; both up and down.	
	O. Preventive maintenance and lubrication category: Subtotal score	

P.	Predictive Maintenance and Condition Monitoring Technology Application	
Item No.	Benchmark Item Rating: Excellent—10, Very good—9, Good—8, Average—7, Below average—6, Poor—5 or less	Rating
1.	Equipment has been evaluated for the application of predictive maintenance (PdM) technology. The scope and frequency of PdM services has been established on all applicable equipment.	
2.	A plan for using current PdM technology is being developed or is now being put in action. PdM techniques used on all applicable equipment = 10; Application of PdM in progress based on plan = 9; PdM plan developed, no progress = 6; No plan for PdM = 5 or less.	
3.	Maintenance, engineering, and others have technical knowledge and necessary skills for using PdM techniques.	
4.	Optimum routes for PdM inspections are established.	
5.	Inspection and testing intervals and procedures are periodically reviewed for changes/improvements and updated as required as part of well-defined PdM change process.	
6.	Planned times are established for all PdM tasks.	
7.	The required level of manpower is being committed to achieve the total scope of PdM services and to react to issues identified from PdM process.	
8.	Goals for increased reliability are established and overall PM/PdM compliance and results are measured against the organization's benchmark.	
9.	All noncompliance to scheduled PdM services is aggressively evaluated and corrected including followup to issues identified from PdM process.	
10.	Maintenance and operations work with close communication, coordination, and cooperation to schedule PdM services.	
11.	The success of PdM is measured based on multiple factors: reduced emergency repairs, increased planned maintenance work, reduced downtime costs, the elimination of the root cause of problems, and improved product throughput and quality, and so on.	
12.	PM/PdM is a highly visible function within maintenance and is well received as an operational strategy.	
13.	Corrective repair work orders are generated and become planned work as a result of PdM inspections and provide one measure of PdM success.	
14.	Critical equipment has been evaluated for the application of continuous monitoring technology.	
15.	A process for evaluating and eliminating the root causes of failure is in place.	
16.	Positive results from PM/PdM and overall reliability improvements are being communicated throughout the entire operation.	
	P. Predictive maintenance and condition monitoring technology application category: Subtotal score	

Q.	Process Control and Instrumentation System Technology	
Item No.	Benchmark Item Rating: Excellent—10, Very good—9, Good—8, Average—7, Below average—6, Poor—5 or less	Rating
1.	List of all instrumentation systems is available and sorted for criticality.	
2.	Instrument systems assigned asset tracking numbers. along with other vital information such as location, contract service vendor, type (process, regulatory compliance, and condition monitoring).	
3.	Operational instructions and data, installation drawings., catalog information, maintenance documents, and parts information on file and indexed as to parent/main system.	
4.	Instrument system components tagged and identified according to accepted asset database standards.	
5.	Calibration procedures and maintenance tasks written and referenced in work-order system.	
6.	Calibration records up-to-date, filed, and indexed. Historical records dated through previous two years or according to insurance carrier and/or legal counsel.	
7.	Critical system operational and emergency procedures written, up-to-date, and available to operators at control panels or other essential locations.	
8.	All hand-held and portable instruments used for operational checks and maintenance functions listed with tracking numbers and with service/calibration schedules entered into work-order system.	
9.	All laboratory measurement equipment identified with tracking numbers., calibration stickers, and calibration requirements. Schedules entered in work-order system.	
	Q. Process control and instrumentation system technology: Subtotal score	

R.	Energy Management and Control	
Item No.	Benchmark Item Rating: Excellent—10, Very good—9, Good—8, Average—7, Below average—6, Poor—5 or less	Rating
1.	Energy costs are reflected in facility budget via monthly tracking of energy usage, demand, and cost by each customer location/building. Designated individual analyzes usage, demand, and costs on a monthly basis to identify and correct problems.	
2.	An energy team (or designated staff position) has been formed and chartered to deal with energy management issues and is actively involved to advise on utility conservation and its recommendations and results published periodically.	
3.	Steam-trap surveys are performed to look at operating status, inlet and outlet pressures, replacement needs, and done on a routine basis as part of the PM program.	
4.	The plant/facility air system has been properly sized and air leaks are routinely corrected.	
5.	Overall systems analysis of the heating and cooling systems of existing facilities is being conducted to optimize the efficiency and operation of the entire heating and cooling system.	
6.	Boilers in the facility are well maintained and periodically inspected and trimmed for maximum efficiency.	
7.	A water management strategy is in place to control process water, monitor usage, and address corrosion.	
8.	A comprehensive energy audit has been conducted for the facility based upon state and federal energy management guidelines.	
9.	Chronic facility breakdowns and problems impacting energy are aggressively investigated to determine root causes.	
10.	New PdM technologies such as infrared analysis is introduced to check the condition of roofs, switch gears, and the like.	
11.	Energy efficient motors, lights, ballasts, and so forth are used throughout the facility or being planned as part of new facilities and major renovations.	
12.	Automatic facility energy control systems are in place and being planned for use in new facilities or as planned retrofits.	
	R. Energy management and control category: Subtotal score	

S.	Maintenance Engineering Support	
Item No.	Benchmark Item Rating: Excellent—10, Very good—9, Good—8, Average—7, Below average—6, Poor—5 or less	Rating
1.	Engineering and maintenance work closely during the design and specification stages to improve equipment reliability and maintainability.	
2.	Purchase of new equipment and modifications to existing equipment is subject to maintenance review prior to final approval.	
3.	Engineering provides key support to maintenance and operations for improving overall equipment effectiveness.	
4.	Engineering provides key support to maintenance during installation and start-up of new equipment to ensure that operating specifications are achieved.	
5.	Engineering supports maintenance as required to troubleshoot chronic equipment problems and to define/eliminate root causes.	
6.	Engineering and maintenance work closely to develop an effective equipment and spare parts standardization program.	
7.	Capital additions, building systems changes, and facility layout changes are subject to maintenance review before final approval.	
8.	Up-to-date prints, parts/service manuals, and other documentation for equipment and facility assets are available to maintenance.	
9.	Engineering coordinates material requisitioning with maintenance for project work, major overhauls, and machine building.	
	S. Maintenance engineering support category: Subtotal score	

T.	Safety and Regulatory Compliance	
Item No.	Benchmark Item Rating: Excellent—10, Very good—9, Good—8, Average—7, Below average—6, Poor—5 or less	Rating
1.	Maintenance leaders have created a broad-based awareness and appreciation for achieving a safe maintenance operation.	
2.	Maintenance employees attend at least one safety meeting per month.	
3.	Maintenance has shown a continual improvement in its safety record over the past five years.	
4.	All permits and safety equipment is available and prescribed for each job that it is required.	
5.	All cranes, hoists, lift trucks, and lifting equipment are inspected as part of the PM program.	
6.	Good housekeeping within maintenance shops and storerooms is a top priority.	
7.	Maintenance tools, equipment, and leftover materials are always removed from the job site after work completion.	
8.	Maintenance continually evaluates areas throughout the operation where safety conditions can be improved.	
9.	The total scope of regulatory compliance issues within the total organization has been defined and a prioritized plan of action established.	
10.	Maintenance responsibilities related to regulatory compliance have been well defined.	
11.	Maintenance has the technical knowledge and experience to support the organization's regulatory compliance action.	
12.	Maintenance works closely with other staff groups in the organization for a totally integrated approach to safety and regulatory compliance.	
	T. Safety and regulatory compliance category: Subtotal score	

U.	Maintenance and Quality Control	
Item No.	Benchmark Item Rating: Excellent—10, Very good—9, Good—8, Average—7, Below average—6, Poor—5 or less	Rating
1.	Quality control has included maintenance processes within its span of factors impacting quality and has included maintenance factors within its baseline measurement of quality improvement.	
2.	Major repairs and set-ups impacting quality have clearly defined procedures and specifications established.	
3.	Documentation of all equipment conditions, factors, and settings that contribute to quality performance is available.	
4.	Quality of maintenance repairs are evaluated and used as a key performance indicator and as a means to validate crafts skills.	
5.	Maintenance and quality work together with close coordination and cooperation to resolve quality issues related to maintenance processes.	
6.	Optimum machine speeds have been established and included in set-up procedures and operator training.	
7.	All machine-related quality defects are aggressively evaluated and corrected.	
8.	Losses due to minor stoppages, idling, and minor equipment failures are addressed by operations and maintenance for corrections.	
9.	Chronic equipment breakdowns and problems are aggressively investigated as to cause.	
	U. Maintenance and quality control category: Subtotal score	

V.	Maintenance Performance Measurement	
Item No.	Benchmark Item Rating: Excellent—10, Very good—9, Good—8, Average—7, Below average—6, Poor—5 or less	Rating
1.	Maintenance performance measurement includes a wide range of performance indicators in order to evaluate the total effectiveness and impact of maintenance service throughout the operation to include craft labor, planning/scheduling, PM, asset reliability, equipment effectiveness, and cost.	
2.	Maintenance labor and material costs are reported monthly and reviewed against previous costs or budgeted costs to evaluate current trends.	
3.	Equipment downtime attributable to maintenance is monitored. The cost of downtime for major pieces of equipment or processes is known and used to measure value of increased equipment up-time.	
4.	Realistic labor performance standards/estimates have been developed and used for all planned work and recurring tasks.	
5.	Maintenance labor performance is reported monthly or weekly to evaluate actual performance against established performance standards.	
6.	The measurement of craft utilization is available from the labor reporting system to evaluate productive hands-on, wrench time versus nonproductive craft time.	
7.	Periodic reviews are done to evaluate the maintenance operation by determining overall craft utilization and the nature of delays and nonproductive time such as waiting for parts, instructions, unbalanced crew, or waiting for equipment.	
8.	The effectiveness of maintenance planning is evaluated by factors such as percent work orders planned versus total work orders, percent workorders completed as planned versus total planned work orders and percent work orders with estimates versus total work orders completed.	
9.	Baseline performance factors and information is available to evaluate all ongoing improvements against past performance. Periodic reports to summarize and highlight the tangible benefits from continuous maintenance improvement are provided.	
10.	A method to measure performance/productivity of contracted services is in place.	
11.	The craft workforce understands the need for improved craft labor productivity and the challenge to remain competitive with contract service providers.	
12.	Maintenance performance measures are linked to operational performance and support total operations success and profit optimization.	
	V. Maintenance performance measurement category: Subtotal score	

W.	Computerized Maintenance Management Systems (CMMS) and Business System	
Item No.	Benchmark Item Rating: Excellent—10, Very good—9, Good—8, Average—7, Below average—6, Poor—5 or less	Rating
1.	The identification of specific CMMS functional requirements has been clearly defined and a complete definition of system capabilities has been determined based on the size and type of maintenance operation.	
2.	Equipment (asset) history data complete and accuracy 95% or better.	
3.	Spare parts inventory master record accuracy 95% or better.	
4.	Bill of materials for critical equipment includes listing of critical spare parts.	
5.	PM tasks/frequencies data complete for 95% of applicable assets	
6.	Direct responsibilities for maintaining parts inventory database is assigned.	
7.	Direct responsibilities for maintaining equipment/asset database is assigned.	
8.	Initial CMMS training for all maintenance employees with ongoing CMMS training program for maintenance and storeroom employees.	
9.	Adequate support from supplier and consultants is budgeted to ensure a successful CMMS start-up.	
10.	Customization of the CMMS is planned to accommodate specific needs for part numbers, equipment numbers, work order and management report formats, and the like.	
11.	Training for CMMS is a top priority and will be established as an ongoing process for new and existing users of the system.	
12.	System outputs have been developed into a maintenance information system that provides management reports to monitor a wide range of factors related to labor, material, equipment costs, and so forth.	
13.	A CMMS systems administrator (and backup) designated and trained.	
14.	Inventory management module fully utilized and integrated with work order module.	
15.	Reorder notification for stock items is generated and used for reorder decisions.	
16.	CMMS provides MTBF, MTTR , failure trends, and other reliability data.	
17.	Engineering changes related to equipment/asset data, drawings, and specifications are effectively implemented.	
18.	Maintenance standard task database available and used for recurring planned jobs.	
	W. Computerized maintenance mnagement systems (CMMS) and business system category: Subtotal score	

X.	Shop Facilities, Equipment, and Tools	
Item No.	Benchmark Item Rating: Excellent—10, Very good—9, Good—8, Average—7, Below average—6, Poor—5 or less	Rating
1.	Maintenance shop facilities are located in an ideal location with adequate space, lighting, and ventilation.	
2.	Standard tools are provided to craft employees and accounted for by a method that ensures good accountability by crafts and cost ontrol of replacements.	
3.	An adequate number of specialty tools and equipment are available and easily checked out through a tool control procedure.	
4.	All personal safety equipment necessary within the operation is provided and used by maintenance employees.	
5.	Safety equipment for special jobs such as confined space entry, and electrical system lock-out is available and used.	
6.	Maintenance achieves a high level of housekeeping in its shop areas.	
7.	Maintenance maintains a broad awareness of new tools and equipment to improve methods, craft safety, and productivity.	
8.	Maintenance continually upgrades tools and equipment to increase craft safety and productivity.	
9.	An effective process to manage special tools and other maintenance equipment inventory is in place.	
	X. Shop facilities, equipment, and tools: Subtotal score	

Y.	Continuous Reliability Improvement	
Item No.	Benchmark Item Rating: Excellent—10, Very good—9, Good—8, Average—7, Below average—6, Poor—5 or less	Rating
1.	Continuous reliability improvement is recognized as an important strategy as evidenced by the current status of maintenance and the ongoing improvement activities witnessed during the benchmark assessment.	
2.	The current level of commitment to continuous reliability improvement is based on results of overall assessment.	
3.	Maintenance improvement opportunities have been identified with potential costs and savings established.	
4.	Improvement priorities have been established based on projected benefits and valid economic justifications.	
5.	Top leaders have reviewed, modified, and/or approved maintenance improvement priorities and have made a commitment to action.	
6.	Sufficient resources (time, dollars, and staff) have been established to address priority improvement areas.	
7.	Implementation plans and leaders for each priority area are established.	
8.	A team-based approach is used to identify and implement practical solutions to maintenance improvement opportunities.	
9.	Continuous reliability improvement for the maintenance resource of physical asset and equipment resource is rated as:	
10.	Continuous reliability improvement for the maintenance resource of craft labor resources is rated as:	
11.	Continuous reliability improvement for the maintenance resource of MRO material resources is rated as:	
12.	Continuous reliability improvement for the maintenance resource of information resources is rated as:	
13.	Continuous reliability improvement for the maintenance resource of technical skills and knowledge is rated as:	
14.	Maintenance employees participate on functional teams within maintenance and on cross-functional teams with other department employees to develop maintenance improvements.	
15.	Written charters are established for each team to outline reasons for the team, process to be used, resources available, constraints, expectations, and results expected.	
	Y. Continuous reliability improvement category: Subtotal score	

Z.	Critical Asset Facilitation and Overall Equipment Effectiveness (OEE)	
Item No.	Benchmark Item Rating: Excellent—10, Very good—9, Good—8, Average—7, Below average—6, Poor—5 or less	Rating
1.	OEE ratings have been established for major equipment assets or processes to provide a baseline measurement of availability, performance, and quality.	
2.	Priorities have been established with a plan of action for improving OEE.	
3.	The OEE factor of availability is being measured and methodology rated as:	
4.	The OEE factor of performance is being measured and methodology is rated as:	
5.	The OEE factor of quality is being measured and methodology is rated as:	
6.	Equipment improvement teams have been established to focus on improving equipment effectiveness based on established priorities for critical equipment.	
7.	Improvements in OEE are evaluated against baseline (OEE) measurements to determine progress.	
8.	Optimum machine speeds have been established and included in set-up procedures and operator training.	
9.	All machine-related quality defects are aggressively evaluated and corrected.	
10.	A process for critical asset facilitation has been implemented.	
11.	Critical asset facilitation for condition-based maintenance includes CBM technologies and hardware requirements, durations and/or hours for each routine, frequencies for each routine, necessary tools and equipment for each routine, and perform/record baseline of test and measures.	
12.	Critical asset facilitation includes developing PM plan for each craft/system, inspections/service requirements in sequence of events, schedules for lube and filter routes, duration for each routine by subsystem level and by shift, necessary parts, tools and equipment for each task, and identification of facilitated asset in CMMS.	
13.	Critical asset facilitation includes a complete review of required documents, manuals, and drawing required to maintaining and operating the asset.	
14.	Critical asset facilitation includes visual management such as labeling of signal or alarm functions/operations, direction of rotation (drives, chains, motors) and replacement parts numbers, proper control adjustments (pressure, temperature, speed, level, voltage, etc.), energy lockouts (electrical, hydraulics, compressed air, etc.), electrical conductors, functions/designations of source on cabinets, panels, boxes, switches, valves, buttons, light, etc., all fixable, adjustable, or critical fasteners, filters functions (hydraulic, lube, air, etc.), and replacement part no., normal operating ranges/levels/readings and label function monitored, lube point with product number, major component functions (coolant pump, exhaust fan, etc.), motor function being powered (pump, axis drive, screw, conveyor, etc.), fluid type/direction of flow/pressure on pipes/hoses/lines.	

Z.	Critical Asset Facilitation and Overall Equipment Effectiveness (OEE)	
Item No.	Benchmark Item Rating: Excellent—10, Very good—9, Good—8, Average—7, Below average—6, Poor—5 or less	Rating
15.	Critical asset facilitation includes facilitated asset documentation material with the following. List of systems, subsystems and components, CBM routines, CBM baselines, PM routines, Lube and filter routes, List of prints, manuals, and so on for PM and CBM routines, visual management digital images, job safety analysis, MSDS, lockout/tag out documentation, craft asset specific training/certification requirements, part list and/or bill of material for PM routines, startup and shutdown procedures, final operator checklist, documented alignment, and test procedures.	
	Z. Critical asset facilitation and overall equipment effectiveness (OEE) category: Subtotal score	

ZZ.	Overall Craft Effectiveness (OCE)	
Item #	Benchmark Item Rating: Excellent—10, Very good—9, Good—8, Average—7, Below average—6, Poor—5 or less	Rating
1.	Improvement of craft productivity and OCE is being established as an important element of the overall maintenance improvement process.	
2.	Priorities have been established with a plan of action for improving the OCE factor for craft utilization (wrench time).	
3.	Priorities have been established with a plan of action for improving the OCE factor for craft performance.	
4.	Priorities have been established with a plan of action for improving the OCE factor for craft service and quality.	
5.	Improvements in OCE and craft productivity are evaluated against baseline measurements to determine progress.	
6.	A team effort is being used that involves the craft workforce to provide a cooperative effort for improving OCE and eliminating a reactive, fire-fighting strategy that wastes valuable craft resources.	
	ZZ. Overall craft effectiveness category: Subtotal score	

Total Point Range	The Scoreboard for Maintenance Excellence: Rating Summary Comments
2700 to 3000 (90–100%)	**Excellent:** Practices and principles in place for achieving effective maintenance and world-class operations performance based on actual results. Reconfirm overall performance measures. Maintain strategy of Continuous Reliability Improvement. Set higher standards for maintenance excellence and continue to measure results.
2400 to 2699 (80– 89%)	**Very good:** Fine tune existing operation and current practices. Reassess progress on planned or ongoing improvement activities. Redefine priorities and renew commitment to Continuous Reliability Improvement. Ensure top leaders see results and reinforce measurement process and use of Maintenance Excellence Index.
2100 to 2399 (70–79%)	**Good:** Reassess priorities and reconfirm commitments at all levels to maintenance process improvement. Evaluate maintenance practices and develop and implement plans for priority improvements. Ensure that measures to evaluate maintenance performance and results are in place. Initiate strategy of Continuous Reliability Improvement.
1800 to 2099 (60–69%)	**Average:** Conduct a complete assessment of total maintenance operation and current practices. Determine total costs/benefits of potential improvements. Develop and initiate strategy of Continuous Reliability Improvement. Define clearly to top leaders where deferred maintenance is increasing current costs and asset life-cycle costs. Gain commitment from top leaders to go beyond maintenance of the status quo.
Less than 1800 (<60%)	**Below average:** Same as for average, plus depending on the level of the rating and the specific major area that is below average. Immediate attention may be needed to correct conditions having an adverse effect on life, health, safety, and regulatory compliance. Place immediate priority and focus toward key issues, major production assets or facility conditions, building systems, and other equipment where increasing costs and deferred maintenance are having a direct impact on the immediate survival of the business. The capabilities for critical assets to perform intended function are being severely limited by current "state of maintenance."

	Total evaluation items	300
	Total possible points	3000

Appendix

B

The CMMS Benchmarking System

Introduction

Today's information technology for computerized maintenance management offers the maintenance leader an exceptional tool for asset management and for managing the maintenance process as an internal business and "profit center." However, maintenance surveys and assessments conducted by The Maintenance Excellence Institute and others validate that poor utilization of existing maintenance software is often a major improvement opportunity.

The Maintenance Excellence Institute is fully committed to ensuring that all clients now using *computerized maintenance management system* (CMMS) gain maximum value from their software investment. Maintenance organizations take an important step when they invest in software to help manage the business of maintenance. Often they do not achieve maximum *return on investment* (ROI) from CMMS software investments. One of the services from The Maintenance Excellence Institute is to provide "Scoreboard for Maintenance Excellence" assessments that includes an independent benchmark evaluation of the current CMMS installation.

Purpose

The CMMS benchmarking system was developed to support getting maximum value from an investment in CMMS by evaluating how well existing CMMS functionality is being used. This benchmarking system provides a methodology for developing an overall benchmark rating of your CMMS installation. It provides you with a baseline for determining how well CMMS is supporting best practices within your total maintenance operation.It can also be used as your baseline to measure the success of a future CMMS that is just being installed.

Benchmarking your CMMS installation

The CMMS benchmarking system provides a means to evaluate and classify your current installation as either "Class A, B, C, or D."A total of nine major categories are included along with 50 specific evaluation items. Each evaluation item that is rated as being accomplished satisfactorily receives a maximum score of 4 points. If an area is currently being "worked on" a score of 1, 2, or 3 points can be assigned based on the level of progress achieved.For example, if spare parts inventory accuracy is at 92 percent compared to the target of 95 percent, a score of 3 points is given. A maximum of 200 points is possible. A benchmark rating of "Class A" is within the 180 to 200 point range. The complete CMMS benchmarking rating scale is given at the completion of the CMMS benchmarking form.

Conducting the CMMS benchmark evaluation

The Maintenance Excellence Institute conducts the CMMS benchmark evaluation to provide an independent and objective evaluation. We do not sell nor even want to get typical overrides offered for recommending a system. This benchmark process is not to make functionality comparisons between different CMMS systems. But it will identify gaps in existing CMMS functionally, best practice needs, and support decisions on upgrades. Results from the benchmark evaluation will establish a baseline classification for your current installation at a point in time. As a key deliverable, we provide a written report with specific recommendations and a plan of action to improve your CMMS installation to the "Class A" level. This is provided along with guidelines for using the CMMS benchmarking system in the future as an internal benchmarking tool.

CMMS Benchmarking Categories and Item Descriptions	Yes (4 points)	No (0 points)	Working On It (1, 2, or 3 points)
A. CMMS Data Integrity			
1. Equipment (asset) history data complete and accuracy 95% or better			
2. Spare parts inventory master record accuracy 95% or better			
3. Bill of materials for critical equipment includes listing of critical spare parts			
4. Preventive maintenance tasks/ frequencies data complete for 95% of applicable assets			
5. Direct responsibilities for maintaining parts inventory database is assigned			
6. Direct responsibilities for maintaining equipment/asset database is assigned			
B. CMMS Education and Training			
7. Initial CMMS training for all maintenance employees			
8. An ongoing CMMS training program for maintenance and storeroom employees			
9. Initial CMMS orientation training for operations employees			
10. A CMMS systems administrator (and backup) designated and trained			
C. Work Control			
11. A work-control function is established or a well-defined documented process is being used for work management			
12. On-line work request (or manual system) used to request work based on priorities			
13. Work-order system used to account for 100% of all craft hours available			
14. Backlog reports are prepared by type of work to include estimated hours required			
15. Well-defined priority system is established based on criticality of equipment, safety factors, cost of downtime, and the like			

CMMS Benchmarking Categories and Item Descriptions	Yes (4 points)	No (0 points)	Working On It (1, 2, or 3 points)
D. Budget and Cost Control			
16. Craft labor, parts, and vendor support costs are charged to work order and accounted for in equipment/asset history file			
17. Budget status on maintenance expenditures by operating departments is available			
18. Cost improvements due to CMMS and best practice implementation have been documented			
19. Deferred maintenance and repairs identified to management during budgeting process			
20. Life-cycle costing is supported by monitoring of repair costs to replacement value			
E. Planning and Scheduling			
21. A documented process for planning and scheduling has been established			
22. The level of proactive, planned work is monitored and documented improvements have occurred			
23. Craft utilization (true wrench time) is measured and documented improvements have occurred			
24. Daily or weekly work schedules are available for planned work			
25. Status of parts on order is available for support to maintenance planning process			
26. Scheduling coordination between maintenance and operations has increased			
27. Emergency repairs, hours, and costs tracked and analyzed for reduction			
F. MRO Materials Management			
28. Inventory management module fully utilized and integrated with work-order module			
29. Inventory cycle counting based on defined criteria is used and inventory accuracy is 95% or better			
30. Parts kiting is available and used for planned jobs			

CMMS Benchmarking Categories and Item Descriptions	Yes (4 points)	No (0 points)	Working On It (1, 2, or 3 points)
F. MRO Materials Management			
31. Electronic requisitioning capability available and used			
32. Critical and/or capital spares are designated in parts inventory master record database			
33. Reorder notification for stock items is generated and used for reorder decisions			
34. Warranty information and status is available as warranty claims get reimbursement			
G. Preventive and Predictive Maintenance (PM/PdM)			
35. PM/PdM change process is in place for continuous review/update of tasks/frequencies			
36. PM/PdM compliance is measured and overall compliance is 98% or better			
37. The long range PM/PdM schedule is available and leveled loaded as needed with CMMS			
38. Lube service specifications, tasks, and frequencies included in database			
39. CMMS provides MTBF, MTTR, failure trends, and other reliability data			
40. PM/PdM task descriptions contain enough information for new craftsperson to perform task			
H. Maintenance Performance Measurement			
41. Downtime (equipment/asset availability) due to maintenance is measured and documented improvements have occurred			
42. Craft performance against estimated repair times is measured and documented improvements have occurred			
43. Maintenance customer service levels are measured and documented improvements have occurred			
44. The maintenance performance process is well established and based on multiple indicators compared to baseline performance values			

CMMS Benchmarking Categories and Item Descriptions	Yes (4 points)	No (0 points)	Working On It (1, 2, or 3 points)
I. Other Uses of CMMS			
45. Maintenance leaders use CMMS to manage maintenance as internal business			
46. Operations staff understands CMMS and uses it for better maintenance service			
47. Engineering changes related to equipment/asset data, drawings, and specifications are effectively implemented			
48. Hierarchies of systems/subsystems used for equipment/asset numbering in database			
49. Failure and repair codes used to track trends for reliability improvement			
50. Maintenance standard task database available and used for recurring planned jobs			
Total CMMS Benchmark Rating Score: ———	—— +	.....	+ ——

SUMMARY OF CMMS BENCHMARKING EVALUATION

- **TOTAL CMMS BENCHMARK RATING SCORE: _______**
- **CURRENT CMMS BENCHMARK RATING: CLASS ____**
- **CMMS SOFTWARE VERSION: ____________________________**
- **DATE OF CMMS INSTALLATION: ______/________/________**
- **BENCHMARK RATING PERFORMED BY: ________________________**
- **LOCATION: ________________________ DATE: ____/_____/______**

CMMS BENCHMARKING RATING SCALE:	**INDIVIDUAL ITEM GRADING:**
CLASS A = 180–200 POINTS (90% +)	**YES = 4 points**
CLASS B = 140–179 POINTS (70%–89%)	**NO = 0 points**
CLASS C = 100–139 POINTS (50%–69%)	**Working On It = 1, 2 or 3 points**
CLASS D = 0–99 POINTS (0–49%)	

EVALUATION COMMENTS:

Appendix

C

CMMS Functionality Rating Checklist

The following is a tool for defining the *computerized maintenance management system* (CMMS) functionality needed for a replacement CMMS. This checklist can also be used to evaluate and rate the functional requirements for competing CMMS systems during the final CMMS selection process. It provides a listing of 120 specific CMMS functional requirements for rating of the following seven major categories.

1. Miscellaneous advanced CMMS functional requirements
2. Preventive, predictive, and *continuous reliability improvement* (CRI) functional requirements
3. Planning and scheduling functional requirements
4. Work-order functional requirements
5. Equipment master database functional requirements
6. MRO materials management functional requirements
7. Accounting, materials transfer, and purchasing functional requirements

The CMMS functionality rating checklist can also be used as an attachment to a formal *request for proposal* (RFP) allowing each vendor to designate "yes or no" as to their system having the designated functionality. In turn, during CMMS final evaluation process, each functionality item can be given an individual rating from 0 to 5 as shown here.

CMMS Rating Values	
Functionality not available	0
Functionality not adequate	1
Functionality not adequate	2
Functionality good	3
Functionality very good	4
Functionality excellent.	5

Item	A. Miscelleneous Advanced CMMS Functional Requirements	CMMS Rating Values X	Y	Z
1.	*Management of change.* Provide central database for maintenance, engineering change control, mechanical integrity data, and equipment/ process specification data. Record all changes in asset/equipment history and statistical process control information and who/when/why/how changes occurred. Provide integration with engineering database to ensure maintenance input on maintainability issues.			
2.	*Craft training database.* Record and track craft training history, training requirements certification, licensing, and/or interface to databases for training. Provide capability to maintain craft training plans, lesson plans, and the like, as required for craft and operator personnel. Also contractor information.			
3.	*Contract management.* Manage system for maintenance contract specifications, work requirements, evaluating contractor response/performance, full contract administration, and progress monitoring. Provide capability to manage performance-based contracts.			
4.	*Compliance regulation library.* Capability to interface with compliance regulations or procedures library for safety. To include cut/paste from current regulations or procedures directly to work order. Print capability for MSDS sheets and specific work permits as required for safety.			
5a.	*Interface with RF technology.* Provide interface for future use of RF technology for online condition monitoring, real time data transmission of work order information and warehouse operation of this technology. (RF—*radio frequency*)			
5b.	*Interface with Speech Recognition technology* for work requests, time reporting, work order closing, work order and project status updates, etc. Includes all next generation data input technologies.			
	A. Rating of miscellaneous advanced CMMS functional requirements			

Item	B. Preventive, Predictive, and Continuous Reliability Improvement Functional	CMMS Rating Values		
		X	Y	Z
6.	*Reliability performance indicators (RPI).* Provide ability to calculate MTBF (mean between failure) and other key reliability indicators as MTTR and MTBR.			
7.	*PM/PdM compliance.* Provide ability to measure *preventive maintenance/predictive maintenance* (PM/PdM) compliance continuously based on PMS generated to schedule versus PMs actually completed (within a specified window of time). Note: PM to include all lube services, calibrations and scheduled quality inspections.			
8.	*PM/PdM results.* Provide capability to track work orders initiated due to results of PM/PdM inspections that eliminated catastrophic failures.			
9.	*PM/PdM costs.* Provide capability to track PM/PdM cost to include inspections and PM/PdM related work and to document PM results and to provide justification support for PM/PdM continuation as a top priority.			
10.	*Continuous reliability improvement database.* Provide access to a centralized reliability database that is maintained locally but integrated/shared globally throughout all sites as needed. (See also No. 1.)			
11.	*Rebuild/reconditioning costing.* Track in-house cost and quality for rebuilt items to compare in-house cost versus vendor cost for items such as relief valves, motors, and special spares.			
12.	*Interface with online condition monitoring.* Capabilities for direct interface with real-time systems such as ENTEK, CSI, VITEK, and SKF via RF technologies or line link.			
13.	*Reliability and FMECA information.* User-defined, time-stamped condition fields to monitor equipment conditions each time a repair task is performed. Define and track process variables, temperature, vibration, and amps/RPM, and analyze and establish trend data. Provide *failure modes, effects, and criticality* (FMECA) analysis information to identify and evaluate what items are expected to fail and the resulting consequences. (See also No. 10.)			
14.	*Equipment usage.* Update the usage factor for an equipment asset (either fixed or mobile type) for scheduling inspections based on usage not fixed intervals. May include tons, linear meters, and so forth.			

Item	B. Preventive, Predictive, and Continuous Reliability Improvement Functional	CMMS Rating Values		
		X	Y	Z
15.	*PM work order closing.* Allow a partial closure of PM work orders (a work order with multiple PMs to do) that would update PMs accomplished and allow those not completed to be generated again for scheduling later.			
16.	*Initiate/deactivate* PM work orders. Initiate/deactivate preventive maintenance work orders.			
17.	*PM work order suspension update.* Suspend/ unsuspend PM work orders by unit, function, or asset.			
	B. Rating of preventive, predictive and continuous reliability improvement functional requirements			

Item	C. Planning and Scheduling Functional Requirements	CMMS Rating Values		
		X	Y	Z
18.	*Daily maintenance schedule.* Produce a daily maintenance schedule by craft area, lines, and so on, including PMs, which can be performed at regular predetermined intervals by operators.			
19.	*Craft capacity leveling.* Schedule work orders against a known labor availability to ensure full effective craft redeployment, labor utilization, and craft leveling to include contractors.			
20.	*Schedule compliance.* To what degree is the schedule being adhered to based on work planned/scheduled versus actual execution of planned work.			
21.	*Client report writer.* Ability to produce tabular and graphical reports effectively at the client level. Specific reports and metrics related to providing a monthly Maintenance Excellence Index.			
22.	*CMMS-project management interface.* Provide interface with project management systems.			
23.	*Job plan view.* View an existing work-order job plan.			
24.	*Work-order job plan entry/update.* Enter a new job plan or update an existing job plan from job plan templates.			
25.	*Work-order status.* Progress work orders that have been started via direct entry of craft labor hours to the work order, via time and attendance system or other work-order progress methods. Show percent completed and progress of planned versus actual. Actual time by craft data must interface with project management system for contractors and in-house employees. (See No.115.)			

Item	C. Planning and Scheduling Functional Requirements	CMMS Rating Values		
		X	Y	Z
26.	*Vendor inquiry.* Inquire work orders for a given vendor/contractor.			
27.	*Job plan print.* Generate a work-order job plan to a local printer.			
28.	*Work-order scheduling.* Work orders created, planned, approved, scheduled, worked, progressed, completed, and transferred to history.			
29.	*Work orders can be scheduled in 15-min intervals.* Quick entry work order.			
30.	*Just-in-time scheduling.* Blanket work orders can be scheduled for a specific time during the day or as needed.			
31.	*Work-order planning notes.* Create a preclosure W/O screen to allow remarks and narrative to be input before the work is completed.			
32.	*Standard maintenance practice library.* Maintain and provide standard maintenance repair practices, preplanned work scope, historical average of previous repair (*mean time to repair*—MTTR).			
33.	*Process project type work orders.* Checks by event for material availability. Commits and prints MRO pick lists by event when material is available. (See also No. 66.)			
	C. Rating of planning and scheduling functional requirements			

Item	D. Work-order Functional Requirements	CMMS Rating Values		
		X	Y	Z
34.	*Work-order master update.* Add, change, or delete workorder master records.			
35.	*Initiate project or shutdown work orders.* Select and initiate project or shut down work orders. Also capital project work orders.			
36.	*Work-order inquiry by equipment, unit, parent, AFE-project ID.* Obtain a display of all open-in-maintenance work orders for a particular equipment/unit/parent/AFE-project ID/subcomponent.			
37.	*Work-order inquiry.* Inquire against work-order master information, events, materials, and costs.			
38.	*Subcomponent inquiry.* Inquiry work orders for a given subcomponent.			
39.	*Close work orders in maintenance.* Close a work order in maintenance and update the appropriate equipment history and SAP cost.			
40.	*Reopen work orders.* Reopen-in-maintenance a closed-in-maintenance work order.			

Item	D. Work-order Functional Requirements	CMMS Rating Values		
		X	Y	Z
41.	*Time and attendance interface.* Time and attendance system at each site must interface with future CMMS to provide immediate and actual craft labor updates to the work-order file.			
42.	*Work-order material status inquiry.* Obtain the material status for a specific work-order event or activity/task.			
43.	*Closed work-order history inquiry.* Inquire the history for a closed work order.			
44.	*Work-order transaction history inquiry.* Inquire the transaction history for a work order.			
45.	*Work-order transaction accounting history.* Inquire the accounting transaction history for a work order			
46.	*"Shopping cart" update/inquiry.* Inquire or delete stock items from the "shopping cart" for a particular work order.			
47.	*Work-order text file.* Narrative text files and/or attachments can be viewed and/or modified while editing the work order. Allow for text members such as checklists and procedures to be attached to a work order.			
48.	*Online/paperless work orders.* Allow work orders to be created, approved, and closed through a paperless system. Dispatch/complete/reassign work orders to maintenance supervisors/ technicians/operations/contractors.			
49.	*Provide online work-order approvals screens to allow for operator performed maintenance.*			
50.	*Quick entry to work order.* Capability to enter all labor and inventory data on one screen.			
51.	*Paperless work order.* Capability to interface with devices such as Apple Newton Message Pad to create, view update work orders anywhere, and input to CMMS via wireless download. Future capability to interface with speech recognition data collection and entry.			
	D. Rating of work order functional requirements			

Item	E. Equipment Master Database Functional Requirements	CMMS Rating Values		
		X	Y	Z
52.	*Work-order "shadowing."* Capability to reduce redundant work orders for PM when the same PM procedures are scheduled at the same time. System should automatically combine tasks to eliminate redundant PM work orders.			
53.	*Equipment master inquiry.* Obtain equipment master descriptive information for a piece of equipment, manufacturer, serial number, spare parts listing, equipment specifications, repairs performed, failure causes, costs, and the like.			
54.	*Equipment spares inquiry.* Obtain stock items from equipment master data that are component parts of an equipment asset. Add, change, or delete items to spares list or copy spares list to another asset.			
55.	*Equipment spares update.* Add, change, or delete equipment spares information via mass updating capability.			
56.	*Interchangeability inquiry.* Obtain a list of all the equipment assets that use a stock item based upon MRO information. Provide cross-reference link to specifications data for the MRO item.			
57.	*Equipment statistics update.* Update historical maintenance costs for a piece of equipment.			
58.	*Equipment function/asset relation.* Attach/detach/reassign an equipment function and a physical equipment item (asset).			
59.	*Equipment imaging interface.* Given an equipment functional/asset review all subassemblies, parts data, exploded diagrams, repair instructions, for example, visual top and side view, zoom, point, and click.			
60.	*Equipment inquiry by parent.* Obtain all subsidiary equipment information for a specified equipment "parent."			
61.	*Equipment history inquiry.* Obtain equipment historical actual costs for a function or a piece of equipment/asset.			
62.	*Equipment manufacturers inquiry.* Inquire equipment based upon manufacturer, serial number, or stock number or target dates.			
63.	*Equipment spares with quantities inquiry.* Similar to No. 54 above except quantities on hand are shown instead of spares statistics.			
64.	*Equipment route function inquiry.* Inquire route functions to which an equipment function has been assigned.			
65.	*Equipment and spares list updates.* Allow for mass additions/deletions of stock items to/from equipment spare list as well as equipment function/asset changes or relocation within the...?			

Item	E. Equipment Master Database Functional Requirements	CMMS Rating Values		
		X	Y	Z
66.	*CMMS-project management system (PMS).* All repairs related to equipment history must be updated to the CMMS database, for example, equipment master file. CMMS-PMS interface must accomplish this activity to support project planning and equipment history.			
67.	*Equipment function update.* Add, change, delete, or reactivate equipment function information and function specifications.			
	E. Rating value of equipment master data base functional requirements			

Item	F. MRO Materials Management Functional Requirements	CMMS Rating Values		
		X	Y	Z
68.	*Equipment asset update.* Add, change, delete, or reactivate equipment asset information and asset specifications.			
69.	*Combined add and attach of equipment function and asset.* Add an equipment function and asset combination and attach them to each other. Option to add equipment function and/or asset specifications.			
70.	*Local stock item master update.* Add, change, delete, reinstate, or inquire master data for an MRO stock item. Provide common stock catalog and common database across sites of the organization. Change the extended description of a regular stock item or direct purchase.			
71.	*Stock item inquiry.* Obtain descriptive, current purchase order, work order, and statistical information about a stock item			
72.	*Stock cycle count manual entry.* Add stock items into the cycle count.			
73.	*Stock cycle count verification.* Input the results of a physical inventory or cycle count and make an inventory/accounting adjustment.			
74.	*Issue material via pick list.* Issue material for a particular work order or a work-order event as the result of a pick list.			
75.	*Electronic data interchange and electronic commerce.* Capability for EDI, fax transmissions, and paperless internal/external requisitioning approvals, and purchase order creation.Capability for electronic commerce technology use.			

Item	F. MRO Materials Management Functional Requirements	CMMS Rating Values		
		X	Y	Z
76.	*Stock item inquiry.* Inquire against stock items based upon stock noun, manufacturer, manufacturer part number, and the like.			
77.	*Stock noun inquiry.* Inquire stock items based on the stock noun and attributes per key words.			
78.	*Adjust deficient quantity.* Adjust the deficient quantity for a stock item on a work order.			
79.	*Stock item master update.* Add, change, delete, or reactivate stock item master information.			
80.	*Direct issues (no pick list).* Record the direct issue of stock items to a work order or work order event.			
81.	*Spare history (deactivated items).* Allow stock warehouse records to be retained for historical purposes when deactivating a stock item.			
82.	*Bar-coding capability.* Provide interface with industry standard bar-coding systems. Print capability for bin locations, work order, asset tags, vehicle tags, and so forth.			
83.	*Receive material from other warehouses.* Enables the receipt of material into a warehouse in the CMMS system from another warehouse.			
	F. MRO materials management functional requirements			

Item	G. MRO Materials Management Functional Requirements	CMMS Rating Values		
		X	Y	Z
84.	*Vendor electronic catalog interface.* Capability to interface directly with selected vendor part information systems and to input data directly to either the inventory or purchasing functions.			
85.	*Stock history inquiry.* Inquire the recent transaction history for a stock item. Inquire stock transaction history by group identification number or commodity code.			
86.	*Transaction inquiry by document.* Inquire stock transaction history by document number.			
87.	*Transaction inquiry by stock item and condition.* Review stock item rebuild condition code transactions and the resulting status of rebuilt items.			
88.	*Quarantine or inspection/testing activity.* Transfer material to and from quarantine/ inspect/testing activities. Track status of inspection/testing either internal or external.			

Item	G. MRO Materials Management Functional Requirements	CMMS Rating Values		
		X	Y	Z
89.	*Stock owner warehouse record update.* Add, change, delete, or reactivate stock owner warehouse records including consignment vendors. Includes bin location, cycle count, and recorder data. Reconcile consignment information for invoicing.			
90.	*Document management system interface.* Direct Interface with document management systems.			
91.	*Inventory management.* Track inventory by LIFO, FIFO, weighted average cost, ABC analysis, and EOQ analysis. Provide capability for min/max, automatic reorder based on reorder point with, or without, review process, and so on, for example, provide state-of-the-art MRO inventory management procedures/control.			
92.	*Bin location inquiry.* Inquire the contents of a particular bin location.			
93.	*MRO materials returns.* Processing issues, receipts, transfers, and returns of stock to include credit to previously closed work orders.			
94.	Replenishing stock according to established stock reordering criteria.			
95.	*JIT purchase requisitions.* Create purchase requisitions directly from work planning for direct-buy items (JIT) and contracted services.			
96.	*MRO materials management performance.* Initiate manual physical counts or automatic cycle counting requirements. Reporting of stock activity, stock-outs, stock status and inventory accuracy, receipts/issues, delivery service (internal) vendor service and quality, and the like.			
97.	*Print receiving worksheet (for purchase order).* A receiving worksheet will be generated.			
98.	*Request for materials or services.* Request materials or services on a particular work order or event. Produces a pick list and/or requisition. Request items not in warehouse.			
99.	*Delivery date update.* Update the delivery date for a particular work-order event. Includes internal or outside JIT deliveries.			
	G. Rating of MRO materials management functional requirements			

Item	H. Accounting, Materials Transfer, and Purchasing Functional Requirements	CMMS Rating Values		
		X	Y	Z
100.	*Request print/reprint of pick list.* Generate/reprint an unissued pick list.			
101.	*Automatic pick list.* Initiates a pick list for materials that are received which were deficient on work orders prior to their receipt.			
102.	*Utilize cross-docking opportunities to reduce handling of parts in receiving.*			
103.	*Reports of material status on work-order job plans.* Provide an online inquiry or a report of material status of work-order job plans by event/task.			
104.	*Purchasing functions.* A client-server CMMS to mainframe SAP purchasing interface that is currently a critical interface requirement.			
105.	*Change stock accounting class.* Change accounting class for a particular stock item.			
106.	*Sell material or return to vendor.* Record the sale of materials or the return of material to a vendor.			
107.	*Transfer between stock items and/or condition codes.* Make material transfers between stock items and/or condition codes.			
108.	*Material receipt with value.* Record a material receipt with value for items purchased outside the purchasing system.			
109.	*Material transfer between owners and/or warehouses.* Record the issue of material between two and/or two warehouses.			
110.	*Direct issue of material (no pick list).* Record the direct issue of material to a work order.			
111.	*Issue of material via pick list.* Record the issue of material for a particular work-order event as the result of a pick list.			
112.	*Material return.* Record the return of material to a warehouse.			
113.	*Material transfer between work orders.* Transfer material from one work order to another.			
114.	*Finalize material pricing.* Finalize material price averages and delayed pricing in material transfers.			
115.	*Labor reporting by skill code* (craft areas). Report labor hours by skill code on work orders.			
116	*Work-order adjustment.* Adjust costs, initiate/update accounting fields, adjust actual material costs, and open or close a particular work order in accounting.			
117.	*Material transfer.* Make material transfers from warehouse to/from surplus/obsolete accounts.			
118.	*Stock value adjustments.* Adjust the value (price) of a particular stock item and update accounting system.			

Item	H. Accounting, Materials Transfer, and Purchasing Functional Requirements	CMMS Rating Values		
		X	Y	Z
119.	*Accounting transaction update.* Update accounting information for a particular transaction.			
120.	*Stock quantity adjustment.* Adjust the quantity (and value) of a particular stock item.			
	H. Rating of accounting, materials transfer, and purchasing functional requirements			

Total Rating Summary of All CMMS Functional Requirements	CMMS Rating Values		
	X	Y	Z
A. Rating of Miscellaneous Advanced CMMS Functional Requirements			
B. Rating of Preventive, Predictive, and Continuous Reliability Improvement Functional Requirements			
C. Rating of Planning and Scheduling Functional Requirements			
D. Rating of Work-order Functional Requirements			
E. Rating Value of Equipment Master Database Functional Requirements			
F. MRO Materials Management Functional Requirements			
G. Rating of MRO Materials Management Functional Requirements			
H. Rating of Accounting, Materials Transfer and Purchasing Functional Requirements			
Overall summary of CMMS Functional requirements evaluation	X	Y CMMS	Z

Appendix

D

Case Study: What Do Your Crafts Think about CMMS?

Introduction

The (*company*) is currently evaluating improvements to its current CMMS system. It is also evaluating other *computerized maintenance management systems* (CMMS) as a future tool to support continued maintenance excellence. Input from all (*company*) personnel who use the existing CMMS is very important to support a valid evaluation and final decision. Your input as a craftsperson is most important in terms of the day-to-day use of the current system.

Purpose

The purpose of this survey is to allow you the opportunity to provide valuable input to (*company*) evaluation of its current CMMS and to help develop plans for the future. Your role as a craftsperson is a critical, mission-essential part of the overall refinery operations. Your input will be very much appreciated and it will be used to help support a final decision.

1. Name (optional): ______________________________
   Position: __________________________
   Company Site: ❑ Location 1 ❑ Location 2 ❑ Location 3 ❑ Location 4
2. Do you have ready access to a computer or terminal at your shop?
   ❑ Yes ❑ No
3. Do you use a computer to find or order parts? ❑ Yes ❑ No

4. Would you like to receive more computer training on CMMS?
   ❑ Yes ❑ No
   If yes, what specific areas?________________________________

5. What are some of the major weaknesses you see with the current CMMS system? Comments: ________________________________

6. What are the strengths of the current CMMS system from your perspective? Comments: ________________________________

7. The best source of reliability information comes directly from the craftsperson after each repair job is completed through input or conditions discovered, causes of failure, and repair actions taken. Do you feel that you need more training on how best to provide reliability information on the completed work order? Please comment. ________________

8. Would you like to receive more training on troubleshooting equipment?
   ❑ Yes ❑ No
   If yes, what type of training would you like to receive? ________________

9. What type of additional equipment history information would help you to do your job better?
   Explain:

10. Do you use the existing repair history information now in CMMS?
    ❑ Frequently ❑ Sometimes ❑ No
    If no, please comment. ________________________________

11. Do you feel the planning time estimates developed by the maintenance coordinator/planner are:

❑ Accurate ❑ Somewhat accurate ❑ Never accurate ❑ Not sure
Comments:

12. Do you find parts descriptions in CMMS complete and accurate?
❑ Yes ❑ No ❑ Somewhat accurate
If no, please comment.

13. Do you find the current spares list accurate and complete?
❑ Yes ❑ No ❑ Somewhat accurate

14. Is the number of regular stock items set up in CMMS adequate for your craft needs or do you use and reuse "XX" numbers? Comments:

15. Would the addition of selected "XX" to the "spare list" for reference only be of value to you?
❑ Yes ❑ No ❑ Not sure

16. Would "on line" access to parts information, drawings, exploded diagrams, P&IDs, and the like help you in your ability to identify and secure parts? ❑ Yes ❑ No Comments:

17. Do you have difficulty in finding Company stock numbers?
❑ Yes ❑ No
If yes, what area? Please comment.
Spare Parts:
Noun and Attributes:
Manufacturer name and part number:

18. How do you feel about working with planners occasionally to develop reasonable planning times for major repair jobs in your craft skill area? Comments:

19. Do you feel you need more CMMS computer-related training or are you comfortable with your current level?
❑ Comfortable
❑ Need more (be specific):

- ❑ Stock information
- ❑ Work order
- ❑ Equipment history
- ❑ Other:

20. What is your estimate of the *total percentage of your time* spent on *all* of the following activities listed below in a typical day? Approximately ____% of a typical day is spent on activities listed below.
    - ■ Parts chasing/securing parts information
    - ■ Receiving job instructions/job preparation
    - ■ Travel to/from job site
    - ■ Waiting on permits
    - ■ Waiting for operations to prepare equipment
    - ■ Troubleshooting

21. On a scale of 1 to 6 (1 being the highest and 6 being the lowest), rank the activities below according to which one takes the most of your time, the next most time, and so on.
    RANK:
    ________ Parts chasing/securing parts information
    ________ Waiting on permits
    ________ Receiving job instructions
    ________ Waiting for operations to prepare equipment
    ________ Travel to/from job site
    ________ Troubleshooting

22. One of (*company's*) goals is to have maintenance and operations work together with a total team effort. How do you feel this is working in your specific situation?
    Comments: ____
    ____
    ____

SUMMARY: Thank you for your valuable input. A copy of the survey results will be provided to you.

Appendix

E

Maintenance Excellence Glossary

Introduction

The following terms and definitions are from the area of maintenance management, maintenance storeroom operations, inventory management, and MRO materials management areas.*

A

ABC inventory policy Collection of prioritizing practices to give varied levels of attention to different classes of inventories. For example, Class A items typically can make up 15 to 25 percent of stock items, but 75 to 85 percent of inventory value. Class C items, in turn, might be 60 percent of the stock items, but only 10 percent of the inventory value. Class B items would be somewhere in between these two. Using ABC policy helps "focus not on the trivial many but the vital few".

ACE team benchmarking system (Also the ACE system) A team-based process for using experienced craftspeople, supervisors, and planners to develop improved repair methods and then a consensus on maintenance repair times and work content/wrench time. It involves *a consensus of experts* (ACEs) for first evaluating a job for improved methods, tools, special equipment, and root cause elimination. Second, this process then develops a consensus on work content and then "benchmark jobs" are arranged on spreadsheets by craft area for work content comparison (slotting). The process provides a methodology for a planner or supervisor to develop reliable planning times for a wide range of

*Thanks to Sandy Dunn, the leader of Assetivity Pvt Ltd in Australia (http://www.assetivity.com.au) and the owner of probably the world's best site for plant maintenance professionals: the Plant Maintenance Resource Center at http//www.Plant-Maintenance. com for many of the maintenance terms included in this appendix.

jobs using a relatively small number of benchmark jobs arranged on spreadsheets by craft areas.

Actuarial analysis Statistical analysis of failure data to determine the age-reliability characteristics of an item.

Adjustments Minor tune-up actions requiring only hand tools, no parts, and usually lasting less than a half hour.

APL See Applications parts list.

Applications parts list A list of all parts required to perform a specific maintenance activity. Typically set up as a standard list attached to a standard job for routine tasks. Not to be confused with a bill of materials.

Apprentice A craftsperson in training, typically following a specifically defined technical training program while gaining hands-on experience in a craft area to gain a specified number of hours experience resulting in one or more craft area competencies.

Area maintenance A method for organizing maintenance operations in which the first-line maintenance leader is responsible for all maintenance crafts within a certain department, area, or location within the facility.

Asset care An alternative term for the maintenance process. A kinder, gentler term, but still pure maintenance and physical asset management.

Assets The physical resources of a business such as plant equipment, facilities, building systems, fleets, or their parts and major components. In the accounting world, *assets* means money and real property. In maintenance, the term *assets* is commonly taken to be any item of physical plant or equipment. Continuous reliability improvement (CRI) goes beyond the physical asset/resources to include five morel asset or resource areas beyond equipment to include; MRO parts and materials, information, people resources, information

Asset list A register of physical assets (equipment, facilities, building systems, and the like usually with information on manufacturer, vendor, specifications, classification, costs, warranty, and tax status.

Asset management The systematic planning and control of a physical asset resource throughout its economic life; the systematic planning and control of a physical resource throughout its life. This may include the specification, design, and construction of the asset, its operation, maintenance and modification while in use, and its disposal when no longer required.

Asset number A unique alphanumerical identification of an asset list, which is used for its management.

Asset register A list of all the assets in a particular workplace, together with information about those assets such as manufacturer, vendor, make, model, and specifications.

Asset utilization The percentage of total time the equipment/asset is running.

Availability The probability that an asset will, when used under specified conditions, operate satisfactorily and effectively. Also, the percentage of time or number of occurrences for which an asset will operate properly when called upon; the proportion of total time that an item of equipment is capable of performing its specified functions, normally expressed as a percentage. It can be calculated by dividing the equipment available hours by the total number of hours in any given period.

Available hours The total number of hours that an item of equipment is capable of performing its specified functions. It is equal to the total hours in any given period, less the downtime hours.

Average life How long, on average, a component will last before it suffers a failure. Commonly measured by *mean time between failures* (MTBF).

B

Backlog, current Work orders planned and prioritized, waiting scheduling and execution: Work which has not been completed by the "required by date." The period for which each work order is overdue is defined as the difference between the current date and the "required by date." All work for which no "required by" date has been specified is generally included on the backlog. Backlog is generally measured in "crew-weeks"; the total number of labor hours represented by the work on the backlog, divided by the number of labor hours available to be worked in an average week by the work crew responsible for completing this work. As such, it is one of the common *key performance indicators* (KPI) used in maintenance.

Backlog, total: Total backlog is total maintenance requirements. It is the total identifiable work that needs to be performed including a) current backlog, b) known deferred maintenance, c) actual yearly PM/PdM tasks (from point in time total backlog has been defined), d) major overhauls, e) shutdown/turnarounds denoted as to planned date at the actual time a total backlog is being defined plus any other work requiring craft skills.

Bar code An identification method using symbols for encoding data using lines of varying thickness, designation alphanumeric characters.

Benchmarking The process of comparing performance or processes with other organizations or standards such as The Scoreboard for Maintenance Excellence, identifying comparatively high-performance organizations or existing universal best practices, and learning as to what they do that allows them to achieve that high level of performance. Acting upon universal best practice to implement and benchmark shop level results.

Bill of materials (BOM) List of components and parts for an asset, usually structured in hierarchical layers from gross assemblies or major end items to minor items down to component parts; a list of all the parts. Also includes list of parts and materials developed by planner, project engineer, estimator and the like for planned major or minor work.

BOM See Bill of materials.

Breakdown Failure to perform to a functional standard; a specific type of failure where an item of plant or facility equipment is completely unable to function.

Breakdown maintenance A maintenance strategy or policy where no maintenance is done until an item fails and no longer meets its functional standard. May be the best strategy when failure consequence is very, very low. May be fatal if failure consequence and risk is extremely high. See No scheduled maintenance.

Built-in test equipment (BITE) Diagnostic and checkout devices integrated into equipment to assist operation, troubleshooting, and service. For example the On Star System for some General Motors vehicles that automatically dial 911 where you are located (per your GPS location) when an airbag deploys and you do not call On Star central dispatch personally from your vehicle.

C

Calibrate To verify the accuracy of equipment and assure performance within tolerance, usually by comparison to a reference standard that can be traced to a primary standard or baseline.

Call-out or call back To summon a tradesperson to the workplace during his normal nonworking time so that he can perform a maintenance activity (normally an emergency maintenance task).

Capital Durable items with long life or high value that necessitates asset control and depreciation under tax guidelines, rather than being expensed immediately for accounting purposes.

Carrying costs Expense of handling space, information, insurance, special conditions, obsolescence, personnel, and the cost of capital or alternative use of funds to keep parts in inventory. Also called holding costs and can be in 25% to 40% range.

CBM (condition-based maintenance) Maintenance based on the measured condition of an asset. testing and/or inspection of characteristics that will warn of pending failure and performance of maintenance after the warning threshold but before total failure. Predictive maintenance technologies such as vibration analysis, thermography/infrared, oil analysis, and ultrasonic provide tools and technology for condition-based maintenance; an equipment maintenance strategy based on measuring the condition of equipment in order to assess whether it will fail during some future period, and then taking appropriate action to avoid the consequences of that failure. The terms condition-based maintenance, on-condition maintenance and predictive maintenance can be used interchangeably. On Star Systems are condition-based sytems that alert driver when oil needs to be changed based upon miles or hours or when a major vehicle system is failing.

Central maintenance A method for organizing maintenance operations in which the maintenance leader is responsible for all maintenance and all craft areas operating on call from a central location to support the entire operation. May also denote a central maintenance support group for area maintenance and include craft support such as a machine, welding, electronics, security/lock shop specialists, equipment operators for special support equipment used by area maintenance and the like not found within area maintenance.

Change out The removal of a component or part and the replacement of it with a new or rebuilt one.

Chief maintenance officer (CMO) The technical leader (or actual leader) of a profit-centered maintenance operation within a large or small corporation or nonmanufacturing organization. First coined, defined, and promoted by staff from The Maintenance Excellence Institute in the late 1990s, the CMO represents the prototype of the new millennium leader for maintenance and physical asset management leading maintenance toward a true profit and customer-centered maintenance operation.

Checkout The determination of the working condition of a system.

Clean To remove all sources of dirt, debris, and contamination for the purpose of inspection and to avoid chronic losses.

CMMS (See Computerized maintenance management systems) Integrated computer system modules such as work orders, equipment, inventory, purchasing, planning, and preventive maintenance that support asset management and overall maintenance management.

CMMS benchmarking system A methodology developed by staff from The Maintenance Excellence Institute to evaluate the effectiveness of a CMMS/EAM installation. A process for evaluating CMMS/EAM implementation progress and full utilization of existing system functionality to enhance best maintenance practices.

CMO See Chief maintenance officer.

Code Symbolic designation used for identification, for example, failure code, repair code, commodity code.

Commodity code Classifications of parts by group and class according to their material content or type of consolidation of procurement, storage, and use.

Component A constituent part of an asset, usually modular and replaceable that may be serialized and interchangeable; a subassembly of an asset, usually removable in one piece and interchangeable with other, standard components (e.g., truck engine).

Component number Designation, usually structured by system, group, or serial number.

Condition monitoring The use of specialist equipment to measure the condition of equipment. Vibration analysis, oil analysis, and thermography are all examples of condition monitoring techniques.

Conditional probability of failure The probability that an item will fail during a particular age interval, given that it survives to enter that age.

Confidence Degree of certainty that something will happen. For example, low confidence of replenishment means repair parts probably will not be readily available and is one reason that maintenance personnel retain excess parts in uncontrolled areas.

Configuration The arrangement and contour of the physical and functional characteristics of systems, equipment, and related items of hardware or software;

the shape of a thing at a given time. The specific parts used to construct a machine.

Consumables Supplies such as fuel, lubricants, gloves, safety glasses, welding rods, paper, printer ribbons, cleaning materials, and forms that are exhausted during use in operation and maintenance.

Contingency Alternate actions that can be taken if the main actions do not work. Something every maintenance leader should have—contingency plan B if plan A is not the right plan.

Continuous reliability improvement A process developed and used by staff from The Maintenance Excellence Institute that goes beyond current *reliability-centered maintenance* (RCM) approaches to outline a continuous, integrated process for improving total reliability of the following key maintenance resources:

- Equipment/facility resources (asset care/management and maximum uptime via RCM techniques)
- Craft and operator resources (recognizing the most important resource: craftspeople and equipment/process operators)
- Maintenance repair operations (MRO) resources (establishing effective storage and materials management processes)
- Maintenance information resources (effective information technology applications for maintenance)
- Maintenance technical knowledge/craft skills base (closing the technical knowledge resource gap with competency based training)
- Synergistic team processes (tapping the value-added resource of effective leadership-driven teams to support total operations success) "Realizing and applying the value from hidden assets"

Contract acceptance sheet A document that is completed by the appropriate contract supervisor and contractor to indicate job completion and acceptance. It also forms part of the appraisal of the contractor's performance and quality.

Coordination Daily adjustment of maintenance activities to achieve the best short-term use of resources or to accommodate changes in needs for service.

Corrective maintenance (CM) Unscheduled maintenance or repair actions, performed as a result of failures or deficiencies, to restore items to a specific condition. Maintenance done to bring an asset back to its standard functional performance: Any maintenance activity which is required to correct a failure that has occurred or is in the process of occurring. This activity may consist of repair, restoration, or replacement of components and can typically be planned, estimated and scheduled proactively.

Craft availability Percentage of time that craft labor is free to perform productive work during a scheduled working period. Not craft utilization or actual craft wrench time.

Craftsperson Alternative to tradesperson. A skilled maintenance worker who has been formally trained through an apprenticeship program.

Craft utilization Percentage of time that craft labor is engaged in productive work, hands-on during a scheduled working period. The actual wrench time which comes out of craft availability is compared to total time paid and becomes the effectiveness factor for overall craft effectiveness (OCE) and overall craft labor producivity.

Critical Describes items especially important to product or service performance and more vital to the operation than noncritical items. For example, critical spares purchased as a true insurance item for assets/operations with extreme profit, service or safety consequences and risks if immediate replacement is not made to regain asset's function.

Critical equipment Items especially important to performance, capacity, and throughput and more vital to the operation than noncritical items.

Criticality The priority rank of a failure mode based on some assessment criteria.

Critical path method (CPM) A logical method of planning and controlling that analyzes events, the time required, and the interactions of the considered activities.

Critical spare (See Insurance item) Parts and materials that are not used often enough to meet detailed stock accounting criteria but are stocked as "insurance items" because of their essentiality or the lead time involved in procuring replacements; similar to safety stocks, except on low-use parts. Storeroom inventory management should denote critical spares as compared to normal stock items. If inventory turns is a metric it should not include critical spares in the calculation; total annual value of normal stock items used divided by total average value of the normally stocked items minus critical spare value.

Crossdocking Term for the function capability of an inventory management module to track high priority in-bound orders into receiving, to initiate immediate delivery, and to receive/process issue transactions with minimal manual effort. Crossdocking provides quick turnaround at receiving without putting items into storage locations and then having to pick for issue later.

Cycle count An inventory accountability strategy where counting and verification of stock item quantities is done continuously based on a predetermined schedule and frequency based on the ABC classification of the item. As opposed to an annual physical inventory, cycle counting allows for continuous counting and reconciliation of inventory discrepancies.

D

Dead stocks Items for which no demand has occurred over a specific period of time.

Defect A condition that causes deviation from design or expected performance. A term typically used in the maintenance of mobile equipment. A defect is typically a potential failure or other condition that will require maintenance attention at some time in the future, but which is not currently preventing the equipment from fulfilling its functions.

Deferred maintenance Maintenance that can be or has been postponed from a schedule.

Deterioration rate The rate at which an item approaches a departure from its functional standard.

Demand Requests and orders for an item. Demands become issues only when a requested part is given from stock.

Direct costs Any expenses that can be associated with a specific product, operation, or service.

Discard task The removal and disposal of items or parts.

Disposal The act of getting rid of excess or surplus property under proper authorization. Such processes as transfer, donation, sale, abandonment, destruction, or recycling may accomplish disposal.

Down Out of service, usually due to breakdown, unsatisfactory condition, or production scheduling.

Downtime (DT) The time that an item of equipment is out of service, as a result of equipment failure. The time that an item of equipment is available, but not utilized is generally not included in the calculation of downtime. But, DT can also in its most strict sense could very well be expressed as time out of service due to failure divided by the time actually available to perform its primary function.

E

Economic life The total length of time that an asset is expected to remain actively in service before it is expected that it would be cheaper to replace the equipment rather than continuing to maintain it. In practice, equipment is more often replaced for other reasons, including: because it no longer meets operational requirements for efficiency, product quality, comfort, and the like, or because newer equipment can provide the same quality and quantity of output more efficiently.

Economic order quantity (EOQ) Amount of an item that should be ordered at one time to get the lowest possible combination of inventory carrying costs and ordering costs.

Economic repair A repair that will restore the item to a sound condition at a cost less than the value of its estimated remaining useful life.

Emergency maintenance A condition requiring immediate corrective action for safety, environmental, or economic risk, caused by equipment breakdown.

Emergency maintenance task A maintenance task carried out in order to avert an immediate safety or environmental hazard, or to correct a failure with significant economic impact.

Engineering change Any design change that will require revision to specifications, drawings, documents, or configurations.

Engineering change notice (ECN) A control document from engineering authorizing changes or modifications to a previous design or configuration.

Engineering work order The prime document used to initiate an engineering investigation, engineering design activity, or engineering modifications to an item of equipment.

Environmental consequences A failure has environmental consequences if it could cause a breach of any known environmental standard or regulation.

Equipment configuration List of assets usually arranged to simulate the process, or functional or sequential flow.

Equipment maintenance strategies The choice of routine maintenance tasks and the timing of those tasks, designed to ensure that an item of equipment continues to fulfill its intended functions.

Equipment repair history A chronological list of defaults, repairs, and costs on key assets so that chronic problems can be identified and corrected and economic decisions made.

Equipment use Accumulated hours, cycles, distance, throughput, or performance.

Estimated plant replacement value The estimated cost of capital required to replace all the existing assets with new assets capable of producing the same quantity and quality of output.

Estimating index The ratio of estimated labor hours required for completing the work specified on work orders to the actual labor hours required to complete the work specified on those work orders, commonly expressed as a percentage. This is also a measure of craft performance, one element of craft labor productivity, particularly when there are well-defined estimating standards.

EWO See Engineering work order.

Examination A comprehensive inspection with measurement and physical testing to determine the condition of an item.

Expediting Special efforts to accelerate a process. An expediter coordinates and assures adequate supplies of parts, materials, and equipment.

Expense Those items that are directly charged as a cost of doing business. They generally have a short, nondurable life. Most nonrepairable repair parts are expensed when installed on equipment.

Expensed inventory Parts written off as a "cost of sales." Material transferred from ledger inventory to expensed inventory is to be used within 12 months.

Expert system Decision support software with some ability to make or evaluate decisions based on rules or experience parameters incorporated in the database; a software-based system which makes or evaluates decisions based on rules established within the software. Typically used for fault diagnosis.

F

Fail-safe An item is fail-safe if, when the item itself incurs a failure, that failure becomes apparent to the operating workforce in the normal course of events.

Failure Termination of the ability of an item to perform its required function to a standard: an item of equipment has suffered a failure when it is no longer capable of fulfilling one or more of its intended functions. Note that an item does not need to be completely unable to function to have suffered a failure. For example, a pump that is still operating, but is not capable of pumping the required flow rate, has failed. In reliability centered maintenance terminology, a failure is often called a functional failure.

Failure analysis The logical, systematic examination of an item or its design, to identify and analyze the probability, causes, and consequences of real or potential malfunction. A study of failures to analyze the root causes, to develop improvements, to eliminate or reduce the occurrence of failures. (See FMECA: failure modes, effects, and criticality analysis.)

Failure cause See Failure mode.

Failure coding Identifying and indexing the causes of equipment failure on which corrective action can be based, for example, lack of lubrication, operator abuse, material fatigue, and the like; a code typically entered against a work order in a CMMS which indicates the cause of failure (e.g., lack of lubrication, and metal fatigue).

Failure consequences A term used in reliability centered maintenance. The consequences of all failures can be classified as being either hidden, safe, environmental, operational, or nonoperational.

Failure effect A description of the events that occur after a failure has occurred as a result of a specific failure mode. Used in reliability centered maintenance, FMEA, and FMECA analyses.

Failure-finding interval The frequency with which a failure-finding task is performed is determined by the frequency of failure of the protective device, and the desired availability required of that protective device.

Failure-finding task Used in reliability centered maintenance terminology. A routine maintenance task, normally an inspection or a testing task, designed to determine, for hidden failures, whether an item or component has failed. A failure finding task should not be confused with an on-condition task, which is intended to determine whether an item is about to fail. Failure-finding tasks are sometimes referred to as functional tests.

Failure mode Any event which causes a failure.

Failure modes, effects, and criticality analysis A structured method of assessing the causes of failures and their effect on production, safety, cost, quality, and so forth.

Failure modes and effects analysis A structured method of determining equipment functions, functional failures, assessing the causes of failures, and their failure effects. The first part of a reliability centered maintenance analysis is a failure modes and effects analysis.

Failure pattern The relationship between the conditional probability of failure of an item, and its age. Failure patterns are generally applied to failure modes. Research in the airline industry established that there are six distinct failure patterns. The type of failure pattern that applies to any given failure mode is of vital importance in determining the most appropriate equipment maintenance strategy. This fact is one of the key principles underlying reliability centered maintenance.

FMECA See Failure modes, effects, and criticality analysis.

FMEA See Failure modes and effects analysis.

Forward workload All known backlog work and work which is due or predicted to become backlog work within a prespecified future time period.

Failure rate The number of failures per unit measure of life (cycles, time, miles, events, and the like) as applicable for the item.

Fault tree analysis (FTA) A review of failures, faults, defects, and shortcomings based on a hierarchy or relationship to find the root cause.

FMECA Failure mode, effect, and criticality analysis, a logical, progressive method used to understand the root causes of failures and their subsequent effect on production, safety, cost, quality, and the like.

Fill rate Service level of a specific stock point. An 85 percent fill rate means that if 100 parts are requested, then 85 of them are available and issued.

First in-first out (FIFO) Use the oldest item in inventory next. FIFO accounting values each item used at the cost of the oldest item in inventory. Contrasts with LIFO (last in-first out).

Forecast To calculate or predict some future event or condition, usually as a result of rational study and analysis of pertinent data. The projection of the most probable: as in forecasting failures and maintenance activities.

Frequency Count of occurrences during each time period or event. A typical frequency chart for inventory plots demand versus days.

Function A separate and distinct action required to achieve a given objective, to be accomplished by the use of hardware, computer programs, personnel, facilities, procedural data, or a combination thereof; or an operation a system must perform to fulfill its mission or reach its objective; the definition of what we want an item of equipment to do, and the level of performance which the users of the equipment require when it does it. An item of equipment can have many functions, commonly split into primary and secondary functions.

Functional failure Used in reliability centered maintenance terminology. The inability of an item of equipment to fulfill one or more of its functions. Interchangeably used with failure.

Functional levels Rankings of the physical hierarchy of a product. Typical levels of significance from the smallest to the largest are part, subassembly, assembly, subsystem, and system.

Functional maintenance structure A method for organizing the maintenance operation where the first-line maintenance leader is responsible for conduction of a specific kind of maintenance, for example, electrical maintenance, pump maintenance, and HVAC maintenance.

G

Gantt chart A bar chart format of scheduled activities showing the duration and sequencing of activities.

General support equipment (GSE) Equipment that has maintenance application to more than a single model or type of equipment.

Go-line Used in relation to mobile equipment. Equipment which is available, but not being utilized is typically parked on the go-line. This term is used interchangeably with ready line.

H

Hardware A physical object or physical objects, as distinguished from capability or function. A generic term dealing with physical items of equipment-tools instruments, components, parts-as opposed to funds, personnel, services, programs, and plans, which are termed "software."

Hidden failure A failure which, on its own, does not become evident to the operating crew under normal circumstances. Typically, protective devices which are not fail-safe (examples could include standby plant and equipment, and emergency systems).

Hold for disposition stock Defective material held at a stock location pending removal for repair or for scrap.

I

Identification Means by which items are named or numbered to indicate that they have a given set of characteristics. Identification may be in terms of name, part number, drawing number, code, stock number, or catalog number. Items may also be identified as part of an assembly, a piece of equipment, or a system.

Indirect costs Expenses not directly associated with specific products, operations, or services; usually considered overhead.

Infant mortality The relatively high conditional probability of failure during the period immediately after an item returns to service.

Inherent reliability A measure of the reliability of an item, in its present operating context, assuming adherence to ideal equipment maintenance strategies.

Inspection A review to determine maintenance needs and priority on equipment. Any task undertaken to determine the condition of equipment, and/or to determine the tools, labor, materials, and equipment required to repair the item.

Insurance items Parts and materials that are considered as critical spares, but not used often enough to meet detailed stock accounting criteria. Insurance items are stocked because of their essentially or the lead time involved in procuring replacements. They may be of high dollar value to classify them as capital spares.

Interchangeable Parts with different configurations and numbers that may be substituted for another part, usually without any modification or different performance, since they have the same form, fit, and function.

Interface A common boundary between two or more items, characteristics, systems, functions, activities, departments, or objectives. That portion impinges upon or directly affects something else.

Interval based Periodic preventive maintenance based on calendar time.

Inventory turnover Ratio of the value of materials and parts issues annually to the value of materials and parts on-hand, expressed as percentage.

Inventory Physical count of all items on hand by number, weight, length, or other measurement; also any items held in anticipation of future use.

Inventory control That phase or function of logistics that includes management, cataloging, requirements, determination, procurement, inspection, storage, distribution, overhaul, and disposal of material. Managing the acquisition, receipt, storing, and issuance of materials and spare parts: managing the investment efficiently of the stores inventory.

Issues Stock consumed through stores.

Item Generic term used to identify a specific entity. Items may be parts, components, assemblies, subassemblies, accessories, groups, parents, components, equipment, or attachments.

Item of supply An article or material that is recurrently purchased, stocked, distributed, used, and is identified by one distinctive set of numbers or letters throughout the organization concerned. It consists of any number of pieces or objects that can be treated as a unit.

J

JIT Just-in-time, buzzword term for proactive planning of many processes such as JIT inventory service, and JIT maintenance services.

K

Keep full Term used for maintaining set levels of shop stock inventory of Class C items. (See also Shop stock.)

Key performance indicators A select number of key measures that enable performance against targets to be monitored.

KPI See Key performance indicators.

L

Last in-first out (LIFO) Use newest inventory next. LIFO accounting values each item used at the cost of the last item added to inventory. Contrasts with FIFO (first in-first out).

LCC See life-cycle costing.

Lead time Allowance made for that amount of time estimated or actually required to accomplish a specific task such as acquiring a part.

Ledger inventory Items carried on the corporate financial balance sheet as material valued at cost.

Level of repair (LOR) Locations and facilities at which items are to be repaired. Typical levels are operator, field technician, bench, and factory.

Level of services (stores) Usually measured as the ratio of stock outs to all stores issues.

Life That strange experience you have all day, every day. In a maintenance context, you may want to look at equipment life.

Life cycle The series of phases or events that constitute the total existence of anything. The entire "womb to tomb" scenario of a product from the time concept planning is started until the product is finally discarded.

Life-cycle cost All costs associated with the items of life cycle including design, manufacture, operation maintenance, and disposal; a process of estimating and assessing the total costs of ownership, operation, and maintenance of an item of equipment during its projected equipment life. Typically used in comparing alternative equipment design or purchase options in order to select the most appropriate option.

Logistics engineering The professional art of applying science to the optimum planning, handling, and implementation of personnel, materials, and facilities including life-cycle designs, procurements, production, maintenance, and supply.

Logistic support analysis (LSA) A methodology for determining the type and quantity of logistic support required for a system over its entire life cycle. Used to determine the cost-effectiveness of asset-based solutions.

LSA See Logistic support analysis.

M

Maintainability The inherent characteristic of a design or installation that determines the ease, economy, safety, and accuracy with which maintenance actions can be performed. Also, the ability to restore a product to service or to perform preventive maintenance within required limits. The rapidity and ease with which maintenance operations can be performed to help prevent malfunctions or correct them if they occur, usually measures as mean time to repair; the ease and speed with which any maintenance activity can be carried out on an item of equipment. May be measured by *mean time to repair*. Is a function of equipment design and maintenance task design (including use of appropriate tools, jigs, work platforms, and the like).

Maintainability engineering The application of applied scientific knowledge, methods, and management skills to the development of equipment, systems, projects, or operations that have the inherent ability of being effectively and efficiently maintained; the set of technical processes that apply maintainability theory to establish system maintainability requirements, allocate these requirements down to system elements, and predict and verify system maintainability performance.

Maintenance The function of keeping items or equipment in, or restoring them to, serviceable condition. It includes servicing, test, inspection, adjustment/alignment, removal, replacement, reinstallation, troubleshooting, calibration, condition determination, repair, modification, overhaul, rebuilding, and reclamation. Maintenance includes both corrective and preventive activities. Any activity carried out to retain an item in, or restore it to, an acceptable condition for use or to meet its functional standard. Related areas may include equipment movement, initial asset installation support, facility system modification per internal resource availability on and in a facilities management operation minor construction may be a key maintenance responsibility.

Maintenance engineering Developing concepts, criteria, and technical requirements for maintenance during the conceptual and acquisition phases of a project. Providing policy guidance for maintenance activities, and exercising technical and management direction and review of maintenance programs. A staff function intended to ensure that maintenance techniques are effective, equipment is designed for optimum maintainability persistent and chronic problems are analyzed, and corrective actions or modifications are made.

Maintenance excellence index (MEI) An essential component to The Maintenance Excellence Institute's implementation of profit-centered maintenance. It is a shop level internal benchmark and a progressive approach to performance measurement within maintenance operations. It is achieved by integrating multiple metrics into a composite total MEI value. It includes the comparison of current baseline performance to both the performance goal and baseline value for each metric selected. This approach is ideally suited to measure progress toward maintenance excellence across multiple sites within an organization.

Maintenance policy A statement of principle used to guide maintenance management decision making. This may also be more detailed to include storeroom operations procedure, planning/scheduling procedures, ACE System estimating procedures, purchasing procedures all as guides to standardization, training, and consistency of best practice application

Maintenance requirements Specific areas defined below sum up a total maintenance backlog; some ready for scheduling from Current backlog, some for upcoming PMPdMs and some identified deferred maintenance needs and the like.

- **Inspection** is a maintenance requirement when the basic objective is to assure that a requisite condition or quality exists. In order to inspect for the desired condition, it may be necessary to remove the item, to gain access by removing other items, or to disassemble partially the item for inspection purposes. In such cases, these associated actions necessary to accomplish the required inspection would be specific tasks.
- **Troubleshooting** is a maintenance operation that involves the logical process (series of tasks) that leads to positive identification, location, and isolation of the cause of a malfunction.
- **Remove** is a maintenance requirement when the basic objective is to separate the item from the next higher assembly. This requirement is usually applied for a configuration change.
- **Remove and replace** is a requirement that constitutes the removal of one item and replacement of it with another like item. Such action can result from a failure or from a scheduled action.
- **Remove and reinstall** is a maintenance requirement when an item is removed for any reason, and the same items reinstalled.

- **Adjustment/alignment** is a maintenance requirement when the primary cause of the maintenance action is to adjust or align, or to verify adjustment/ alignment of specific equipment. Adjustment/alignment accomplished subsequent to repair of a given item is not considered a separate requirement and is included as a task in the repair requirement.
- **Functional test** constitutes a system or subsystem operational checkout either as a condition verification after the accomplishment of corrective maintenance action or as a scheduled requirement on a periodic basis.
- **Conditioning** is a maintenance requirement whenever an item is completely disassembled, refurbished, tested, and returned to a serviceable condition, meeting all requirements set forth is applicable specifications. It may result from either a scheduled or unscheduled requirement and is generally accomplished at the depot/factory level of maintenance

Maintenance schedule A comprehensive list of planned maintenance and its sequence of occurrence based on priority in a designated period of time; a list of *planned maintenance* tasks to be performed during a given time period, together with the expected start times and durations of each of these tasks. Schedules can apply to different time periods (e.g., daily schedule and weekly schedule).

Maintenance strategy Principles and strategies for guiding decisions for maintenance management; a long-term plan, covering all aspects of maintenance management which sets the direction for maintenance management, and contains firm action plans for achieving a desired future state for the maintenance function.

Maintenance shutdown A period of time during which a plant, department, process, or asset is removed from service, specifically for maintenance.

Maintenance task routing file A computer file containing skills, hours, and descriptions to perform standard maintenance tasks.

Mean time between failures A measure of equipment reliability. Equal to the number of failures in a given time period, divided by the total equipment uptime in that period.

Mean time to repair A measure of maintainability. Equal to the total equipment downtime in a given time period, divided by the number of failures in that period.

MIL-STD U.S. Military Standard.

Model work order A work order stored in the CMMS which contains all the necessary information required to perform a maintenance task. (See also Standard job.plan)

Modification Change in configuration; any activity carried out on an asset which increases the capability of that asset to perform its required functions.

Modularization Separation of components of a product or equipment into physically and functionally distinct entities to facilitate identification, removal, and replacement unitization.

MRO Term for maintenance repair operations and generally used as MRO items referring to parts, materials, tools, and equipment used in the maintenance process.

MRO materials management The overall management of the process for requisitioning, storage/warehousing, purchasing, inventory management, and issue of MRO type items used in the maintenance process. A modern term for the mission-essential area of logistics for executing the maintenance process.

MTBF See Mean time between failures.

MTTR See Mean time to repair.

N

NDT Nondestructive testing of equipment to detect abnormalities in physical, chemical, or electrical characteristics using such technologies as ultrasonic (thickness), liquid dye penetrates (cracks), x-ray (weld discontinuities), and voltage generators (resistance).

Nonoperational consequences A failure has nonoperational consequences if the only impact of the failure is the direct cost of the repair (plus any secondary damage caused to other equipment as a result of the failure.

Nonrepairable Parts or items that are discarded upon failure for technical or economic reasons.

Nonroutine maintenance Maintenance performed at irregular intervals, with each job unique, and based on inspection, failure, or condition. Any maintenance task which is not performed at a regular, predetermined frequency.

No scheduled maintenance An equipment maintenance strategy, where no routine maintenance tasks are performed on the equipment. The only maintenance performed on the equipment is corrective maintenance, and then only after the equipment has suffered a failure. Also described as a run-to-failure strategy.

O

Obsolescence Decrease in value or use of items that have been superseded by superior items.

Obsolete Designation of an item for which there is no replacement. The part has probably become unnecessary as a result of design change.

OBM Operator-based maintenance—a maintenance excellence strategy where equipment or process operators are trained and accountable for selected maintenance tasks. Also autonomous maintenance.

Oil analysis See Teratology.

On-condition maintenance See Condition-based maintenance.

Operating context The operational situation within which an asset operates. For example, is it a stand-alone piece of plant, or is it one of a duty-standby pair? Is it part of a batch manufacturing process or a continuous production process? What is the impact of failure of this item of equipment on the remainder of the production process? The operating context has enormous influence over the choice of appropriate equipment maintenance strategies for any asset.

Operating hours The length of time that an item of equipment is actually operating.

Operational consequences A failure has operational consequences if it has a direct adverse impact on operational capability (lost production, increased production costs, loss of product quality, or reduced customer service).

Operational productivity of a physical asset Used in the calculation of overall equipment effectiveness, which is physical asset productivity as compared to overall craft effectiveness which is craft labor productivity. The actual output produced from an asset in a given time period divided by the output that would have been produced from that asset in that period, had it produced at its rated capacity. Normally expressed as a percentage.

Order point Quantity of parts at which an order will be placed when usage reduces stock to that level, also called *reorder point* (ROP).

Order quantity Number of items demanded. The *economic order quantity* (EOQ), also called minimum cost quantity, is a specific number; but the actual order quantity may vary as a result of cost, transportation, discounts, or extraordinary demand.

Outage A term used in some industries, for example, electrical power generation, transmission, and distribution. Outage denotes that an individual asset or system is not in use.

Overall craft effectiveness (OCE) The OCE factor is a method developed by The Maintenance Excellence Institute to measure craft labor productivity that combines three key elements: craft utilization, craft performance, and craft service quality. Typically, the OCE factor is determined by only the two elements: percent craft utilization × craft performance. Compares to OEE in basic concepts but applies directly to productivity of craft labor assets.

Overall equipment effectiveness (OEE) The OEE factor is a method to measure overall physical asset productivity first used in the USA and later adopted by others claiming origination. The OEE factor combines three key elements: equipment availability, performance, and quality measurement into a common metric that reflects key elements of the manufacturing environment. The OEE factor equals percent availability × percent performance × percent quality. Compares to OCE in the basic concept but applies directly to the productivity of physical assets.

Overhaul A comprehensive examination and restoration of an item to an acceptable condition.

P

Pareto's principle Critical few, often about 20 percent, of parts or people or users that should receive attention before the insignificant many, which are usually about 80 percent. Named after Alfredo Pareto an Italian intellectual with common sense, uncommon for some of his time in history.

Part numbers Unique identifying numbers and letters that denote each specific part configuration; also called stock numbers or item numbers.

PCM See Profit-centered maintenance.

PCCM See Profit- and Customer-Centered Maintenance.

PdM See Predictive maintenance.

Periodic maintenance Cyclic maintenance actions carried out at regular intervals, based on repair history data, use or elapsed time.

Percent planned work The percentage of total work (in labor hours) performed in a given time period which has been planned in advance.

PERT chart See Project evaluation and review technique (PERT) chart.

P-F interval A term used in reliability centered maintenance. The time from when a potential failure can first be detected on an asset or component using a selected predictive maintenance task, until the asset or component has failed. Reliability centered maintenance principles state that the frequency with which a predictive maintenance task should be performed is determined solely by the P-F interval.

Pick list A selection of required store items for a work order or task.

Planned maintenance Maintenance carried out according to a documented plan of tasks, skills, and resources; any maintenance activity for which a predetermined job procedure has been documented, for which all labor, materials, tools, and equipment required to carry out the task have been estimated, and their availability assured before commencement of the task.

Plant engineering A staff function whose prime responsibility is to ensure that maintenance techniques are effective, that equipment is designed and modified to improve maintainability, that on-going maintenance technical problems are investigated, and appropriate corrective and improvement actions are taken. Used interchangeably with maintenance engineering and reliability engineering.

PM See Preventive maintenance.

Potential failure A term used in reliability centered maintenance. An identifiable condition which indicates that a functional failure is either about to occur, or is in the process of occurring.

PRA See Probabilistic risk assessment.

Predictive maintenance (PdM) Use of measured physical parameters against known engineering limits for detecting, analyzing, and correcting equipment problems before a failure occurs; examples include vibration analysis, sonic testing, dye testing, infrared testing, thermal testing, coolant analysis, teratology, and equipment history analysis. Subset of preventive maintenance that uses nondestructive testing such as spectral oil analysis, vibration evaluation, and ultrasonic with statistics and probabilities to predict when and what maintenance should be done to prevent failures; an equipment maintenance strategy based on measuring the condition of equipment in order to assess whether it will fail during some future period, and then taking appropriate action to avoid the consequences of that failure. The terms condition-based maintenance, on-condition maintenance, and predictive maintenance can be used interchangeably.

Preventive maintenance (PM) Maintenance carried out at predetermined intervals, or to other prescribed criteria, and intended to reduce the likelihood of a functional failure. Actions performed in an attempt to keep an item in a specific operating condition by means of systematic inspection, detection, and prevention of incipient failure; an equipment maintenance strategy based on replacing, overhauling, or remanufacturing an item at a fixed interval, regardless of its condition at the time. Scheduled restoration tasks and scheduled discard tasks are both examples of preventive maintenance tasks. See also scheduled maintenance.

PRIDE-in-Maintenance Coined originally in 1981 as the theme for a presentation to the craft workforce at a manufacturing plant in Greenville, Mississippi. It is about changing the hearts, minds, and attitudes about the profession and practice of maintenance. It is about PRIDE. and *people really interested in developing excellence* in maintenance operations of all types. Its foundation starts with the most important maintenance resource, the crafts workforce. The goal is to achieve *pride* in maintenance from within the crafts workforce and among their maintenance leaders and to have top leaders realize the true value of their total maintenance operation and take positive action.

Priority The relative importance of a single job in relationship to other jobs, operational needs, safety, and so on, and the time within which the job should be done: used for scheduling work orders.

Proactive maintenance A maintenance strategy that is anticipatory and includes a level of planning; any tasks used to predict or prevent equipment failures.

Probabilistic risk assessment A "top-down" approach used to apportion risk to individual areas of plant and equipment, and possibly to individual assets so as to achieve an overall target level of risk for a plant, site, or organization. These levels of risk are then used in risk-based techniques, such as reliability centered maintenance to assist in the development of appropriate equipment maintenance strategies, and to identify required equipment modifications.

Probabilistic safety assessment Similar to Probabilistic risk assessment, except focused solely on safety-related risks.

Procurement Process of obtaining persons, services, supplies, facilities, materials, or equipment. It may include the function of design, standards determination, specification development, and selection of suppliers, financing, contract administration, and other related functions.

Project evaluation and review technique (PERT) chart Scheduling tool which shows in flowchart format the interdependencies between project activities.

Profit-centered maintenance (PCM) A value adding business approach to the leadership and management of maintenance and physical asset management Simply stated it asks the question: "If I owned this maintenance operation as a business to make a profit, what would I do differently?" On a broader scope it is the application of world-class maintenance practices, attitudes, and leadership principles. When applied, it makes an in-house maintenance operation equivalent to a profit center with both a financial system and performance measurement process in place to validate results.

Profit- and customer-centered maintenance (PCCM) On a broader scope it combines the philosophies of profit centered with customer centered into management and leadership of all types of maintenance processes. It is the application of maintenance best practices, attitudes, and leadership principles to both profit and maintenance customer service. When applied, it makes an in-house maintenance operation equivalent to a profit center when both a financial system and performance measurement process is in place to validate results.

Protective device Devices and assets intended to eliminate or reduce the consequences of equipment failure. Some examples include standby plant and equipment, emergency systems, safety valves, alarms, trip devices, and guards.

Provisioning Process of determining and selecting the varieties and quantities of repair parts, spares, special tools, and test and support equipment that should be procured and stocked to sustain and maintain equipment for specified periods of time. It includes identification of items of supply, establishing data for catalogs, technical manuals and allowance lists, and providing instructions and schedules for delivery of provisioned items.

PSA See Probabilistic safety assessment.

Purchase requisition The prime document raised by user departments authorizing the purchase of specific materials, parts, supplies, equipment, or services from external suppliers.

Purchase order The prime document created by an organization, and issued to an external supplier, ordering specific materials, parts, supplies, equipment, or services.

Q

Quality rate Used in the calculation of overall equipment effectiveness. The proportion of the output from a machine or process which meets required product quality standards. Normally specified as a percentage.

R

RCM See Reliability centered maintenance.

Reaction time/response time The time required between the receipt of an order or impulse triggering some action and the initiation of the action.

Ready line Used in relation to mobile equipment. Equipment which is available, but not being utilized is typically parked on the ready line. This term is used interchangeably with go-line.

Rebuild Restore an item to an acceptable condition in accordance with the original design specifications.

Rebuild/recondition Total teardown and reassembly of a product, usually to the latest configuration.

Redesign A term which, in reliability centered maintenance, means any one-off intervention to enhance the capability of a piece of equipment, a job procedure, a management system, or people's skills.

Redundancy Two or more parts, components, or systems joined functionally so that if one fails, some or all of the remaining components are capable of continuing with function accomplishment; fail-safe backup.

Refurbish Clean and replace worn parts on a selective basis to make the product usable to a customer. Less involved than rebuild.

Reliability The probability that an item will perform its intended function without failure for a specified time period under specified conditions. The ability of an item to perform a required function under stated conditions for a stated period of time is usually expressed as the mean time between failures. Normally measured by mean time between failures.

Reliability analysis The process of identifying maintenance of significant items and classifying them with respect to malfunction on safety environmental, operational, and economic consequences. Possible failure mode of an item is identified and an appropriate maintenance policy is assigned to counter it. Subsets are *failure mode, effect, and criticality analysis* (FMECA), *fault tree analysis* (FTA), risk analysis, and HAZOP (hazardous operations) analysis.

Reliability centered maintenance (RCM) Optimizing maintenance intervention and tactics to meet predetermined reliability goals. A structured process, originally developed in the airline industry, but now commonly used in all industries to determine the equipment maintenance strategies required for any physical asset to ensure that it continues to fulfill its intended functions in its present operating context. A number of books have been written on the subject.

Reliability engineering A staff function whose prime responsibility is to ensure that maintenance techniques are effective, that equipment is designed and modified to improve maintainability, that ongoing maintenance technical problems are investigated, and appropriate corrective and improvement actions are taken. Used interchangeably with plant engineering and maintenance engineering.

Reorder point (ROP) Minimum quantity, established by economic calculation and management direction, which triggers the ordering of more items.

Repair To restore an item to an acceptable condition by the renewal, replacement, or mending of worn or damaged parts. Restoration or replacement of parts or components as necessitated by wear, tear, damage, or failure; to return the facility, equipment, or part to efficient operating condition; any activity which returns the capability of an asset that has failed to a level of performance equal to, or greater than, that specified by its Functions, but not greater than its original maximum capability. An activity which increases the maximum capability of an asset is a modification.

Repair parts Individual parts or assemblies required for the maintenance or repair of equipment, systems, or spares. Such repair parts may also be repairable or nonrepairable assemblies, or one-piece items. Consumable supplies used in maintenance or repair such as wiping rags, solvents, and lubricants, are not considered repair parts. Repair parts are also service parts.

Repairable Parts or items that are technically and economically repairable. A repairable part, upon becoming defective, is subject to return to the repair point for repair action.

Replaceable item Hardware that is functionally interchangeable with another item but differs physically from the original part to the extent that installation of the replacement requires such operations as drilling, reaming, cutting, filling, or shimming in addition to normal attachment or installation operations.

Restoration Any activity which returns the capability of an asset that has not failed to a level of performance equal to, or greater than, that specified by its functions, but not greater than its original maximum capability. Not to be confused with a modification or a repair.

Return on assets An accounting term. Let us not get into a lengthy discussion of the relative merits of various accounting standards, how assets should be valued (book value, replacement value, depreciation rates, and methods), and differences between tangible and intangible assets. This is the stuff that accountants have wet dreams over, but not maintenance engineers. In practical terms, as it impacts on maintenance, return on assets is the profit attributable to a particular plant or factory, divided by the amount of money invested in plant and equipment at that plant or factory. It is normally expressed as a percentage. As such, it is roughly equivalent (in principle—please excuse the pun!) to the interest rate that you get on money invested in the bank, except that in this case the money is invested in plant and equipment.

RIME A maintenance priority methodology entitled the "ranking index for maintenance expenditures." Provides method to include a ranking of equipment/asset criticality combined with the repair work classification ranking to produce a priority index value.

Risk The potential for the realization of the unwanted, negative consequences of an event. The product of conditional probability of an event, and the event outcomes.

Risk management The strategy that an organization deploys to identify, prioritize, manage and implement plans to lessen or eliminate risks of all types.

Rotable A term often used in the maintenance of heavy mobile equipment. A rotable component is one which, when it has failed, or is about to fail, is removed from the asset and a replacement component is installed in its place. The component that has been removed is then repaired or restored and placed back in the maintenance store or warehouse, ready for reissue.

RPI Reliability performance indicators or key metrics that relate to the measurement of asset reliability. Examples include:

- *Maximum corrective time* (MCT) and *maximum preventive time* (MPT). The most time expected for maintenance, usually specified at 95 percent confidence level.

- *Mean active maintenance time* (MAMT). Weighed average of mean corrective time and mean preventive time, but excluding administrative and logistics support time.
- *Mean downtime* (MDT). Average time a system cannot perform its mission; including response time, active maintenance, supply time, and administrative and logistics support time.
- *Mean time between failure* (MTBF). The average time/distance/events a product or equipment process delivers between breakdowns.
- *Mean time between maintenance* (MTBM). The average time between corrective and preventive actions.
- *Mean time to repair* (MTTR). The average time it takes to fix a failed item.

Routine maintenance task Any maintenance task that is performed at a regular, predefined interval.

Running maintenance Maintenance that can be done while the asset is in service.

Run-to-failure No scheduled maintenance—an equipment maintenance strategy, where no routine maintenance tasks are performed on the equipment. The only maintenance performed on the equipment is corrective maintenance, and then only after the equipment has suffered a failure. Also described as a no-scheduled maintenance strategy.

S

Safety consequences A failure has safety consequences if it causes a loss of function or other damage that could hurt or kill someone.

Safety stock Quantity of an item, in addition to the normal level of supply, required to be on, had to permit continuing operation with a specific level of confidence if the supply is late or demand suddenly increases.

Salvage The saving of reuse of condemned, discarded, or abandoned property, and of materials contained therein for reuse or scrapping. As a noun, it refers to property that has some value in excess of its basic material content, but is in such condition that it has no reasonable prospect of original use, and its repair or rehabilitation is clearly not practical.

Schedule compliance The number of scheduled jobs actually accomplished during the period covered by an approved schedule; also the number of scheduled labor hours actually worked against a planned number of scheduled labor hours, expressed as percentage; one of the key performance indicators often used to monitor and control maintenance. It is defined as the number of scheduled work orders completed in a given time period (normally one week), divided by the total number of scheduled work orders that should have been completed during that period, according to the approved maintenance schedule for that period. It is normally expressed as a percentage, and will always be less than or equal to 100 percent. The closer to 100 percent, the better the performance for that time period.

Scheduled discard task Replacement of an item at a fixed, predetermined interval, regardless of its current condition; a maintenance task to replace a component with a new component at a specified, predetermined frequency, regardless of the condition of the component at the time of its replacement. An example would be the routine replacement of the oil filter on a motor vehicle every 6000 miles. The frequency with which a scheduled discard task should be performed is determined by the useful life of the component.

Scheduled maintenance (SM) Preplanned actions performed to keep an item in specified operating condition by means of systematic inspection, detection, and prevention of incipient failure. Sometimes called preventive maintenance, but actually a subset of PM.

Scheduled operating time The time during which an asset is scheduled to be operating, according to a long-term production schedule.

Scheduled restoration task A maintenance task to restore a component at a specified, predetermined frequency, regardless of the condition of the component at the time of its replacement. An example would be the routine overhaul of a slurry pump every 1000 operating hours. The frequency with which a scheduled restoration task should be performed is determined by the useful life of the component.

Scheduled work order A work order that has been planned and included on an approved maintenance schedule.

Scoping A planning activity which outlines the extent/scope and detail of work to be done, and defines the resources needed.

Scoreboard for excellence Baseline for today's most comprehensive benchmarking guides for maintenance operations. Developed initially in 1981 and enhanced into its present format of five different "Scoreboard for Excellence" versions. See Scoreboard for facilities management excellence and Scoreboard for maintenance excellence for manufacturing operations in Appendix A.

Scoreboard for facilities management excellence Today's most comprehensive benchmarking guide for facilities management and maintenance operations. Developed along the same format as the "Scoreboard for Maintenance Excellence," the new millennium version includes 27 evaluation (best practice) categories and 300 evaluation categories. An excellent benchmarking guide for physical asset management operations within large physical plant and facilities complexes such as universities, state and municipal building complexes, healthcare facilities, secondary school complexes, and retail organizations with nationwide system of company-owned retail sites. Provides an essential benchmarking guide where results become an important external benchmark against recognized best practices and also the user's baseline for continuous reliability improvement.

Scoreboard for maintenance excellence Today's most comprehensive benchmarking guide for plant maintenance operations. Developed initially in 1982 and enhanced into its present new millennium format of 27 evaluation (best practice) categories and 300 evaluation categories. Used by over 4000 organizations and for benchmarking all types of maintenance operations. Provides an essential benchmarking guide where results become an important external benchmark against recognized best practices and also the user's baseline for continuous reliability improvement.

Secondary damage Any additional damage to equipment, above and beyond the initial failure mode, that occurs as a direct consequence of the initial failure mode.

Secondary failures Malfunctions that are caused by failures of another item.

Secondary function A term used in reliability centered maintenance. The secondary functionality required of an asset-generally not associated with the reason for acquiring the asset, but now that the asset has been acquired, the asset is now required to provide this functionality. For example a secondary function of a pump may be to ensure that all of the liquid that is pumped is contained within the pump (i.e., the pump does not leak). An asset may have tens or hundreds of secondary functions associated with it.

Serial number Number or letters that uniquely identify an item.

Service contract Contract calling directly for a contractor's time and effort rather than for a specific end product.

Service level Frequency usually expressed as a percentage, with which a repair part demand can be filled through a particular service stock echelon. A 95 percent level of service means that 95 out of 100 demands are properly issued. If viewed from the end-customer service technician perspective, the service level is the percent of parts received out of those requested, from all levels of the support system.

Serviceability Characteristics of an item, equipment, or system that make it easy to maintain after it is put into operation. Similar to maintainability.

Servicing The replenishment of consumables needed to keep an item in operating condition.

Shelf life The period of time during which an item can remain unused in proper storage without significant deterioration.

Shop stock Self-service items such as Class C SKUs such as nuts, bolts, and fitting that are stored directly in the shop work area. May be on consignment directly from the vendor or vendor may inventory and "keep full" as needed without significant paperwork requirement.

Shutdown That period of time when equipment is out of service.

Shutdown maintenance Maintenance done while the asset is out of service, as in the annual plant shutdown.

SKU Stock keeping unit—a warehouse inventory management term for individual stock items carried in inventory.

Specifications Physical, chemical, or performance characteristics of equipment, parts, or work required to meet minimum acceptable standards.

Standard item Part, component, material, subassembly, assembly, or equipment that is identified or described accurately by a standard document or drawing.

Standardization Process of establishing the greatest practical uniformity of items and of practices to assure the minimum feasible variety of such items and practices to optimum interchangeability.

Standard job A work order stored in the CMMS which contains all the necessary information required to perform a maintenance task and can be used as a possible template for other similar types of work. (See also Model work order.)

Standby Assets installed or available but not in use.

Standing work order A work order that is left open either indefinitely or for a predetermined period of time for the purpose of collecting labor hours, costs, and/or history for tasks for which it has been decided that individual work orders should not be raised. Examples would include standing work orders raised to collect time spent at safety meetings, or in general housekeeping activities; a work order that remains open, usually for the annual budget cycle, to accommodate information, small jobs, or for specific tasks.

Stock number Number assigned by the stocking organization to each group of articles or material, which are then treated as if identical within the using supply system also called part number, item number, or part identifier.

Stock out Indicates that all quantities of a part normally on hand have been used, so that the items are not presently available. Demand for a nonstock part is usually treated as a separate situation.

Stores issue The issue and/or delivery of parts and materials from the store or warehouse.

Stores requisition The prime document raised by user departments authorizing the issue of specific materials, parts, supplies, or equipment from the store or warehouse.

Supply Procurement, storage, and distribution of items.

Support equipment Items required to maintain systems in effective operating condition under various environments. Support equipment includes general and special-purpose vehicles, power units, stands, test equipment, tools, or test benches needed to facilitate or sustain maintenance action, to detect or diagnose malfunctions, or to monitor the operational status of equipment and systems.

T

Technical data and information Includes, but is not limited to, production and engineering data, prints and drawings, documents such as standards, specifications, technical manuals, changes in modifications, inspection and testing procedures, and performance and failure data.

Terotechnology An integration of management, financial, engineering, operating maintenance, and other practices applied to physical assets in pursuit of an economical life cycle; the application of managerial, financial, engineering, and other skills to extend the operational life of, and increase the efficiency of, equipment and machinery.

Test and support equipment All special tools and checkout equipment, metrology and calibrations equipment, maintenance stands, and handling equipment required for maintenance. Includes external and *built-in test equipment* (BITE) considered part of the supported system or equipment.

Thermography The process of monitoring the condition of equipment through the measurement and analysis of heat. Typically conducted through the use of infrared cameras and associated software. Commonly used for monitoring the condition of high-voltage insulators and electrical connections, as well as for monitoring the condition of refractory in furnaces and boilers, among other applications.

Throwaway maintenance Maintenance performed by discarding used parts rather than attempting to repair them.

Total asset management An integrated approach (yet to be developed!) to asset management which incorporates elements such as reliability centered maintenance, total productive maintenance, design for maintainability, design for reliability, value engineering, life-cycle costing, probabalistic risk assessment, and others to arrive at the optimum cost-benefit-risk asset solution to meet any given production requirements.

Total productive maintenance (TPM) A maintenance strategy first used in the USA for equipment maintenance management program that emphasizes operator involvement in equipment maintenance, continuous improvement in equipment productivity, and measurement of *overall equipment effectiveness* (OEE).

TPM See Total productive maintenance.

Tradesperson Alternative to craftsperson. A skilled maintenance worker who has been formally trained through an apprenticeship program.

Teratology The process of monitoring the condition of equipment through the analysis of properties of its lubricating and other oils. Typically conducted through the measurement of particulates in the oil, or the measurement of the chemical composition of the oil (spectrographic oil analysis). Commonly used for monitoring the condition of large gearboxes, engines, and transformers among other applications.

ToSS See Total system support.

Total system support (ToSS) The composite of all considerations needed to assure the effective and economical support of a system throughout its programmed life cycle.

Troubleshooting Locating or isolating and identifying discrepancies or malfunctions of equipment and determining the corrective action required. Should be measured/planned and considered as part of wrenchtime.

Turnaround time Interval between the time a repairable item is removed from use and the time it is again available in full serviceable condition.

Turnover Measurement on either numbers of parts or on monetary value that evaluates how often a part is demanded versus the average number kept in inventory. For example, if two widgets are kept in inventory and eight are used each year, then the turnover is 8/2 = 4X per year. In monetary terms turnover is cost of inventory sold/average cost of inventory carried.

U

Unplanned maintenance Maintenance done without planning or scheduling could be related to a breakdown, running repair, or corrective work: any maintenance activity for which a predetermined job procedure has not been documented, or for which all labor, materials, tools, and equipment required to carry out the task have been not been estimated, and their availability assured before commencement of the task.

Unscheduled maintenance (UM) *Emergency maintenance* (EM) or *corrective maintenance* (CM) to restore a failed item to usable condition.

Up In a condition suitable for use.

Uptime It is defined as being the time that an item of equipment is in service and operating.

Usage Quantity of items consumed or necessary for product support. Usage is generally greater than the technical failure rate.

Useful life The maximum length of time that a component can be left in service, before it will start to experience a rapidly increasing probability of failure. The useful life determines the frequency with which a scheduled restoration or a scheduled discard task should be performed.

Utilization The proportion of available time that an item of equipment is operating. Calculated by dividing equipment operating hours by equipment available hours. Generally expressed as a percentage.

V

Value engineering A systematic approach to assessing and analyzing the user's requirements of a new asset, and ensuring that those requirements are met, but not exceeded. Consists primarily of eliminating perceived "non-value-adding" features of new equipment.

Variance analysis Interpretation of the causes for a difference between actual and some norm, budget, or estimate.

Vibration analysis The process of monitoring the condition of equipment, and the diagnosis of faults in equipment through the measurement and analysis of vibration within that equipment. Typically, conducted through hand-held or permanently positioned accelerometers placed on key measurement points on the equipment. Commonly used on most large items of rotating equipment such as turbines, centrifugal pumps, motors, and gearboxes.

W

Warranty Guarantee that an item will perform as specified for at least a specified time, or will be repaired or replaced at no cost to the user.

Wear-out Deterioration as a result of age, corrosion, temperature, or friction that generally increases the failure rate over time.

Workload The number of labor hours needed to carry out a maintenance program, including all scheduled and unscheduled work and maintenance support of project work.

Work order (WO) A unique control document that comprehensively describes the job to be done; may include formal requisition for maintenance, authorization, and charge codes, as well as what actually was done. The prime document used by the maintenance function to manage maintenance tasks. It may include such information as a description of the work required, the task priority, the job procedure to be followed, the parts, materials, tools and equipment required to complete the job, the labor hours, costs and materials consumed in completing the task, as well as key information on failure causes, and what work was performed.

Work request The initial request for maintenance service or work usually as a statement of the problem. The work request provides the preliminary information for creation of the work order. Depending on the cost and scope of a work request an approval process may be required before the work order is created, planned, and scheduled. The prime document raised by user departments requesting the initiation of a maintenance task. This is usually converted to a work order after the work request has been authorized for completion.

Appendix F

Maintenance Excellence Strategy Team Charter Example

XYZ Company Leadership—Preliminary Review Date: ___________
XYZ Company Leadership—Final Approval Date: __________
XYZ Company Maintenance Excellence Strategy Team Review Date: ________
XYZ Company Maintenance Excellence Strategy Team Acceptance Date: _______

I. OPPORTUNITY: (*What is the reason this team exists?*)

To provide technical leadership, direction, and support to maintenance best practice development and *computerized maintenance management system* (CMMS) implementation at all XYZ Company sites. To support each plant in conducting a "Scoreboard for Maintenance Excellence" assessment and developing a prioritized plan of action for improvements. To ensure that the strategic maintenance plans for each site are integrated with the XYZ Company business planning process. To ensure that both internal and external resources for implementation of improvements are allocated.

- Serve as the XYZ Company facilitation process for sharing of maintenance best practices across all current and future sites.
- Serve as the steering team for CMMS implementation throughout current and future XYZ Company sites.
- Provide approval and support allocation of internal and external resources to implement strategic maintenance plans developed for each site.
- Support a corporatewide culture change at the operator/maintenance craft level for the process of operator-based maintenance.

- Support evolution toward a "pride in ownership" philosophy for care and operation of physical assets at each plant.

II. PROCESS: (*What are the steps to be followed, and what are the questions to be answered by this team?*)

a. Team orientation and charter review.
b. Ensure that XYZ Company leadership is fully committed to the maintenance strategy now being developed, understands team objectives, and agrees on what needs to be achieved for maintenance excellence throughout XYZ Company.
c. Understand current constraints to existing information systems and business systems.
d. Understand current recommendations for maintenance/engineering improvements presented in the initial maintenance assessments.
e. Supports finalization of CMMS decision, develop the recommended XYZ Company CMMS strategy, and define support for CMMS path forward at all plants.
f. Present or support presentations to XYZ Company leadership with costs/benefits for best practice implementations as they evolve.
g. Review, approve, and support implementations of best practices as they evolve in all plants.
h. Ensure implementation of the CMMS benchmarking system for all XYZ Company plants to evaluate long-term success of CMMS investments.
i. Develop and approve charters for site level CMMS implementation teams and any teams for PM/PdM review, equipment numbering, parts numbering, and inventory database review.
j. Provide consensus on companywide metrics to be used to validate implementation results via the XYZ Company *Maintenance Excellence Index* (MEI).
k. Ensure that all plant-level teams are established and that necessary resources are allocated to meet the CMMS implementation schedule.

III. EVIDENCE OF SUCCESS: (*What results are expected, in what time frames, for this team to be successful?*)

a. The XYZ Company MEI is to be fully implemented by ______.
b. The new CMMS is installed and "goes live" by ________.
c. The XYZ Company CMMS benchmarking system is established to evaluate progress toward a "Class A" status.
d. Positive trends on all XYZ Company MEI metrics are being achieved (baseline metric values compared to performance goals) three months after the actual MEI implementation date.

e. Projected savings and benefits are being achieved based on the savings time line for each plant.

IV. RESOURCES: (*Who are the team members, team leader, team facilitator; who will support the team if needed; how much time should be spent both in meetings and outside of meetings?*)

a. This team will be a cross-functional team and have functional representation from the following areas:
- VP of operations: team leader
- Engineering representative
- Corporate finance/accounting representative
- Corporate information services representative
- Maintenance manager representative
- Maintenance supervisor representative
- Crafts representatives
- Human resource representative
- Team resources: consulting staff

b. The kick-off meeting for the XYZ Company maintenance excellence strategy team will be for approximately 1 h at (time) ______on (date)______ in (location)______. The team shall meet as frequently as defined by the team and the team leader with maximum use of telephone conferencing for both XYZ Company and consulting staff. This team's activities and success will be considered as part of each team member's job.

c. The consulting project team will provide support for team orientation; facilitation of charter approval and any JIT training of this team as it is required in the future.

V. CONSTRAINTS: (*What authority does the team have, what items are outside the scope of the team, and what budget does the team have?*)

a. Recommendations are to be approved by XYZ Company leadership and all capital requests will follow current XYZ Company procedures.

b. This team must obtain buy-in from XYZ Company plant operations to overcome a reactive maintenance strategy and change to a proactive, continuous reliability-based strategy for maintenance excellence. Also buy-in to greater accountability for assets used by operations and greater "pride in ownership" by equipment operators.

c. Travel costs to be accounted for and incurred by the respective plants and corporate team members.

VI. EXPECTATIONS: (*What are the outputs, when are they expected, and to whom should they be given?*)

a. Support and enable implementation of the strategic maintenance plans at each plant and ensure that maintenance plan of actions are fully integrated with the XYZ Company business plan.
b. Evaluate the CMMS strategy being developed by each plant, determine best path forward, support presentation of CMMS team recommendations to XYZ Company leaders for resource funding.
c. Support allocation of in-house resources to implement CMMS related teams at each plant.
d. Support and enable well-defined CMMS implementation plan of actions at each plant.
e. Ensure implementation of the XYZ Company CMMS benchmarking system as a long-term metric to ensure full CMMS utilization and to validate benefits of the CMMS investment.
f. Provide final review approval and ensure that the XYZ Company MEI is implemented and remains a long-term measure of maintenance's contribution to XYZ Company's total operations success and increased profit.
g. Ensure positive interaction between maintenance/engineering and corporate information services for full integration of CMMS with current and/or future business systems.
h. Minutes are to be completed for all team meetings, sent to XYZ Company leadership as required.
i. Present team findings, recommendations, and results to XYZ Company leadership as required.
j. Provide information to corporate communications staff as required.

Appendix

G

ACE Team Benchmarking Team Charter Example

Charter For The ACE Team At ——————
Maintenance Strategy Team—Preliminary Review Date ——————
Maintenance Strategy Team—Final Review ——————
ACE Team Review Date ——————
ACE Team Acceptance Date ——————

I. OPPORTUNITY: (*What is the reason this team exists?*)

The ACE team exists to provide a well-qualified team of experienced craftspeople, technicians, and supervisors to establish benchmark repairs jobs and work content time for these jobs. The ACE team is chartered to help develop the ACE system for establishing maintenance performance standards at each XYZ site.

II. PROCESS: (*What are the steps to be followed, and what are the questions to be answered by this team?*)

a. Orientation, charter review, and charter acceptance or modification.
b. Ensure that all team members understand team objectives and agree on what needs to be achieved and the criticality of this initiative to the planning and scheduling process.
c. Understand the current concepts of the ACE System as defined in the XYZ Company maintenance planning and scheduling SOP.
d. Understand the basics of the new XYZ system for CMMS/EAM, the characteristics, functionality, and performance.
e. Determine critical repair jobs that should be used as representative benchmark jobs; define key steps and elements for each benchmark job; and define any special tools, safety requirements, and other special requirements for the job.

f. Determine ways to improve doing the jobs being analyzed as benchmark jobs considering better tools, equipment, skills, and even better PM/PdM techniques to avoid this type failure of problem.
g. Conduct the ACE team process as outlined in the 10-step approach from the SOP.
h. Develop a team consensus on work content times for all the benchmark jobs selected.
i. Continuously improve the ACE team process within the XYZ organization as an element of our *continuous reliability improvement* (CRI) efforts.

III. EVIDENCE OF SUCCESS: (*What results are expected, in what periods, for this team to be successful?*)

a. A sufficient number of benchmark jobs will be developed as to individual tasks and steps along with estimated work content times to complete the site's ACE team spreadsheets.
b. The actual period to complete the initial spreadsheets will depend on time allocated by the ACE team at each site.
c. ACE teams from one site are expected to share their results with the other sites. Due to the similar nature of equipment the sharing of benchmark job write-ups and even work content times that ACE teams have developed can be shared throughout the operation.
d. Overall success will be determined by each planner having adequate spreadsheets that cover all construct areas as well as types of crafts work (mechanical, electrical, and so on) so that planning times can eventually be established for 80 percent or more of the available craft hours.

IV. RESOURCES: (*Who are the team members, team leaders, team facilitators, who will support the team if needed; how much time should be spent both in meetings and outside of meetings?*)

The ACE team should consist of the following representatives:

- One maintenance planner: team leader
- One maintenance supervisor
- Two to three crafts representatives from Area 1
- Two to three crafts representatives from Area 2
- Two to three craft representative from Area 3

Note: Crafts representatives should rotate periodically and sufficient numbers designated so as to have at least two to three representatives from each craft areas when benchmark jobs from these areas are being reviewed for job steps and estimated for work content time.

An initial ACE team meeting will be for 1 h on _______. The team shall meet initially for at least 3 h each week. This team's activities and success will be considered as part of each team member's job.

V. CONSTRAINTS: (*What authority does the team have, what items are outside the scope of the team, and what budget does the team have?*)

a. No changes to organization structure are anticipated.
b. Benchmark job plans are to be reviewed and approved by the maintenance manager.
c. Each team must obtain buy-in and overcome concerns from the other crafts on their estimates for benchmark jobs and repair methods recommended for each benchmark job.
d. Team presents implementation status reports as required and any additional recommendations to the XYZ maintenance excellence strategy team.
e. The ACE team has the authority to recommend new and improved repair methods, new tools to help craft productivity and safety, and to recommend other improvements to improve asset reliability as developed during the ACE team process.

VI. EXPECTATIONS: (*What are the outputs, when are they expected, and to whom should they be given?*)

a. Spreadsheets for the site that cover all crafts areas and construct types are completed by__________.
b. Reliable planning times are provided for benchmark jobs so that effective planning, performance measurement, backlog control, and level of PM work can be established with a high level of confidence.
c. ACE team provides a steady source of continuous improvement ideas to make repair jobs safer and easier.
d. Minutes are to be completed for all team meetings and sent to the maintenance manager and the XYZ maintenance excellence strategy team.

Appendix

H

Ace Team Forms

<table>
<tr><td colspan="7">ACE Team Benchmark Job Analysis
Benchmark Job Description</td></tr>
<tr><td colspan="3"></td><td colspan="4">Benchmark job no:</td></tr>
<tr><td colspan="3"></td><td colspan="4">Craft</td></tr>
<tr><td colspan="3"></td><td colspan="4">Ref. drawing:</td></tr>
<tr><td colspan="3"></td><td colspan="4">No. of crafts:</td></tr>
<tr><td colspan="3"></td><td colspan="4">Analyst:</td></tr>
<tr><td colspan="3"></td><td colspan="4">Date:</td></tr>
<tr><td>Line No.</td><td>No. of Crafts</td><td>Operation Description</td><td>Ref. Code</td><td>Unit Time</td><td>Freq</td><td>Total Time</td></tr>
<tr><td colspan="3">Notes</td><td colspan="4">Benchmark time for work content</td></tr>
<tr><td colspan="3"></td><td colspan="4">Standard work group</td></tr>
</table>

ACE Team Benchmatking System: Work Groups and Time Ranges

Work Groups	Ranges From (h)	Benchmark Time (Slot Time)	Up To (h)
A	0.0	0.1	0.15
B	0.15	0.2	0.25
C	0.25	0.4	0.5
D	0.5	0.7	0.9
E	0.9	1.2	1.5
F	1.5	2.0	2.5
G	2.5	3.0	3.5
H	3.5	4.0	4.5
I	4.5	5.0	5.5
J	5.5	6.0	6.5
K	6.5	7.3	8.0
L	8.0	9.0	10.0
M	10.0	11.0	12.0
N	12.0	13.0	14.0
O	14.0	15.0	16.0
P	16.0	17.0	18.0
Q	18.0	19.0	20.0
R	20.0	22.0	24.0
S	24.0	26.0	28.0
T	28.0	30.0	32.0

ACE Team Spreadsheet for Work Groups A, B, C, and D Craft: ______________ Code: ________________			
Task Area			
Group A: 0.1 h (0.0+) (0.15)	Group B: 0.2 h (0.15+) (0.25)	Group C: 0.4 h (0.25+) (0.5)	Group D: 0.7 h (0.5+) (0.9)

ACE Team Spreadsheet for Work Groups E, F, G, and H			
Craft: ____________ Code: ______________			
Task Area			
Group E: 1.2 h (0.9+) (1.5)	Group F: 2.0 h (1.5+) (2.5)	Group G: 3.0 h (2.5+) (3.5)	Group H: 4.0 (3.5+) (4.5)

ACE Team Spreadsheet for Work Groups I, J, K, and L			
Craft: ______________ Code: ________________			
Task Area			
Group I: 5.0 h (4.5+) (5.5)	Group J: 6.0 h (5.5+) (6.5)	Group K: 7.3 h (6.5+) (8.0)	Group L: 9.0 h (8.0+) (10.0)

ACE Team Spreadsheet for Work Groups M, N, O, and P			
Craft: ______________ Code: ________________			
Task Area			
Group M: 11.0 h (10.0+) (12.0)	Group N: 13.0 h (12.0+) (14.0)	Group O: 15.0 h (14.0+) (16.0)	Group P: 17.0 h (16.0+) (18.0)

ACE Team Spreadsheet for Work Groups Q, R, S, and T			
Craft: ______________ Code: ________________			
Task Area			
Group Q: 19.0 h (18.0+) (20.0)	Group R: 22.0 h (20.0+) (24.0)	Group S: 26.0 h (24.0+) (28.0)	Group T: 30.0 h (28.0+) (32.0)

Appendix

I

OEE and Maintenance 5S Forms

5 S Levels of Maintenance Excellence

Maintenance leader ____________ Shop __________
Date _________

Level No.	Sorting	Date Achieved
1.	Necessary and unnecessary items are mixed together in the work area.	
2.	Necessary and unnecessary items separated (includes excess inventory).	
3.	Unnecessary items have been removed from work area.	
4.	A method has been established to maintain the work area free of unnecessary items.	
5.	Employees continuously seeking improvement opportunities.	

Level No.	Simplifying	Date Achieved
1.	Shop tools supplies and materials randomly located.	
2.	A designated location has been established for all items.	
3.	Designated locations are marked to make organization more visible.	
4.	A method has been established to recognize with a visual sweep if items are out of place or exceed quantity limits.	
5.	A method has been developed to provide continual evaluation, and a process is in place to implement improvements.	

Level No.	Sweeping	Date Achieved
1.	Shop/office and special tools and equipment are dirty and disorganized.	
2.	Work/break areas are cleaned on a regularly scheduled basis.	
3.	Shop, break areas, special tools, and equipment are cleaned daily.	
4.	Housekeeping tasks are understood and practiced continually.	
5.	Area employees have devised method of preventive shop cleaning and maintenance.	

Level No.	Standardizing	Date Achieved
1.	No attempt is being made to document or improve current process.	
2.	Methods are being improved but changes have not been documented.	
3.	Changes are being incorporated and documented.	
4.	Information on process improvements and reliable methods is shared.	
5.	Employees are continually seeking the elimination of waste with changes documented and information shared.	

Level No.	Self-Discipline	Date Achieved
1.	Minimal attention spent on housekeeping and safety.	
2.	A recognizable effort has been made to improve the condition of the work/shop area.	
3.	Housekeeping and safety practices have been developed and are utilized.	
4.	Follow through of housekeeping and safety practices is evident.	
5.	There is a general appearance of a confident understanding of, and adherence to the 5 S program.	

Overall Equipment Effectiveness (OEE) Form

Machine:____________________ Date:__________

Product:_______________ Shift:__________

OEE %

Asset Availability

A. Total available time:	If 8-h shift = 480 min, if 10-h shift = 600 min		Min
B. Planned downtime:	Time for meetings, cleaning, breaks, PM, set-ups		Min
C. Run time:	Total available time – planned downtime	A – B	Min
D. Unplanned downtime	a. Breakdown minutes b. Changeover minutes c. Minor stoppages	a + b + c	Min
E. Net operating time:	Run time – unplanned downtime	C – D	Min
F. Asset availability	Net operating time ÷ run time × 100	E ÷ C × 100	%

Asset Performance

G. Processed amount	(All units run per shift: good and bad)		Total Units
H. Design cycle time:	Total minutes required to process each production unit per machine design speed		Min/ Unit
I. Asset performance:	(Design cycle time × processed amount) ÷ net operating time	[(H × G) ÷ E] × 100	%

Asset Quality Output		
J. Total rejects or off quality	(machine-related defects)	Units
K. Asset quality output:	Processed amount –total rejects ÷ processed amount [(G – J) ÷ G] × 100	%
Overall Craft Effectiveness		
L. Overall equipment effectiveness	(Availability % × performance % × quality %) × 100 = OEE%	%

Appendix

J

Customer Service Pocket Pal Example

This appendix provides a proven means to simplify craft time reporting to work orders and to capture nonwrench time in a quick and easy method by the crafts workforce. It also provides a process:

1. To report failure codes
2. To report parts and materials used
3. To report nonwrench time and provide capability to measure craft utilization and productivity
4. To report work accomplished as it occurs daily for input to CMMS by crafts or by clerical staff
5. To enable direct data entry by crafts using handheld PDA or to data entry device

How To Use The Customer Service Pocket Pal

The Pocket Pal serves many important purposes and can be used to make notes of work being performed and report of all work performed against an established work order or job order. It can then be entered from your Pocket Pal notes to the system by the controllers.

Note: The "How to Use The Customer Service Pocket Pal" instructions are included following a preview of both the front and back of this time- and material-accountability form.

Customer Service Pocket Pal (Front Side)						
Crafts Name:____________ Craft ID # _________ Date_______ Shift___ Shop_____						
W.O No.	Asset No.	Brief Description of Work Performed or Service Provided to Customer	Start Time	Stop Time	Total Time	√ if Parts Used
Nonwrench time codes: WP-WE-WO-TT-TS			Total time reported today on pocket pal →→→→→			

Customer Service Pocket Pal Example: (Back Side)			
Use This Back Side For Your Notes on Parts, Comments on Failures and Other Notes or Comments		Turn in at End of Your Work Day	
W.O. NO. or Asset No.	Failure Code	Part Number and Quantity Needed for Repair Work	IF PART USED √ OFF
Important repair notes:			
YES √ I returned all unused parts to storage area: Name_________Date_________			

Use the Pocket Pal form each day to report all work. There may not be a work order established before some jobs are started.

This may include areas such as:

- Any type of on-call support to operations where you providing coverage.
- Operations and services type of work is recorded on the pocket pal.

- Reporting time to all emergency jobs when called off of a planned job.
- Reporting time to small convenience jobs when called off a planned job.
- Service covered by a standing work order.
- Service in keeping areas clean and safe.
- Reporting of craft training time to a standing work order.
- Reporting of meeting time to a standing work order.

Nonwrench time

The Pocket Pal will be used for the reporting of all nonwrench time over 15 min (or 0.25 h) to an established work order. This is very important. All delays that keep you from completing a planned job within the estimated time should be reported. They should be noted on the Pocket Pal for the work order it applies to. The codes for reporting of nonwrench time of over 0.25 h (15 min) are:

a. WP—*waiting parts* includes searching for parts and parts information.

b. WE—*waiting equipment* includes waiting for equipment and waiting for customer to release area for work.

c. WO—*waiting other* to include instructions, getting prints, manuals necessary for doing your job.

d. TT—*travel time* to and from job site, storeroom, and all travel associated with a work order.

e. TS—*troubleshooting*—this is a very important time to define the problem, scope of work, and parts/materials required.

Additional work

The Pocket Pal will also be used to report extra work above the planned scope of work for your job order. If you have been authorized to perform work over and above the planned scope of work, note this when reporting your time to the assigned workorder.

a. *For example.* You have a planned PM for a generator. You are performing the PM as scheduled and find a need for another repair that should be done. You have the necessary parts and are given the okay to go ahead and do the extra work. The extra work does not delay the schedule.

b. *Next steps.* You then enter the work order no. or asset no. on your Pocket Pal, a brief description of work performed, start/stop time, and total time for the extra work. You will then check (√) that parts were used and continue to complete the planned PM under its assigned WO number.

c. *But remember this*! You must also issue the parts used to a PM job order or to the asset no. where parts were used for additional work beyond the PM work.

d. *If parts are not available for the additional work.* If parts were not available, you should notify your supervisor and create a work request to have the job scheduled later when parts are available. You should also create a parts request so that the necessary parts can be ordered and the work scheduled at a later time.

e *You are the key to a successful accountability and accuracy of authorized shop stock/bench stock inventory for your shop.*

Appendix

K

A 1999 Vision of Maintenance at the Millennium: Revisited in 2006

In August 1999, the editor of *Industrial Maintenance and Plant Operations* (IMPO) magazine asked me for some comments on the topic: "Maintenance at the Millennium" for the October 1999 issue of IMPO. As a member of the editorial advisory board then, I was more than glad to share some of my thoughts and visions for the future of the maintenance process and the profession. Jim's questions and my responses are included here.

As you have the chance to reflect on my thoughts, think about your vision for maintenance excellence during the millennium. When you read this it will probably be during the new millennium. As a maintenance leader, regardless of your title or position, you must be prepared. Do you have a vision for continuous maintenance improvement, continuous technology advancement, and the continuous challenges to planning and executing the maintenance process?

It does not matter whether it is a fleet, facilities, or plant maintenance operation in either the private or public sector. Physical asset management and the mission-essential maintenance process are forever. This is because, like gravity, extinction, and taxes, maintenance is also forever! You can and will be a key contributor to the total operation success of your operation in the new millennium if you start the journey now.

Here are my comments to some questions (six years ago) that you must also consider.

1. What do you think has been the most important development in plant maintenance in the past 20 years? (Think in terms of how you have taken advantage of these developments.)
   - Growth of reliability improvement technologies: The most important technology development has been the growth of new reliability improvement technologies combined with best practice strategies for enhancing what I call *continuous reliability improvement* (CRI). Very briefly, CRI addresses improvement toward more than just the physical asset. CRI definitely includes Reliability Centered Maintenance (RCM) techniques but goes well beyond the scope of RCM to include all assets and the total business process of asset care/maintenance. CRI involves continuous improvement processes for:
     - Physical assets
     - MRO items, parts and material assets
     - People assets
     - Craft skills base, core competency assets
     - Information assets
     - Team processes that multiply people assets
   - New technologies plus the synergistic power of people assets achieve continuous reliability improvement. The evolution of reliability improvement technologies has given the maintenance leader a wide range of tools to identify, predict, diagnose, detect, and continuously monitor the physical asset for opportunities to increase uptime. The advancements in computerized maintenance management systems (CMMS) as information technology for enterprise asset management (EAM) also belong in the reliability-improvement technology category. CMMS/EAM is a reliability tool that helps to maximize information assets when used properly.

     Concurrent with the evolution of reliability technology tools, the growth of techniques and best practice strategies to eliminate the root cause of problems has also occurred during the 20 years from 1980 to 2000. People assets using CRI, RCM, TPM, and Six Sigma techniques along with the new reliability improvement technologies eliminate root cause and increase uptime. The advancement of technologies to identify problems combined with the techniques to eliminate causes of physical asset problems has been a very significant development that will continue to grow.
2. What types of advances do you anticipate in the next 20 years? (Think about this; "It's almost 21st Century, where is my CMO?")
   - The Chief Maintenance Officer (CMO) will emerge as a recognized corporate leadership position. Based upon observations during my last thirty years of experience in the 20th Century, a CMO (or an

equivalent) will be absolutely essential for long-term profitability in the new millennium. Organizationally, I think the Chief Maintenance Officer (CMO) will eventually evolve into a recognized corporate position. We all know that the CFO watches the money assets, the CIO manages the information assets, and the CPO procures and manages material assets. People assets fall under the scope of human relations (HR). Team processes to multiply people assets and support business process continuous improvement may be sponsored at various levels.

The CEO/COO that does not have a CMO accountable for physical asset management will be gambling with stockholder's equity against a "stacked deck." The CEO/COO must understand the "state of maintenance" in their operation and the physical asset management process. The future capable company will require a proactive, capable CMO just as they need a CFO, CIO, and CEO. I will bet on this one!

- Reliability improvement technologies will continue to evolve. Greater use of radio frequency (RF) technology to support condition-based monitoring will occur. RF technology will enable mobile, timely, and vital communications with the mobile workforce of craftspeople. It is happening now with over 12,000 Sears technicians. BellSouth has plans for some 15,000 technicians to use their TechPlus mobile computing platform. Process control systems will become more integrated with condition-based monitoring systems that in turn link to CMMS/EAM systems or even back to the condition-based equipment suppliers for real-time trouble shooting.
- The use of the Internet will expand exponentially. Condition based monitoring systems will eventually link data collection directly back to suppliers such as CSI, ENTEK, SKF, and others via the Internet or RF. Reliability data analysis by reliability experts providing contract services will be as close as e-mail. E-commerce will continue to expand and direct links to the overall supply chain will enhance procurement of MRO parts/materials and services.
- Information technology for the "life-cycle information loop" will be available. I like to refer to it as the "life-cycle information loop" now possible via IT advances. Life-cycle information for critical assets will be facilitated by the Internet, intranets, and real-time data collection resulting in "real information" on the shop floor. This will take life-cycle costing, equipment design/redesign, reliability improvement and execution of maintenance to new levels. The craft person doing the repair will be in the life-cycle information loop with the OEM, subsystem and MRO providers, the asset designers, local engineering, the local asset documentation sources, and the customers of maintenance. The IT technology is available now and

cost for data collection and integration will eventually be a non-issue for maintenance leaders in the new millennium.

- Contract maintenance will continue to grow. As I have stated in this article and many times before in other publications, "maintenance is forever." Whether we call it asset care or physical asset management, effective maintenance of our minds, bodies, souls, cars, homes, and the physical assets that produce goods and services will be necessary. Due to downsizing and "dumbsizing" (By the way I think Walt Flannery at Chevron and on IMPO's editorial board coined the term "dumbsizing" and deserves credit) of maintenance along with the lack of internal leadership many operations will continue to lose their core competencies to do maintenance. The core requirement for maintenance will be even more important in the next millennium due to technology advances.

  I could possibly be wrong, but TVM (total virtual maintenance) will not come to pass. Profit-centered contract maintenance providers will consume internal maintenance operations that continue a cost-centered approach. In house maintenance operations will continue to lose when they are unable to replenish and/or maintain their core competencies in maintenance. Stock in good contract maintenance providers will still be a good investment in the new millennium.
- The contribution of maintenance to total operations success and profitability will finally be recognized. Maintenance and physical asset management will be more closely linked to enterprisewide performance and profitability. Successful Maintenance leaders will think "profit-centered," establish strategic maintenance plans that are integrated with the business plan, and validate return on investment with effective measurement processes. The emerging CMO with "profitability" and effective leadership and technical skills will facilitate this process in larger multisite operations. Smaller maintenance operations will also have their CMO equivalents too. This may be the enlightened supervisor/plant engineer/multipurpose person who runs an 8 to 15 person within a small plant operation.
- Maintenance excellence processes will become standardized. We will learn from practices that have been a part of supply and maintenance in our military services for many, many years. Standards that encompass all required practices to maximize total financial and functional value over the complete life cycle of the physical asset will evolve. Such practices will include asset design/redesign, process modifications, engineering change control, capital procurement, installation and commissioning, staffing, MRO materials management, storeroom operations, maintenance, and reliability

business practices, life-cycle costing, full utilization of CMMS/EAS, and much more.

This will include standardization and integration of physical asset management technologies in the form of software, hardware, and associated information technology practices. The CMO will ensure this standardization occurs across a multisite operation but be receptive to change, promote continuous reliability improvement, and the sharing of best practices.

- Certification requirements and technical competencies of maintenance leaders will become more not less important. Certification and self-governing of the practices of maintenance and the physical asset management profession similar to the medical, legal, financial, and engineering design professions will increase. The importance of certifications offered, for example, by the Association of Facilities Engineering, the Vibration Institute, and Professional Engineering will grow. As catastrophic examples of "bad maintenance" continue to take their toll in lives, I think that the media will eventually elevate this to the political arena.

  When the high cost of deferred maintenance (infrastructure, pubic facilities, roads, etc., is fully understood by politicians and the public and becomes a real political agenda, there will be actions to manage and maintain our public and private resources much better. Maintenance is not a hot media topic unless a plane crashes or the Maytag repairman is a trendy advertising strategy. If the taxpayers could get the truth on the high cost of deferred maintenance to schools, roads, bridges, and public property by the media the level of concern could create more cost-effective political action.
- Stronger academic curriculum will evolve to meet the demand for greater technical competencies. Four-year educational institutions will enhance their engineering curriculum to include maintenance best-practice topics accompanied by in-the-field internships. Two-year technical colleges also will supplement maintenance management topics and intensify their technical hands-on training. Technical training developers/vendors will also continue to prosper as a supplement to an educational system that fails to prepare students with job skills in trade's areas.

Some other comments on maintenance at the millennium include:

- Measurement of maintenance performance via internal benchmarking will take on renewed importance as profit-centered maintenance becomes a standard practice.
- Real total operations consulting organizations will understand maintenance and provide total solutions linked to the total operation.

- ISO/QS requirements will specifically refer maintenance processes that are needed for a quality management system.
- Harvard Business Review will finally use the word "maintenance" in an article title or "asset management" to mean something other than money assets.

As a special note I included a quote from one of my previous articles.

> *"It is very interesting to note that maintenance is still available as a hot new business topic for the Harvard Business Review. Currently, there is not even one single reference to the word maintenance in article titles from July 1995 to October 1999. Back in 1991 there was one entitled "Northwest Airlines Financing Maintenance Facility in Minnesota." There's a good article (for sale) on "How High Is Your Return on Management" . . . but nothing on return on maintenance ... of all the physical assets that produce all goods and services. There are of course no references at all (now) to a CMO."*

3. In what ways do you feel technology will change your job in the next decade? (I had to think about this question from my personal perspective; here is what I sent to IMPO.)
   - All the advancements that I forecasted so to speak in Question #2 will probably occur from 2000 to 2010. From the perspective of total operations consulting company, we must stay abreast of all the advancements in maintenance that can help serve our growing client base. If, I personally do not keep up with this technology leap, my value to current and future clients will diminish very quickly.
   - The advancements in information technology have and will continue to help improve my personal productivity. However, no matter how quick and powerful the laptop, my service to clients must result in successful implementations of best practices. There must be measurable, bottom line results to each client or my job as a consultant will quickly change to being my wife's full-time yardman and driver.
4. What single aspect of plant maintenance or management do you think will represent the biggest challenge to plant operations in the next decade?
   - Unleashing the hidden asset from people at all levels in an operation will continue as a big challenge and opportunity. Developing maintenance leaders, rather than perpetuating maintenance managers of the status quo will be essential to the success of multiplying the latent hidden value of people assets. I think that the technical aspects of maintenance sometimes are the easiest to transfer when compared to the people side of business.

     Getting 21st Century craftspeople to work together for a common goal requires leadership and an effective teaming process.

Maintenance business process continuous process improvement (BPCI) must be fully linked to the organization's BPCI efforts and the quality excellence process. Technical and personal leadership development to unleash hidden people assets is required. This will remain a very real challenge to fleet, facility, and plant maintenance operations if there is not a true maintenance champion in the form of a CMO.

I have always been a big, big fan of Alvin Tofler since I read "Future Shock" back in 1970, "The Third Wave" in 1980, and his third book in 1990, which I forget the name of. Personal adaptability to change was one key Tofler point that has stuck with me since reading his works. Adaptability is important to all of us when it comes to unleashing the hidden people assets. There is a very bright future for the profession of maintenance and physical asset management in the new millennium. My comments really just scratch the surface because of the proven fact that "whatever the mind of man can conceive, it can achieve."

5. What role will the Internet play in your job in the next decade?
   - Previous comments touched upon this one pretty well from my perspective. It is obvious that the future use of the Internet and intranets will be exponential not linear.
6. In what ways will educational requirements need to change in order for you and your peers to cope with increasing technological advances?
   - Previous comments also address my thoughts on this one. But, if we think craft skills training is expensive, try staying ignorant of Electro-mechanical principles!
7. What advice would you give to a student interested in pursuing a career in the maintenance profession?
   - As I said before, "Maintenance is forever" and there will be a continuous need for maintenance technical resources at all levels and in all types of operations. Craftspeople are becoming a scarce resource and their value and their so called status (and pay) in the workforce will only go up. From my perspective, craftspeople with a life-long learning goal, rate extremely high on the ladder of career paths.

     I think that a student is wise to choose a hands-on career in maintenance or to pursue an engineering field that prepares one for more technical roles in engineering and maintenance. I do not see many young MBA's serving as CMO's but the successful CMO will know grass roots maintenance, maintenance best practices, finance, and effective business practices along with the technical skills.

8. In what ways do you think your job will become easier/harder in the next 10 years?
   - I personally don't think that my career in the maintenance and total operations consulting arena will become easier. My personal productivity may improve via information technology advances. Maintenance and total operations consulting is grass roots; down in the trenches, results-based consulting which is fun and personally very rewarding. When attitudes about maintenance eventually change, selling the investment and service may become easier but the consulting work itself will remain a challenge. The hard part will be staying abreast of new technologies and their proper application. It will be a difficult time for us all if we are adaptable and are not totally committed to life-long learning and professional development.

The last question is for you the reader, "What are your thoughts on maintenance at the millennium right now?" Your positive thoughts and vision for maintenance excellence when put into action will make a difference in your operation!

Appendix

L

Comparison of Scoreboards

Comparison of the Scoreboard for Manufacturing Plant Maintenance and the Scoreboard for Pure Facilities Maintenance Type Operations

CAT	**The Scoreboard for Maintenance Excellence** **Benchmark Category Descriptions (Part I)**	Items	Total Points	CAT	**The Scoreboard for Facilities Management Excellence** **Benchmark Category Descriptions (Part I)**	Items	Total Points
A.	The Organizational Culture and PRIDE in Maintenance	6	60	A.	The Organizational Culture and PRIDE in Maintenance	5	50
B.	Maintenance Organization, Administration, and Human Resources	12	120	B.	Facilities Organization, Administration, and Human Resources	10	100
C.	Craft Skills Development and PRIDE in Maintenance	12	120	C.	Craft Skills Development and PRIDE in Maintenance	10	100
D.	Operator Based Maintenance and PRIDE in Ownership	6	60	D.	Facilities Management Supervision/Leadership	10	100
E.	Maintenance Supervision/Leadership	9	90	E.	Business Operations, Budget, and Cost Control	15	150
F.	Maintenance Business Operations, Budget, and Cost Control	12	120	F.	Work Management and Control: Maintenance and Repair (M/R)	10	100
G.	Work Management and Control: Maintenance and Repair (M/R)	12	120	G.	Work Management and Control: **Construction and Renovation (C/R)**	5	50
H.	Work Management and Control: **Shutdowns and Major Overhauls**	6	60	H.	Facilities Maintenance and Repair Planning and Scheduling	15	150
I.	Shop Level Planning and Scheduling	18	180	I.	Facilities Construction and Renovation Planning and Scheduling	10	100
J.	Shutdown and Major Planning/Scheduling and Project Management	9	90	J.	**Facilities Planning and Property Management**	10	100
K.	Manufacturing Facilities Planning and Property Management	9	90	K.	***Facilities Condition Evaluation Program**	**5**	**50**
L.	Production Asset and Facilities Condition Evaluation Program	6	60	L.	Facilities Storeroom Operations and Internal MRO Customer Service	15	150

M.	Storeroom Operations and Internal MRO Customer Service	12	120	M.	MRO Materials Management and Procurement	10	100
N.	MRO Materials Management and Procurement	12	120	N.	Preventive Maintenance and Lubrication	20	200
O.	Preventive Maintenance and Lubrication	18	180	O.	Predictive Maintenance and Condition Monitoring Technology Applications	10	100
P.	Predictive Maintenance and Condition Monitoring Technology Applications	15	150	**P.**	***Building Automation and Control Systems Technology**	**5**	**50**
Q.	Process Control, Building Automation, and Instrumentation Systems Technology	9	90	**Q.**	***Utilities Systems Management**	**10**	**100**
R.	Energy Management and Control	12	120	R.	Energy Management and Control	10	100
S.	Maintenance Engineering Support	9	90	S.	Facilities Engineering Support	10	100
T.	Safety and Regulatory Compliance	12	120	T.	Safety and Regulatory Compliance	15	150
U.	Maintenance and Quality Control	9	90	**U.**	***Security Systems and Access Control**	**10**	100
V.	Maintenance Performance Measurement	12	120	V.	Facilities Management Performance Measurement	15	150
W.	Computerized Maintenance Management System (CMMS/EAM) and Business System	18	180	W.	Facilities Maintenance Management System (FMMS) and Business System	15	150
X.	Shop Facilities, Equipment, and Tools	9	90	X.	Shop Facilities, Equipment, and Tools	10	100
Y.	Continuous Reliability Improvement	15	150	Y.	Continuous Reliability Improvement	10	100
Z.	Asset Facilitation and Overall Equipment Effectiveness	15	150	**Z.**	***Grounds and Landscape Operations**	**15**	**150**
ZZ.	Overall Craft Effectiveness (OCE)	6	60	**ZZ.**	***Housekeeping Service Operations**	**15**	**150**
The Scoreboard for Maintenance Excellence		**300**	**3000**	**The Scoreboard for Facilities Management Excellence**		**300**	**3000**
Total Evaluation Items and Points				**Total Evaluation Items and Points**			

Index

ABOUT THE AUTHOR

RALPH W. "PETE" PETERS is founder of The Maintenance Excellence Institute, focused on Worldwide Services with Shop Level Results. He is an engineer with over 36 years of practical experience in manufacturing operations and direct leadership of maintenance for plant, facilities, fleet, and road maintenance operations. Mr. Peters is a recognized leader in the areas of implementing maintenance best practices, developing effective productivity measurement/improvement, and initiating long-term operational improvement processes within both the public and private sectors. His value as a consultant is enhanced through his direct manufacturing plant management experience, his direct maintenance management, and leadership of large fleet and facilities maintenance operations.

He has helped operations such as UNC-Chapel Hill, the U.S. Air Force's Air Combat Command, Atomic Energy Canada Ltd., Boeing Commercial Airplane Group, Rocketdyne, Caterpillar, Ford, Honda, Polaroid, Lucent, Heinz, General Foods, Biglots Stores, Sheetz Inc., Marathon Ashland Oil, Great River Energy, Wyeth-Ayerst (US & IR), Cooper Industries, National Gypsum, Carolinas Medical Center, North Carolina Department of Transportation, and the U.S. Army Corps of Engineers achieve success in plant, fleet, and facilities maintenance operations.

He is the author of over 200 articles and publications and is a frequent speaker and trainer who has delivered speeches and workshops on maintenance and operations improvement topics worldwide in over 15 countries. He received both his BSIE and MIE from North Carolina State University and is a graduate of the U.S. Army Command and General Staff Course, the Engineer Officers Advanced and Basic Courses, the Military Police Officers Course, and the Civil Affairs Officers Course.

His definitive program for profit- and customer-centered maintenance was created by focusing on the basics and successful implementation of practical solutions. Pete is also the author of maintenance chapters in four major books, a CMMS living book, two E-Books, and over 200 articles and white papers. He has helped hundreds of organizations and individuals apply knowledge for solutions that provide measurable results. He lives in the United States of America in Raleigh and Oak Island, North Carolina with his wife and the familes of two sons and their three grandchildren.